T0317536

AN INTRODUCTION TO HIGH CONTENT SCREENING

AN INTRODUCTION TO HIGH CONTENT SCREENING

Imaging Technology, Assay Development, and Data Analysis in Biology and Drug Discovery

Editors

STEVEN A. HANEY
DOUGLAS BOWMAN
ARIJIT CHAKRAVARTY

Associate Editors

ANTHONY DAVIES
CAROLINE SHAMU

Published by John Wiley & Sons, Inc., Hoboken, New Jersey.
Published simultaneously in Canada.

For general information on our other products and services or for technical support, please contact our Customer Care Department within the United States at (800) 762-2974, outside the United States at (317) 572-3993 or fax (317) 572-4002.

Wiley also publishes its books in a variety of electronic formats. Some content that appears in print may not be available in electronic formats. For more information about Wiley products, visit our web site at www.wiley.com.

Library of Congress Cataloging-in-Publication Data:

An introduction to high content screening : imaging technology, assay development, and data analysis in biology and drug discovery / editors, Steven A. Haney and Doug Bowman ; associate editors, Arijit Chakravarty, Anthony Davies, and Caroline Shamu.
   1 online resource.
  Includes bibliographical references and index.
  Description based on print version record and CIP data provided by publisher; resource not viewed.
  ISBN 978-1-118-85941-4 (ePub) – ISBN 978-1-118-85947-6 (Adobe PDF) – ISBN 978-0-470-62456-2 (cloth)
  I. Haney, Steven A., editor.   II. Bowman, Doug, active 2014, editor.   III. Chakravarty, Arijit, editor. IV. Davies, Anthony, active 2014, editor.   V. Shamu, Caroline, editor.
  [DNLM: 1. Drug Design. 2. Systems Biology–methods. 3. Data Interpretation, Statistical. 4. Drug Evaluation, Preclinical–methods. 5. Image Processing, Computer-Assisted–methods. 6. Models, Biological. QV 26.5]
  RM301.25
  615.1′9–dc23

2014027680

10 9 8 7 6 5 4 3 2 1

CONTENTS

12 Automation and Screening   181

John Ringeling, John Donovan, Arijit Chakravarty, Anthony Davies,
Steven A. Haney, Douglas Bowman, and Ben Knight

15 Supervised Machine Learning   231

Jeff Palmer and Arijit Chakravarty

Appendix A   Websites and Additional Information on Instruments, Reagents, and Instruction   247

Appendix B   A Few Words About One Letter: Using R to Quickly Analyze HCS Data   249

Steven A. Haney

PREFACE

We have been living in the "information age" for over a generation now, and although the term itself has lost much of its cachet, it is more true than ever. As we discuss teaching high content screening (HCS) specifically, there are many excellent options available for obtaining information. Study guides, protocols, and online tutorials are plentiful and are typically carefully written by seasoned practitioners. Why then would one consider using a book, particularly when so many of the other options are free? We have discussed this issue at length, and undertook this project because we believe there are several strong arguments in its favor:

1. A book allows comprehensive discussion of a highly complex system. There are many excellent protocols available for specific problems. However, these protocols tend to be case studies and specific solutions rather than a presentation of a series of options and how to integrate them.

2. A book is better suited for basic principles, rather than teaching a particular system or platform. By emphasizing principles and drawing upon examples from several of the available platforms, we have designed this text to grow with scientists as they gain more experience with HCS. Platforms change frequently, if we were to emphasize how to use one or more systems, the utility of the book would fade as specific upgrades are introduced. However, all platform-specific upgrades are built around adapting the platform to better meet these fundamental principles. A discussion of principles enables a deeper understanding of how each platform approaches the problem of robustly capturing cellular information, and to highlight advantages of one approach over another as examples are considered.

3. A book that has been collectively edited can provide a sustained discussion across many chapters. One challenge for some methods books is that topics are

presented as part of a collection of independently written essays; each author or group of authors makes assumptions about what will be presented elsewhere, and these assumptions do not always hold true. As such, in many methods volumes, significant information can be presented in each chapter about one topic, but important information related to that topic may be missing or covered thinly in other chapters.

These reasons are particularly relevant to a discussion of HCS. To begin such a discussion, it is helpful to recognize that learning HCS can be made more difficult because HCS itself can mean different things to different people, and such distinctions may not be obvious. In many cases, an HCS assay could mean one that measures the localization of a transcription factor to the nucleus using a "canned" algorithm. Such algorithms provided by the instrument manufacturer can simplify analysis, but occasionally can also hide some of the data processing steps from the experimenter. In other cases, phenotypic profiling (a measure of changes to a cell through the integration of many morphological features) or the quantification of rare events may introduce specialized data analysis methods that will make an experiment much more difficult to perform, and yet these types of studies may be relatively uncommon for many laboratories.

This book was conceived and written under the philosophy that HCS is not truly challenging on a technical level, but that it requires a good understanding of several distinct areas of biological and data sciences and how to integrate them to develop a functioning HCS laboratory. As such, learning HCS requires building an understanding of basic biology, immunofluorescence, image processing, and statistics. This is covered in more detail in Chapter 1. In addition, every instrument vendor provides training on their instruments, and in several cases, this training can be excellent, but in each case, there are common principles that each platform needs to address—it is the focus of this book to treat the principles of imaging and data analysis directly.

We have endeavored to create a tone for the presentation of this material that is direct but not overly technical. We feel that this helps maintain the principles-based approach and keep the text light when we can. We do not shy away from technical terms, they are important, and in fact we strive to define a handful of new terms where the discussion can be difficult to follow without appropriate distinctions.

We have many people to thank. We thank our editor, Jonathan Rose, for many things but mostly for his forbearance. Keeping organized during the collaborative effort, where multiple editors worked on each chapter, was a logistical nightmare and added several years to the completion of this book. Lin Guey is thanked for helpful discussions beyond her direct contributions to the data analysis chapters. Lifei Liu and Karen Britt are thanked for contributing images used to illustrate some of the concepts we discuss.

STEVEN A. HANEY
DOUGLAS BOWMAN
ARIJIT CHAKRAVARTY
ANTHONY DAVIES
CAROLINE SHAMU

CONTRIBUTORS

Douglas Bowman, Molecular and Cellular Oncology, Takeda Pharmaceuticals International Co, Cambridge, MA 02139

John Bradley, Molecular and Cellular Oncology, Takeda Pharmaceuticals International Co, Cambridge, MA 02139

Kristine Burke, Molecular and Cellular Oncology, Takeda Pharmaceuticals International Co, Cambridge, MA 02139

Jay Copeland, Department of Systems Biology, Harvard Medical School, Boston, MA 02115

Arijit Chakravarty, Drug Metabolism and Pharmacokinetics, Takeda Pharmaceuticals International Co, Cambridge, MA 02139

Anthony Davies, Translational Cell Imaging, Queensland University Of Technology, Brisbane, Australia

John Donovan, Lead Discovery, Takeda Pharmaceuticals International Co, Cambridge, MA 02139

Craig Furman, Research Technology Center, Pfizer, Cambridge, MA 02139

Lin T. Guey, Biostatistics, Shire Human Genetic Therapies, Lexington, MA 02421

Steven A. Haney, Quantitative Biology, Eli Lilly and Company, Indianapolis, IN 46285

Ben Knight, Lead Discovery, Takeda Pharmaceuticals International Co, Cambridge, MA 02139

Alice McDonald, Translational Biomarkers, Epizyme, Cambridge, MA 02139

Jeffrey Palmer, Biostatistics, Genzyme, Cambridge, MA 02139

John Ringeling, Lead Discovery, Takeda Pharmaceuticals International Co, Cambridge, MA 02139

Caroline Shamu, ICCB-Longwood Screening Facility, Harvard Medical School, Boston, MA 02115

Vaishali Shinde, Molecular Pathology, Takeda Pharmaceuticals International Co, Cambridge, MA 02139

1

INTRODUCTION

Steven A. Haney

1.1 THE BEGINNING OF HIGH CONTENT SCREENING

Microscopy has historically been inherently a descriptive endeavor and in fact it is frequently described as an art as well as a science. It is also becoming increasingly recognized that image-based scoring needs to be standardized for numerous medical applications. For example, for medical diagnoses, interpretation of medical images has been used since the 1950s to distinguish disorders such as cervical dysplasias and karyotyping [1]. Cameras used in microscopes during this era were able to capture an image, reduce the image data to a grid that was printed on a dot-matrix printer and integrated regional intensities to interpret shapes and features. In essence, these principles have not changed in 50 years, but the sophistication and throughput with which it is done has increased with advances in microscope and camera design and computational power. In the early 1990s, these advances were realized as automated acquisition and analysis of biological assays became more common.

Advances in automated microscopy, namely the automated movement of slides on the stage, focusing, changing fluorophore filters, and setting proper image exposure times, were also essential to standardizing and improving biomedical imaging. Automated microscopy was necessary to reduce the amount of time required of laboratory personnel to produce these images, which was a bottleneck for these studies, especially medical diagnoses. A team of scientists from Boston and Cambridge, Massachusetts described an automated microscope in 1976 that directly anticipated its use in subcellular microscopy and image analysis [2]. The microscope, and a processed image of a promyelocyte captured using the instrument, are shown in Figure 1.1.

An Introduction to High Content Screening: Imaging Technology, Assay Development, and Data Analysis in Biology and Drug Discovery, First Edition. Edited by Steven A. Haney, Douglas Bowman, and Arijit Chakravarty.
© 2015 John Wiley & Sons, Inc. Published 2015 by John Wiley & Sons, Inc.

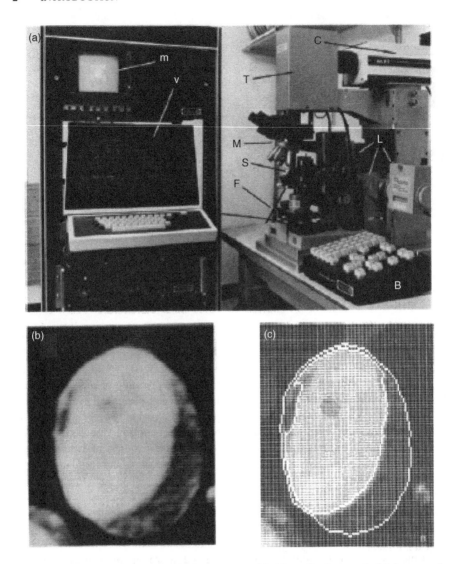

FIGURE 1.1   *An early automated microscope used in biomedical research.* (a) An example of an automated fluorescence microscope. Letters inside the figure are from the original source. The system is outfitted with controlled stage and filter movements (S and F), a push-button console for manual movements (B), a television camera and monitor (T and m) and a video terminal for digitizing video images (v). (b) A video image of a promyelocyte and (c) image analysis of (b), showing, an outline of the nucleus and cell borders, which can be used in automated cell type recognition. Reproduced with permission from [2]. Copyright 1974 John Wiley & Sons.

Until the mid-1990s, automated microscopy was applied in basic research to address areas of high technical difficulty, where rigorous measurements of subtle cellular events (such as textural changes) were needed, events that took place over long time periods or were rare (which made it challenging to acquire sufficient numbers of images of each event). In medicine, automated imaging was used to standardize the interpretation of assay results, such as for the diagnosis of disease from histological samples (where it was notoriously difficult to achieve concordance among clinical pathologists). Adapting quantitative imaging assays into a screening context was first described by Lansing Taylor and colleagues [3], who commercialized an automated microscope capable of screening samples in multiwell plates (a format that had emerged as an industry standard during this time period). The term "high content" was coined to contrast the low throughput screening in these imaging assays with the increasing scale of high throughput primary drug discovery screens. Many groups have since demonstrated the usefulness of automated microscopy in drug discovery [4, 5] and basic research [6, 7]. During this phase (the early 2000s), data acquisition, image analysis, and data management still imposed limits on image-based screening, but it did find an important place in the pharmaceutical industry, where expensive, labor-intensive assays critical for late-stage drug development were a bottleneck. One example is the micronucleus assay, an assay that measures the teratogenicity of novel therapeutics through counting the number of micronuclei (small nonnuclear chromosomal fragments that result from dysregulation of mitosis). An increase in the number of cells that contain micronuclei is indicative of genotoxicity, so this assay is frequently part of a screening program to make a go/no go decision on clinical development [8]. The assay requires finding binucleate cells and checking for a nearby micronucleus. For each compound assayed, a single technician might spend many hours in front of a microscope searching and counting nuclei. Automation of image capture and analysis not only reduced the work burden of researchers, but it also made the analysis itself more robust [9]. Similar applications were found in the field of cell biology, where automated microscopy was utilized to collect and analyze large data sets [10, 11].

Following from these early implementations, high content screening (HCS) has been widely adopted across many fields as the technology has improved and more instruments are available commercially. The speed at which images can be analyzed is limited by computer power, as more advanced computer technology has been developed, the scale at which samples can be analyzed has improved. Faster computers also mean that more measurements per cell can be made; shapes of cells and subcellular structures can be analyzed as well as probe intensities within regions of interest. This has led to the quantification of subtle morphological changes as assay endpoints. A widely used application of this approach has been receptor internalization assays, such as the Transfluor^{TM} assay to measure the activation of GPCRs through changes in the pattern of receptor staining, from even staining over the surface of the cells to dense puncta following internalization of the activated receptors through vesicle formation [12]. Concomitant with the increase in the sophistication of the assays themselves, improvements in the mechanical process of screening samples has also fed the growth of HCS. Gross-level changes, such as integrating plate handling robotics and fine-level changes, such as improvements in sample detection and autofocusing, have improved the scale of HCS to the point where image-based readouts are possible for true high throughput screens (screens of greater than 100,000 compounds) [5].

HCS has a strong presence in basic biological studies as well. The most widely recognized applications are similar to screening for drug candidates, including siRNA screening to identify genes that control a biological process, and chemical genetics, the identification of small molecules that perturb a specific cellular protein or process. While operationally similar to drug screening, they seek to explain and study biological questions rather than lead to therapeutics explicitly. Additional uses of HCS in basic science include the study of model organisms. Finally, the use of multiparametric single cell measurements has extended our understanding of pathway signaling in novel ways [11].

1.2   SIX SKILL SETS ESSENTIAL FOR RUNNING HCS EXPERIMENTS

At this point we want to touch on the fundamental skill sets required to successfully set up and use an HCS system to address a biological problem, and how responsibilities might be divided up in different settings. The six major skill sets required to develop and run an HCS project are shown in Figure 1.2. Each area is distinct enough as to be a full-fledged area of expertise (hence introducing these areas as "skill sets"), but

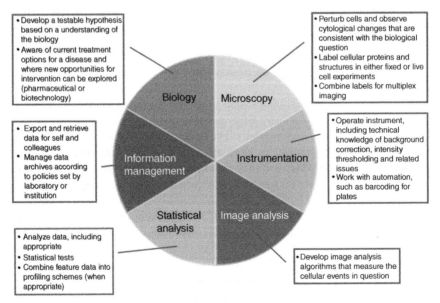

FIGURE 1.2   *The basic skill sets essential for establishing and running HCS experiments.* Skills noted in the figure are discussed in detail in the text.

typically a person is competent in more than one area. It is rare that all roles can been successfully filled by one person. Therefore, the ability to develop a collaborative team is essential to HCS. It is also very important to understand that these roles vary between groups, and this can cause problems when people move between groups or as groups change in size. The skill sets are the following.

1.2.1 Biology

The biologist develops the question that needs to be answered experimentally. In academia, the biologist is typically a cell biologist and oftentimes is also capable of collecting images by HCS as well. In industrial circles (pharma and biotech), a therapeutic team may be led by a biochemist or *in vivo* pharmacologist, who may have little training in fluorescence microscopy. The key area of expertise here is an appreciation of these problems and an ability to formulate strategies (experimental systems and assays) to address them. There is also a significant understanding of how cellular models in the laboratory relate to the biology *in vivo*. In addition to understanding the fundamental biological question, understanding how to establish a cellular model that incorporates relevant aspects of the biological environment is important.

1.2.2 Microscopy

Although many of the HCS systems are sold as turnkey "black-boxes," it is important to have a good understanding of fundamental microscopy components (staining techniques, reagents, and optics) as each has a significant impact on the quality of data generated by the instruments. For example, the choice of illumination system and filter sets determine which fluorescence wavelengths (fluorophores) you can use to stain specific cellular compartments. Other microscope objective characteristics (numerical aperture, magnification, and working distance) also impact both the types of samples one can image as well as the spatial resolution of the resulting images. More information on these topics are covered in Chapters 3 and 7. If the biological question is posed by someone who is not a trained microscopist, then it is important to discuss technical aspects with someone who has such training, which is why these skills are frequently part of the Platform Manager responsibilities (see below), particularly when the HCS instrument is used in a core facility.

1.2.3 HCS Instrumentation (Platform Manager)

The platform manager focuses on the hardware and software needed to keep an HCS facility running smoothly. Much of the information needed to run a particular HCS instrument is obtained from the instrument vendor, including instrument operation and, in some cases, strategies for data analysis; the Platform Manager will be the one who interacts with the vendors directly, particularly for handling challenging problems with the instrumentation or for scheduling updates to the instrument. Although the image acquisition configuration is often simplified by the user interface software, a solid understanding of imaging hardware (laser autofocus, CCD camera exposure time, PMT amplifier gain) is needed for optimal use of the instrument. In addition, it is important to know how to integrate the HCS instrument with other laboratory automation instruments to increase overall throughput. Relevant automation instrumentation includes robotic plate handling devices (robot arms with plate "stackers" or "hotels") that serve plates to the HCS instrument one at a time and store plates after they are read, automated tissue culture incubators to store plate used for live cell imaging, and plate barcoding equipment. We go into more detail on these in Chapters 5 and 12.

1.2.4   Image Analysis

The identity of the person that contribute to the image analysis can be very fluid. In many cases, this position functions as an extension of the microscopy skill set, but it is also becoming a more specialized position, particularly as HCS experiments grow in complexity or subtlety, such as spheroids and primary cells (which may be plated at confluence) require more work to develop an algorithm that is suitable. Many of the instruments include "canned" algorithms that can be applied to a range of assays (cell counting, mitotic index, cell health, neurite outgrowth, etc.). There are also third-party applications, such as the open-source CellProfiler^{TM} and the commercial analytical package Definiens^{TM} that are compatible with all of the common platform image set files and require more effort to understand how to use them than the "shrink-wrapped" algorithms. The skills are covered in more detail in Chapters 4, 5, and 15.

1.2.5   Statistical Analysis

The data analyst needs to understand the design and objectives of the experiment and apply the appropriate statistical tests to characterize any conclusions. The scope of the data analysis needs can vary greatly depending on the user needs: whether you are using the instrument for single experiments or you are running a screen with hundreds of compounds. Depending on the experiment, analysis can be straightforward, even routine, or it can be complex with a potential for making false conclusions if the proper statistical tests are not used.

HCS in a screening environment typically means that an assay is being run to identify hits, and the methods for determining how robust a screen is can be evaluated by someone with good screening experience, but not necessarily having a rigorous background in statistics. Assays used for HTS are typically well validated and use a limited set of well-characterized cell lines. Positive and negative controls produce visually distinct results, and are therefore easy to measure. As such, few images need to be obtained (as few as one per well) and compound (or RNAi reagent) effects are evaluated relative to these controls. Analysis of the data typically involves measures of automation and cell culture performance, in the form of heatmaps of individual plates to locate patterns that suggest systematic or spurious problems, followed by normalizations of the data across plates and an evaluation of each treatment according to a single measure, typically a Z-score (see Chapter 9). Such an analytical stream is fairly straightforward. Phenotypic patterns, such as composite measures of multiple features, require testing of the metrics themselves and a scheme for integrating features. Features can be measured according to Z-scores, much like in a single endpoint HTS assay. However, using multiple morphological features to evaluate the effect treatment effects can lead to false conclusions, because testing a wide number of potential features as potential assay endpoints leads to spurious associations. The latter is a subtle statistical problem, and because of pitfalls such as this, analyzing such data requires stronger training or experience than is typical for a bench scientist [13]. In addition to Chapter 9, data analysis is covered in Chapters 8–13, which cover the concept of assay metrics in HCS and progress through statistical analysis of screening data and multivariate methods.

1.2.6   Information Technology Support

Information technology (IT) expertise is needed to implement a data management solution according to mandates from the research team and their institution. HCS generates unprecedented volumes of data. Storing and retrieving data in a stable, validated IT environment that conforms to institutional guidelines requires both a thorough understanding of how to manage data, but also an understanding of the needs of different scientists who use the HCS instrument. HCS vendors are well aware of the data management requirements, but rarely provide complete solutions. There may also be need to integrate with user databases, such as linking the plate barcodes with a compound library as described above. Further details related to informatics requirements are described in Chapter 6.

1.3   INTEGRATING SKILL SETS INTO A TEAM

While skill sets were delineated above, it does not necessarily take six people to run an HCS experiment. Two or three functions might be covered by a single person, but rarely more. Therefore, HCS is a collaborative endeavor, and instead of challenging oneself with learning several complex roles, it is most productive to consider what roles each person can carry well and what skill sets need to be filled. The ability to play more than one role is influenced by the scientific and business environment. A large pharmaceutical company typically has institutional policies that mandate an IS/IT group implement data management policies that insure the security and preservation of data. This group will typically identify existing server space or install local servers for image and data storage, and will establish procedures for backing-up and archiving data. An academic lab or a small biotech company using HCS may have lesser needs for dedicated IT support since many image or data analysts (particularly those that have experience in high volume data-driven projects such as transcriptional profiling or proteomics) will be able to set up a fileserver and help to organize data. In such a case, the roles of image/data analyst and IT manager might be combined.

Most commonly, the roles of biologist and microscopist will be combined. Sometimes, a biologist who is not trained in cell biology might articulate an important question that can be addressed by HCS, but that person may not have sufficient microscopy experience to establish a robust HCS assay. In such a case, a scientist should not be dissuaded from proposing a high content experiment, but needs to collaborate with a microscopist. The roles of HCS instrumentation specialist, image and data analyst can be combined in drug screening groups. A high throughput screening (HTS) group in a pharmaceutical company typically manages many instruments, and works with biologists to adapt assays for HTS, but this can come at the expense of flexibility, as rapid progression through many screens can limit the time that can be spent on cell lines or imaging challenges for a particular project.

Who, then, develops the image analysis algorithms? This is probably the most collaborative piece of running an HCS experiment. Certainly, the person running the HCS instrument should have experience using the standard image analysis tools packaged with most HCS instruments, but some assays might require help from an expert image analyst to write custom image analysis algorithms, using software such as MATLAB^{TM}, FIJI^{TM}, or CellProfiler^{TM}, that are not packaged with the HCS instrument. As noted above, this is becoming common in larger facilities. The biologist/microscopist functions are also intrinsically involved in developing image analysis algorithms as the process is iterative: an algorithm is developed and tested on control samples, the results are evaluated by the microscopist/biologist to confirm that they capture physiologically and experimentally relevant parameters, the algorithm is then improved further and re-evaluated, until an optimal and robust image algorithm is produced.

1.4 A FEW WORDS ON EXPERIMENTAL DESIGN

Finally, it is worth a few minutes to discuss HCS experiments from a broader perspective. Researchers beginning to appreciate the power of HCS can become overwhelmed. One common assumption for users embarking on HCS is that obtaining a meaningful result requires imaging many, many cells at high magnification using the most sensitive camera and analyzed using the most complex imaging algorithms. In truth, even the most basic HCS experiments are substantially richer in data and more statistically significant than traditional cell biological and biochemical assays (Western blotting and ELISA technology) that measure responses of cell populations only and do not provide information about individual cells. So much so, in fact, that it is worth taking some time to consider how much (or rather, how little) sophistication is necessary to answer the scientifically relevant question. As an example, a dose response curve typically requires at least three replicates per dose, and 5 to 12 compound concentrations to determine the potency of a small molecule. In many cases, more than triplicate values are used per dose. The additional replicates are necessary because dose–response curves are typically quite noisy near the IC50. In contrast, an HCS translocation assay at a moderate magnification will determine the extent of activity of the compound on 30–150 cells per field. As such a single field will capture enough that truly spurious noise is not a problem. Such assays still require replicates, due to systematic or experimenter error, but the historical problem of noise and scatter are handled much better by imaging technologies, a detailed treatise on this is presented in Chapter 9. Lastly, the sensitivity of an HCS imager is very high, and it can measure very subtle changes. As such, it is common that a low magnification objective is usually sufficient to observe the change as a function of compound dose, and using a lower magnification objective means faster acquisition times and fewer images that need to be collected.

HCS has found a place in the highly rigorous and standardized discipline of HTS. HTS in modern drug discovery relies on the ability to robustly make singular measurements over very many samples (upward of 1 million in large-pharma HTS campaigns), and HCS accomplishes this by capturing a single image per well. At the other end of the continuum, there are approaches to drug development and basic biology that leverage the sensitivity of HCS to integrate many (largely) independent effects of a perturbation to determine the extent and similarity of it to other perturbations. These approaches do in fact benefit from better imaging and large numbers of cells, but they are far less common that the simpler HCS assays.

1.5 CONCLUSIONS

HCS is not a single technological innovation, but the aggregation of a handful of independent technologies that give a highly flexible approach to quantitative cell biology. No single aspect of HCS is truly difficult to learn, but pulling together an understanding of each of the core technologies takes time. Most vendors of HCS instruments commit a lot of effort to training their users, and these efforts are essential to becoming fluent with their instruments.

There is greater variability of educational opportunities for learning the complete set of skills that contribute to HCS, in part because there are many places where HCS is used. Screening laboratories will place a premium on reliability and minimizing day-to-day variability. Drug discovery laboratories (and many academic laboratories that study cellular signaling pathways) will focus on the flexibility of the platform, and the capability of measuring the activity of a large number of signaling pathways and functional outcomes. Systems biology, tissue biology, pharmacology, and other disciplines also make use of the unique capabilities of HCS. All of these will be discussed in this book. In each case, the core process of HCS will be described but linking it to the needs of the laboratory will depend on the HCS team.

KEY POINTS

1. HCS represents the integration of diverse skills. In general, scientists working in HCS will have a high level of expertise in a few areas, but will rely on others with complementary expertise to form a robust team.

2. An appreciation of the power of HCS is invaluable, ironically because there are many occasions where a simple and efficient assay is optimal. Such cases are common and will not call on all of the experimental and analytical power of HCS that is available, just a clear vision of the problem and how it can be solved.

FURTHER READING

There are many review articles available that discuss the role of HCS in biology and drug discovery. In addition, the following books are multi-author efforts that present many perspectives and case studies in the practice of HCS.

Haney, S. (ed.). *High Content Screening: Science, Techniques and Applications*. John Wiley and Sons, Hoboken, NJ, 2008.

Inglese, J. Measuring biological responses with automated microscopy. *Methods in Enzymology*, 2006, **414**: 348–363. Academic Press, New York, NY.

Taylor, D.L. et al. *High Content Screening: A Powerful Approach to Systems Cell Biology and Drug Discovery*. Humana Press, New York, NY, 2006.

REFERENCES

1. Eaves, G.N. Image processing in the biomedical sciences. *Computers and Biomedical Research*, 1967, **1**(2): 112–123.

2. Brenner, J.F. et al. An automated microscope for cytologic research: a preliminary evaluation. *Journal of Histochemistry and Cytochemistry*, 1976, **24**: 100–111.

3. Giuliano, K. et al. High-content screening: a new approach to easing key bottlenecks in the drug discovery process. *Journal of Biomolecular Screening*, 1997, **2**: 249–259.

4. Haney, S.A. et al. High content screening moves to the front of the line. *Drug Discovery Today*, 2006, **11**: 889–894.

5. Hoffman, A.F. and Garippa, R.J. A pharmaceutical company user's perspective on the potential of high content screening in drug discovery. *Methods in Molecular Biology*, 2006, **356**: 19–31.

6. Abraham, V.C., Taylor, D.L., and Haskins, J.R. High content screening applied to large-scale cell biology. *Trends Biotechnology*, 2004, **22**(1): 15–22.

7. Evans, J.G. and Matsudaira, P. Linking microscopy and high content screening in large-scale biomedical research. *Methods in Molecular Biology*, 2007, **356**: 33–38.

8. Ramos-Remus, C. et al. Genotoxicity assessment using micronuclei assay in rheumatoid arthritis patients. *Clinical and Experimental Rheumatology*, 2002, **20**(2): 208–212.

9. Smolewski, P. et al. Micronuclei assay by laser scanning cytometry. *Cytometry*, 2001, **45**(1): 19–26.

10. Feng, Y. et al. Exo1: a new chemical inhibitor of the exocytic pathway. *Proceedings of the National Academy of Sciences*, 2003, **100**(11): 6469.

11. Perlman, Z.E. et al. Multidimensional drug profiling by automated microscopy. *Science*, 2004, **306**(5699): 1194–1198.

12. Oakley, R.H. et al. The cellular distribution of fluorescently labeled arrestins provides a robust, sensitive, and universal assay for screening G protein-coupled receptors. *Assay and Drug Development Technologies*, 2002, **1**(1 Pt 1): 21–30.

13. Malo, N. et al. Statistical practice in high-throughput screening data analysis. *Nature Biotechnology*, 2006, **24**(2): 167–175.

SECTION I

FIRST PRINCIPLES

As we get started, we begin with some discussion of the basics of image capture. These include the tools for labeling and visualizing cells, the mechanics of image capture, and the transition to a digital record. For most cell biologists, Chapters 1 and 2 will be pure review, although an effort is made to bring forth some important concepts and properties that can be missed by those with practical but not formal training (i.e., learning microscopy through getting checked out on the lab scope by another student).

The image processing discussion brings out the highly integrative nature of HCS. Understanding the digital nature of an image, and the general concepts behind taking this digital information and reconstructing the shape and texture of the cell are critical events that define the adaptation of images to quantitative cellular measurements. These chapters are presented first, because they cover material that is less protocol based, but serves as parts of the conceptual material upon which the subsequent chapters are based. Of particular importance is the recognition that HCS tracks cells as individual objects and records many facets of each cell's structure. This can be new territory for an assay development scientist, but this is a property of HCS that will receive a lot of attention in the data analysis chapters.

An Introduction to High Content Screening: Imaging Technology, Assay Development, and Data Analysis in Biology and Drug Discovery, First Edition. Edited by Steven A. Haney, Douglas Bowman, and Arijit Chakravarty.
© 2015 John Wiley & Sons, Inc. Published 2015 by John Wiley & Sons, Inc.

SECTION 1

FIRST PRINCIPLES

2

FLUORESCENCE AND CELL LABELING

Anthony Davies and Steven A. Haney

2.1 INTRODUCTION

The high content screening (HCS) process is centered around the fluorescence micro-scope or similar imaging technologies. The fluorescence microscope is a mature and trusted technology that has been used by cell biologists for decades, permitting the study of complex biological processes at the cellular and subcellular levels. With the advent of new generations of probes, markers, and dyes, the biologist now has access to a rich and diverse toolbox offering the capability of visualizing specific cellular and subcellular entities such as organelles and proteins. The use of these fluorescently labeled probes and markers has consistently grown with time. The reasons for this are clear, they offer excellent signal to noise characteristics and can be designed with well-defined excitation and emission characteristics. In addition to this, their toxicity to living cells and tissues is generally low. Fluorescence enables the detection of even low abundance cellular targets as well as offering the capability to multiplex (i.e., use several probes simultaneously) in both living and fixed cells [1,2]. The key advantage of the multiplexing approach is that it allows the biologist to simultaneously monitor the temporal and spatial relationships between multiple cellular targets in an intact biological system [3]. In this chapter we will be dealing with the use of fluorescence in high content analysis and for the purposes of simplicity we will refer to these fluorescent labels and dyes as fluorescent probes.

An Introduction to High Content Screening: Imaging Technology, Assay Development, and Data Analysis in Biology and Drug Discovery, First Edition. Edited by Steven A. Haney, Douglas Bowman, and Arijit Chakravarty.
© 2015 John Wiley & Sons, Inc. Published 2015 by John Wiley & Sons, Inc.

2.2 ANATOMY OF FLUORESCENT PROBES, LABELS, AND DYES

Broadly speaking, a fluorescent probe comprises two separate functional components (graphically diagrammed in Figure 2.1). The first is the fluorophore, a molecule that emits a light signal of defined wavelength(s) when excited by incident light. Fluorophores are characterized by functional groups or chemical moieties that absorb light energy over a narrow range of wavelengths and then reemit energy at longer (slightly redder) wavelengths. The spectral absorption and emission properties of some commonly used fluorophores are shown in Figure 2.2. The amount of light absorbed and reemitted and the spectral characteristics of these compounds depend on both the chemical and physical properties of the fluorophore and also the chemical environment in which these molecules are placed. The second component is the targeting or localizer moiety, that is, a molecule that is bound to the fluorophore which causes the probe to localize to discrete macromolecules or cellular compartments (Figure 2.1). Fluorophores can be attached to a variety of probes or targeting or localizing region such as monoclonal antibodies, ligands, and peptides, all of which can be engineered to bind to specific biological targets.

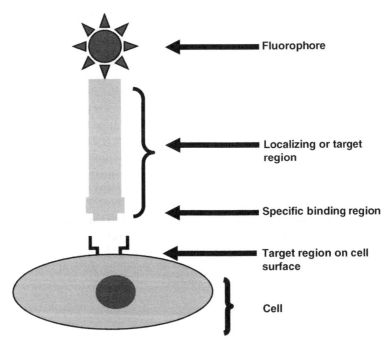

FIGURE 2.1 *Anatomy of a fluorescent dye.* A diagrammatic representation of an example of a fluorescent probe showing the fluorophore, and the interactions between the localizer (with specific binding region (key)) and the target region on the cell (lock).

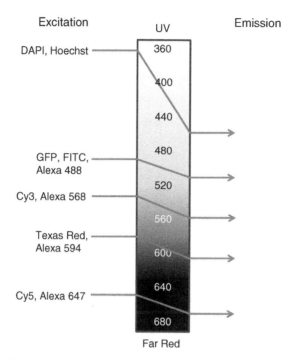

FIGURE 2.2 *Change in excitation and emission spectra of common fluorophores.* A graphical representation of the spectral changes of common dyes is shown in the figure. The greater the slope, the larger the Stokes' shift of the fluorophore. Dyes shown in the figure are generalized to group each by spectral family, which impose restrictions on multiplexing due to these groupings that are based on overlapping excitation and emission wavelengths.

2.3 STOKES' SHIFT AND BIOLOGICAL FLUOROPHORES

Fluorescence was first described by Sir William Stokes in 1852. He observed that the light emitted from minerals such as fluorspar was of a longer wavelength than that of the incident light illuminating these minerals. The term Stokes' shift was subsequently coined in his honor to describe this phenomena [4]. At the atomic level, this shift from wavelength to longer wavelengths occurs when electrons are energized or excited from their ground state by incident light rays. Eventually, these electrons drop back to their stable ground state and rerelease or emit their excess energy as light of a longer wavelength [5]. As the value for Stokes' shift (i.e., the difference between the excitation and emission wavelength) increases, it becomes easier to separate the excitation from the emission light using fluorescence filter combinations (see Chapter 3).

An example of how Stokesian fluorescence is employed is the use of Fluorescein, which is an example of a fluorophore that has been commonly used in biology for many years. Fluorescein is excited at a wavelength of 495 nm and emits light at a wavelength of 517 nm. To achieve the maximum fluorescence intensity and hence the

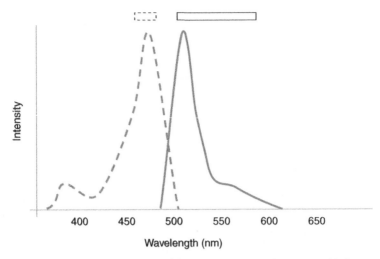

FIGURE 2.3 *Schematic representation of the absorbance and emission of light energy by a fluorescent molecule.* Wavelengths that are captured by a fluorescent molecule, and the extent to which the energy is captured, is shown by the dashed line. Emission of the energy in the form of lower energy, longer wavelength light is shown by the solid line. Boxed regions denote optimized bandpass filter regions for excitation (dashed box) and emission (solid box) regions that maximize the collection of fluorescent light.

best signal possible, it is necessary to excite a fluorophore at the wavelength closest to the peak of its excitation curve. To further enhance signal strength it is also necessary to capture emitted light from the broadest possible range of emission wavelengths including the peak emission for that fluorophore (see Figure 2.3).

Crucially, while the Stokes' shifts (diagramed in Figures 2.2 and 2.3) are peak values, the actual range of excitation and emission wavelengths are broadly distributed (as shown in Figure 2.3). The long tails in these distributions exemplify one of the great limitations to fluorescence microscopy, that relatively few fluorophores can be combined before they interfere with each other (four fluorophores are the maximum that can be combined with the same localization). On the other hand, the spread of a single fluorophore's emission wavelengths and the excitation range of another fluorophore enables some sophisticated, but powerful, fluorescence methods such as Förster resonance energy transfer (FRET), where one fluorophore can excite another when the two are in close proximity (within 10 nm, the distance of two proteins that form a complex). We will introduce this and related methods later in this chapter. Excitation and emission spectra for some common fluorophores are shown in Table 2.1.

2.4 FLUOROPHORE PROPERTIES

To effectively use and troubleshoot fluorescent reagents in high content biology, it is important to understand the basic properties of a fluorophore. In many basic assays, these decisions will not come into play, but as one starts to develop new assays, multiplex several fluorophores, or starts working with the many new variants of fluorescent proteins, these factors become important [3]. Apart from the already mentioned absorption and emission characteristics of a given fluorophore, other properties of fluorophores that should be considered are as follows.

2.4.1   The Extinction Coefficient (Efficiency of Absorption)

In simple terms the extinction coefficient is the efficiency at which a fluorophore absorbs excitation light at a given wavelength. The likelihood of light being absorbed by a fluorophore is proportional to the cross-sectional area of that molecule. Therefore the greater the value for this coefficient the higher the probability that light of a given wavelength is absorbed.

2.4.2   Quantum Yield

The quantum yield of a given fluorophore is defined as the ratio of the number of photons emitted to those absorbed by fluorescent molecules and is expressed in a value range between 0.1 and 1.0. A quantum yield value below 1 indicates that energy is lost or converted into other forms of energy such as molecular vibrations, heat or photochemical reactions, rather than reradiated as light.

At first glance it may seem that the properties mentioned above (namely extinction coefficient and quantum yield), could almost be considered as abstract theoretical concepts with no real relevance in the practical world of cellular imaging. However knowing the numerical values for these for any given dye will permit the side by side comparison of the "brightness" of a given fluorophore, since the *fluorescence intensity* is proportional to the product of the *extinction coefficient* and *quantum yield*.

In short, by applying the simple equation shown below, the HCS biologist can make informed decisions when selecting dyes for their *fluorescence intensity or "brightness"*.

$$\text{Extinction coefficient} \times \text{Quantum yield} = \text{Brightness}$$

2.4.3   Fluorescence Lifetime

The fluorescence lifetime can be defined as the mean time after excitation that molecules stay in their excited state before returning to their ground state. The more commonly used fluorophores transit from the excited to the ground state in approximately 0.5 to 20 nanoseconds. Dyes can be measured for their decay times instead of their Stokes' shifts. Interest in fluorescence lifetime imaging is growing steadily within the high content imaging field, as it offers the capability of visualizing the changes in the fluorescence lifetime properties of dyes and endogenous fluorophores within cells and tissues. This can provide valuable insights into the biochemical environment. In addition, the fluorescent lifetime of a dye is also an important characteristic when considering applications such as fluorescence resonance energy transfer (FRET, discussed below).

2.4.4   Loss of Signal (Fading or Signal Degradation)

Signal loss or fading is a common problem when using fluorescent markers and probes, and it is usually caused by the following.

2.4.4.1   Quenching   The loss of signal due to short-range interactions between the fluorophore and the local molecular environment such as low pH, or in the presence of halogen compounds or other fluorophores. The energy is lost as fluorescence because it is transferred to other molecules that do not produce a signal that can be detected by the microscope. When the transfer occurs to a reagent that can be detected by the microscope, this becomes resonance energy transfer, which is the basis of FRET microscopy, discussed below. It can be possible to reduce quenching through changing the solution composition of the sample.

2.4.4.2   Photobleaching   This is the destruction of the fluorophore due to the presence of reactive oxygen species (ROS). The phenomenon of photobleaching is widely exploited for studying the diffusion and motion of biological molecules in a method known as Fluorescence recovery after photobleaching (FRAP). Aside from its role in specialized microscopy methods, photobleaching can pose serious problems during image acquisition. Either the signal can diminish to a point where detection is difficult or can lead to misleading data because of changes in signal intensity. Photobleaching can be avoided in several ways. The first is to choose a fade-resistant dye such as from the Alexa^{TM} and Dyelite^{TM} series. If the use of a fade-resistant dye is not possible, or proves to be ineffective, the following steps can be taken. (i) Increase sensitivity or gain of microscope detector and reduce proportionally the time that the sample is exposed to the light source, or (ii) reduce or eliminate free radical production within the sample by means of antioxidants or free radical scavengers such as carotenoids [6, 7]. In fixed end-point assays, photobleaching is normally not a problem because each sample is acquired in a single acquisition scan.

2.5   LOCALIZATION OF FLUOROPHORES WITHIN CELLS

As mentioned earlier, a vast number of fluorescent dyes, probes, and markers are currently available. For the sake of simplicity, we will only cover some the main application areas and the most commonly used fluorescent probes.

2.5.1   Nuclear Stains

DNA-binding dyes are almost always used in high content imaging as a means of contrasting cell nuclei. Nuclear staining is essential as an assay endpoint, in addition to its common function as a means for identifying cells during image analysis. Indeed nuclear morphology and the staining intensity of DNA are the mainstay of many classical HCS assays and are universally recognized as sensitive physiological markers. Nuclear stains are unique, in that the chemical structure that is responsible

TABLE 2.1   Commonly Used Fluorophores and Their Excitation and Emission Spectral Characteristics

Fluorochrome	Excitation (nm)	Emission (nm)
DAPI	368	461
Hoechst[a]	350	461
Fluorescein	495	525
Rhodamine	552	570
Texas Red	596	620
Cy2	492	510
Cy3	550	570
Cy5	650	670

These dyes are still in use, but are becoming historical references to describe the properties of newer fluorophores and of filter sets to use them.

[a]The DNA intercalating dye produced by Hoechst that is used most commonly is typically abbreviated with just the company name, but each compound synthesized by the company was numbered sequentially, and the complete name of this dye is Hoechst 33258. Other Hoechst dyes are also used in fluorescence microscopy.

for its fluorescence is also critical for its binding to nucleic acids. In many cases, the interaction alters the fluorescence properties of the fluorophore. For the most useful nuclear dyes, fluorescence increases as the dye complexes with DNA, reducing the need to wash out the dye before imaging.

The nucleus of eukaryotic cells contains both DNA, which is contained within the chromosomes, and RNA which is mainly concentrated within the nucleolus. There are a number of DNA-binding dyes used for imaging, the uses and properties of which are detailed in Table 2.2. In terms of selection of these dyes for HCS, it is important to consider the following.

- Excitation and emission properties of the dye: as the nuclear stain is often the mainstay of most assays particular attention should be paid to this as issues such as spectral bleed-through may lead to errors when performing image analysis. This of course holds for all dyes and stains used in HCS but the nucleus is the most commonly stained cellular compartment. Nuclei stain brightly, and are often the organelles on which images are focused. Counting nuclei is often carried out as a surrogate for counting cells.

- The ability of the dye to pass through the cell membrane: cell permeability is often a desirable feature of any dye as their use is not dependent on permeabilization of the cell prior to administration of the dye. Also, cell-permeable dyes can be used in live cell assays. Conversely, cell-impermeable dyes are also extremely useful as they can be used to assess the health of a cellular population. One feature of acute cellular toxicity is an increase in plasma membrane permeability which can be monitored by measuring uptake of impermeable membrane dyes. Examples of such dyes are TOTO 3, Propidium Iodide and Draq7 (see Table 2.2).

TABLE 2.2   **Common Nuclear Fluorophores**

Dye name	Exc (nm)	Emm (nm)	Cell permeant	Binding mode	Live/fixed applications
Hoechst 33342	350	461	Permeant	Minor groove	Live and fixed cell-based assays
DAPI	358	461	Semi permeant	AT-selective	Fixed cell-based assays
Propidium iodide	530	625	Impermeant	Intercalator	Fixed, permeabilzed, identification of dead cells
Draq5	647	670	Permeant	Minor grove	Fixed, permeabilized and live cell based assays
Draq7	633	694	Impermeant	Minor grove	Fixed, Permeabilzed, identification of dead cells
TOTO3	640	660	Impermeant	Bis-intercalator	Fixed, permeabilzed, identification of dead cells

- Finally, a very important factor when selecting a nuclear stain is its target within the nucleus. Most nuclear stains (but not all) bind specifically to DNA (see Table 2.2); the region of the DNA molecule to which the stain binds can be of critical importance depending on the assay application. The binding mechanism of any dye only becomes an issue if other competing dyes or molecules are also used in the experiment. For example, care should be taken when using drugs such as anthracyclines [8] (e.g., doxorubicin) and dyes such as Hoechst 33342, as they both preferentially occupy the minor groove of the DNA molecule which can result in one molecule displacing another. This can result in dimly stained nuclei for samples where a high concentration of doxorubicin is used.

2.5.2   Fluorescent Proteins

Fluorescent proteins (FP) such as GFP have been widely used for imaging biological processes that occur at the cellular and subcellular levels for many years. GFP was first isolated in the early 1960s where it was purified from the jelly fish *Aequorea victoria* or more commonly known as the Crystal Jelly. In 1992, the potential of GFP as a tool for biomedical research was first realized when Douglas Prasher [9] reported cloning and nucleotide sequence of wt GFP. Soon after, wt GFP was expressed in bacteria and *Caenorhabditis elegans* where the properties of this protein were first characterized. It was found that the GFP molecule folded and was fluorescent at room temperature, and more importantly without the need for exogenous cofactors specific to the jellyfish. Despite the success of these studies it was discovered that using wt GFP was not without technical drawbacks, which included dual peaked excitation spectra, pH sensitivity, chloride sensitivity, poor fluorescence quantum yield, poor photostability and poor folding at 37°C. Suffice it to say that many of the problems associated with the use of fluorescent proteins have now been resolved and indeed the colour palate of these proteins has expanded tremendously (see Table 2.3 and [10] for examples).

The advantage of using fluorescent proteins over the more conventional dyes and stains is that they can be introduced and expressed by living cells and tagged onto functional proteins. This capability allows for the real-time monitoring of functional changes in the spatial distribution and expression activity of specific targets. Fluorescent proteins allow, for example, monitoring the turnover of specific cellular proteins over a wide range of times and experimental conditions, offering more informative data regarding the physiological role of a given target molecule(s). Technically, these proteins can be used both as markers and specific gene reporters, with the added benefit that once transfection has been achieved very little is required in the way of assay preparation. Indeed, fluorescent proteins can be used both in live cell and fixed end-point assays [11]. Intrinsically fluorescent proteins are most common, but some proteins are ligand-binding domains that can complex with fluorescent ligands. These ligands can be added to cells during the course of the experiment or during sample preparation as other fluorescent reagents would be, giving some greater experimental options.

While the differences between fluorescent proteins are complex and subtle, there are some general guidelines in considering the selection of one or more fluorescent proteins. The first consideration for choosing a fluorescent protein is not that different than one would consider when choosing a fluorescent label: excitation and emission characteristics of the protein in question. As seen in Table 2.3, there is a wide selection of fluorescent proteins available extending beyond the visible spectrum. The next consideration would be the brightness (see the definition earlier) and photostability. One factor that is peculiar to fluorescent proteins is that some will have issues surrounding oligomerization and its potential for inducing toxicity of the fusion protein. In many cases, the wild type fluorescent proteins naturally aggregate into dimers or tetramers the consequence of this is that proteins in this state may have a tendency to form inclusion bodies within the cell. The disadvantage of this is that the protein aggregates will not function as desired and the precipitated proteins will exert toxic effects. Thankfully many of the commercially available fluorescent proteins have been engineered to prevent formation of these protein dimer or tetramer complexes (see Table 2.3).

For many fluorescent stains and dyes, performance is often dependent on the environmental conditions. One significant environmental factor is local pH. Many fluorescent proteins are sensitive to low pH. Care should be taken when selecting fluorescent proteins for use in cellular compartments where pH is typically low such

TABLE 2.3 Examples of Fluorescent Proteins and Their Physical Properties

Spectral classification	Protein name	EAcitation (nm)	Emmission (nm)	% Relative brightness in spectral class[a]	% Relative photostability[b]	>pKa	Oligomerization
Far Red	m-Plum	590	649	5.942	101.9231	<4.5	Monomer
Red	m-Cherry	587	610	23.1884	1846.154	<4.5	Monomer
Orange	m-Orange	548	562	71.0145	173.0769	6.5	Monomer
Yellow-green	EYFP	514	527	73.913	1153.846	6.9	Weak dimer
Green	EGFP	488	507	49.2754	3346.154	6.0	Weak dimer
Cyan	CyPet	435	477	26.087	1134.615	5.0	Weak dimer
UV-excitable green	T-Sapphire	399	511	37.6812	480.7692	4.9	Weak dimer

[a]Percentange brightness.

[b]Stability of proteins was calculated from intensity and f_{time} values (f = photobleaching) derived from free fluorescein which we have given an arbitrary value of 100%.

as the lumen of lysosomes or in secretory granules. In these situations the general rule of thumb would be to use fluorescent proteins with a relatively low PKa value such as m-plum or m-cherry (which have PKa below 4.5) and avoid those with higher PKa such as YFP or m-orange [16]. One drawback of fluorescent proteins is that appropriate expression of these proteins is a prerequisite for their use. This can often result in lengthy transfection optimization steps which can be compounded by variable cellular expression. High content applications can mitigate some of these difficulties due to the capability of this technology to enable subpopulation analysis (i.e., to isolate those cells which are expressing fluorescent proteins at appropriate levels).

2.5.3 Localization Agents

As mentioned earlier, broadly speaking, fluorescent probes comprise of two distinct functional units: the fluorophore and the localising or targeting regions, see below for examples of these. Nucleic-acid-binding fluorophores are the exception to this differentiation between moieties within the molecule. In most other cases, the fluorophore must be linked to a localization agent. The most common localization agents are the following.

2.5.3.1 AM Ester Groups   These molecules are uncharged lipophilic molecules that can easily permeate cell membranes. Once inside the cell, the lipophilic groups become modified by nonspecific cellular esterases that cleave these molecules. This cleavage results in a polarized form of the molecule that is less able to traverse the plasma membrane again. Hence the molecule leaks out of the cell at a vastly slower rate. These types of dyes are effective for labelling the cytoplasm of the cell.

2.5.3.2 AM Ester Groups and Their Use with Ion Indicators   One area where the use of AM esters has been extremely important is with the use of intracellular ion indicators such as Fluo4. With these dyes, this mechanism of hydrolysis of the esterified groups can be employed to both increase the polarity of the molecules, to ensure retention within the cellular cytoplasm, and to activate the probe molecule such that it becomes responsive to its target ion for binding. An example of such a molecule is calcein AM, which contains an acetomethoxy group which masks the part of the molecule that chelates divalent cations such as Ca^{2+}, Mg^{2+}, Zn^{2+}. Once the acetomethoxy group becomes cleaved, the dye becomes fluorescent, and the intensity of fluorescence is directly related to the concentration of its target ions. Hence the AM molecule allows the dye to be sequestered within the cell but only the activated form of the dye will fluoresce strongly, reducing potential background fluorescence from the free dye contained within the fluids bathing the cells.

2.5.3.3 Ligands, Including Toxins   A number of highly specific ligands have been used as localization agents. Generally, these are toxins that act as inhibitors.

Probably the most well-known agent in this class is phalloidin, a fungal toxin that binds to actin [12]. Phalloidin-conjugated fluorescent reagents are used extensively in imaging studies, including a counterstain to provide a cellular context for other dyes and in phenotypic or cytological profiling studies, as actin morphology is affected by many signalling pathways. Other localization agents include ligands for cell surface receptors, GPCRs in particular [20]. Dyes conjugated to these ligands can be used to study the organization of receptors on cell surfaces and how the receptor responds to activation.

2.5.3.4 Antibodies Antibodies are one of the most commonly used molecular targeting systems. These proteins can be large, compared to their targets, as they consist of heavy and light chains which combine to form the antigen-binding sites. In general, the structure of all antibodies are very similar, apart from the antigen-binding site which is known as the hypervariable region. It is the variability found in these regions that allows for the production of antibodies with millions of different antigen-binding characteristics. The region of the antigen that is recognized by the antibody is called the epitope, and can be as short as six amino acid residues. These epitopes are selectively bound by antibodies which carry complimentary binding motif in their hypervariable region in a highly specific interaction which is termed an induced fit [14].

Antibodies provide an excellent means of labeling cells and tissues as they bind stably to their target. Immunofluorescence is a technique that involves the labeling of cellular proteins with primary antibodies and a fluorochrome-conjugated secondary antibody (indirect method) or a directly conjugated primary antibody (direct method) [14]. Indirect immunofluorescence can use an antibody against the target protein, or against an epitope that has been added to the target protein sequence as a fusion protein.

2.5.3.5 Sequence-Specific DNA Hybridization One application that is particularly well suited to for use with HCS technologies is the cytogenetic technique Fluorescence *in situ* hybridization (FISH). This method is used widely in the study and detection and localization of specific DNA nucleotide sequences [15]. FISH is a standard assay for detecting chromosomal abnormalities and gene amplifications [16]. This technique is based on the detection and localization of exogenously applied fluorescently labeled DNA probes which carry nucleotide sequences complimentary those of interest and hence specifically bind to these regions.

2.5.4   Issues that Affect Fluorescent Reagent Choice

For the most part, commercial reagents are highly robust, and these considerations become more important when "going off the ranch," such as selecting dyes to complement a specialized assay, such as a FRET pair.

2.5.4.1 Dye Stability (also see Section 2.4.4.2) Photostability of the dye (i.e., how resistant is the dye to photobleaching), should always be considered when developing an HCS assay. However, it should be noted that many of the newer dyes on the market are now quite stable and remarkably resistant to bleaching and fading (e.g., the Alexa^{TM} and DyLight^{TM} fluorophores). An example where photostability of a dye is critical would be during a live cell experiment where it may be necessary to acquire multiple images of the same cell. Also during larger screens if it is often necessary to store your stained cells for appreciable amounts of time before acquiring images (e.g., if you have large numbers of plates to be scanned).

The second factor is, for want of a better phrase, diffusivity. Typically, diffusion of a dye may result from either a breakdown of the probe and fluorophore complex (e.g., florescent conjugate separating from the antibody), resulting in the diffusion of the dye away from the target into the cell or the whole dye/probe complex detaching for its target. Diffusion of dyes can lead to significant experimental error especially in experiment where the spatial distribution of the dye needs to be quantified (e.g., a cytoplasm to nucleus translocation assay). Again, this is typically not a problem for commercial reagents, but will be a concern in cases where a custom fluorescent ligand is being synthesized.

2.5.4.2 Toxicity of the Dye or Probe Factors

Toxicity is an issue if using a fluorescent dye with live cells, general toxicity would (as the name suggests) be anything that adversely effects the health of the cell. Toxicity can result from properties of the fluorophore or of the localization agent. Dye toxicities can be placed into two categories; the first we can describe as chemical toxicity. An example of a chemical toxicant would be phalloidin which is a commonly used reagent for staining cytoskeleton. The toxicity of this dye emanates from the ability of this molecule to bind irreversibly to filamentous actin (f-actin), hence effectively blocking the cells ability to modify or remodel its cytoskeleton (a process necessary for maintaining normal cell function). Nonspecific dyes, such as those that form esters to cellular proteins generally (such as CMFDA) will also affect cellular functions. These are very valuable methods for labeling cells; rather than avoid using them, it is instead most beneficial to titrate the labeling treatments to the minimal concentration or time necessary to achieve good labeling.

The second type of toxicity seen when using dyes with live cells is phototoxicity. This occurs when the florescent dye or probe reacts with light to form a toxic chemical species such as oxygen free radicals. Examples of this are the membrane dyes PKH2 and PKH26 [17]. It should also be noted that these effects are often cell type specific. Dye or label toxicity is a constant feature when performing live cell assays, experimental strategies will be discussed in greater detail in Chapter 11.

2.5.4.3 Ability of Molecule to Reach Its Target

Typically, this is only an issue if the target of interest lies within the cell itself and in live cell work where the cell membrane must remain intact during the period when the cells need to be labeled. When considering the ability of a molecule to reach its target within the living cell, one should be aware of both the physical and chemical characteristics. At first glance it would seem that the larger the size of a dye or probe, the less easily that molecule will pass into a cell by passive diffusion. However, it is not a linear relationship, and other factors are important.

(i) The 3D shape and conformational flexibility of the molecule can have a dramatic effect on how easily it will pass though the lipid bilayer of the cell. For instance, it is known that molecules which have an elongated chain-like structure diffuse much faster through the cell membrane than a spherical-shaped molecule of a similar size.

(ii) In addition to the passive movement of a dye or probe into the cell by simple diffusion down its concentration gradient, dyes and probes can also be taken up by the cells own transport mechanisms, for example, cytosis (phagocytosis and/or pinocytosis), active transport (e.g., primary and secondary), and facilitated diffusion. This process can often be facilitated by conjugating dyes to plasma membrane transport substrates such as hexose sugars, peptides, etc., which then facilitate dye uptake by means of the cells own endogenous transport mechanisms.

(iii) Also the chemical makeup of the dye strongly affects the ease with which a dye or probe will penetrate the lipid membranes of cells, for example, non-polar molecules in general are much more lipophilic that polar molecules. Interestingly, the lipophilicity of a dye molecule can be modulated or *switched* effectively trapping it internally within the cell. A good example of how a switch in charge can be utilized is the dye calcein AM (see above).

These are properties that can be researched or candidate labeling reagents can be tested in the lab to determine their suitability.

2.6   MULTIPLEXING FLUORESCENT REAGENTS

Multiplexed fluorescence labeling is one of the most important features in HCS. This not only provides context for a specific target within the cell, by defining it relationship to the nucleus and cytoplasm or cell membrane, but provides the capability of simultaneous monitoring of multiple cellular targets using discrete spectral channels or signals emanating from the same biological test sample. It these cases, activation of one pathway or response can be correlated with another, and on a cell-by-cell level, enabling the characterization of correlated or mutually exclusive events.

A multiplexed fluorescence assay involves the use of multiple fluorescent dyes with distinct spectra, which are excited and detected independently. Multiplexing offers many advantages—it allows for the conservation of sample material as several independent assays can be performed on the same sample. In addition, multiplexing improves the quality and the relevance of information gathered from cellular assays, allowing for a better understanding of complex biological processes under study.

Ultimately, multiplexing offers advantages in both the quantity and quality of data, as well as confidence in the results. This is because two independent assays can be run in the same cell, permitting the use of Boolean criteria (e.g., the agent must activate the STAT3 pathway and not activate the NF-κB pathway). Multiplexing a complex phenotypic response with an assay for a specific pathway can also be used to screen for an integrated response rather than a single event.

On occasion, creating a multiplexed assay can be labor intensive, but this it typically the case only when one of the reagents is of poor quality (such as a primary antibody that binds weakly) or a target protein that is expressed at very low levels. Otherwise, HCS platforms can readily accommodate three separate assays (with the fourth typically being a nuclear stain that is essential for most image analysis algorithms). This includes separate optimization steps for each assay/channel, including sample exposure, segmentation, and quantification. How the data from these assays is handled will be discussed later, but sticking to the discussion of combining separate fluorescence reagents, if the two or three assays are robust when prepared individually, they can typically be combined without presenting a serious challenge. The major exception to this is when two primary antibodies are generated in the same species. If the antibodies are monoclonal, then checking the isotype of the antibody is important, as isotype-specific secondary antibodies are commercially available. If this is not a possibility, then labeling one or both of the primary antibodies may be required. If only one is labeled, the unlabeled antibody will need to be added first to the sample, then the secondary antibody is added. Each step will need to be washed well, then the labeled primary antibody is added, with a final washing step. This type of protocol suggests that there are alternatives to logistical conflicts that require a little bit of creativity.

The upside of all of this troubleshooting is that the power of HCS itself becomes amplified. By combining multiple fluorescent reagents, the capacity to define individual cells by several properties (be they coordinated or discrete) adds significant context to cellular responses. The generally heterogeneous behavior of cells in culture can be better appreciated (see Figure 2.4). Such methods open the door to monitoring multiple pathways, understanding the coordination of a single signaling pathway with phenotypic responses such as differentiation or apoptosis, as well as multiparametric profiling and clustering approaches, all discussed in later chapters.

2.7   SPECIALIZED IMAGING APPLICATIONS DERIVED FROM COMPLEX PROPERTIES OF FLUORESCENCE

Given the discussion of fluorophores and their properties, it is worth taking a moment to briefly describe a couple of unique imaging applications that leverage these properties quite elegantly, and to significant scientific gains. Förster resonance energy transfer (FRET) and fluorescence lifetime imaging (FLIM) are two techniques employed to study protein–protein interaction. Although they have the reputation of being

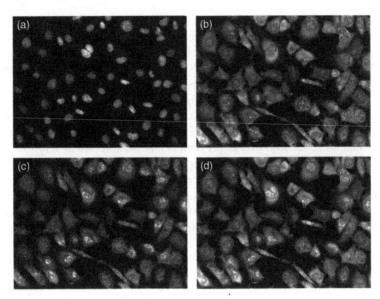

FIGURE 2.4 *Multiplexing of fluorescent reagents.* (a). Nuclear staining. (b) and (c). Immunostaining for two different cellular proteins. (d). A composite image of the three previous frames. The additive nature of the individual images can be appreciated in the monochrome composite, but in an image analysis application, each individual image would be used to create a profile of each individual cell and the composite image can be represented in three colors to highlight the individual phenotypes of the individual cells.

technically advanced and difficult to implement, many investigators use these techniques routinely [18, 19], although FLIM does require some nonstandard hardware, as discussed below.

2.7.1   Förster Resonance Energy Transfer

FRET (although also described as fluorescence resonance energy transfer), describes a process of energy transference between two fluorophores. These fluorophore (or FRET) pairs can be divided into two classes of molecules: the donor and the acceptor. The donor is excited at its optimal wavelength. If FRET occurs, some of the fluorescence energy is captured by the acceptor protein, and emits light at its emission wavelengths, as diagrammed in Figure 2.5. For this to occur, the two molecules must be very close to each other, and therefore, FRET is an excellent means of determining the proximity and interactions of biomolecules within cells.

For energy transfer to occur the donor and acceptor must be within a radius of less than 10 nm of each other. If the donor and acceptor are greater than 10 nm apart, there is no sufficient energy transfer and the acceptor will not fluoresce. The advantages of FRET are clear when one considers the resolving power of the light microscope which is approximately 200 nm [20]. In order for energy transfer to occur it is necessary that the emission spectrum of the donor fluorophore overlaps

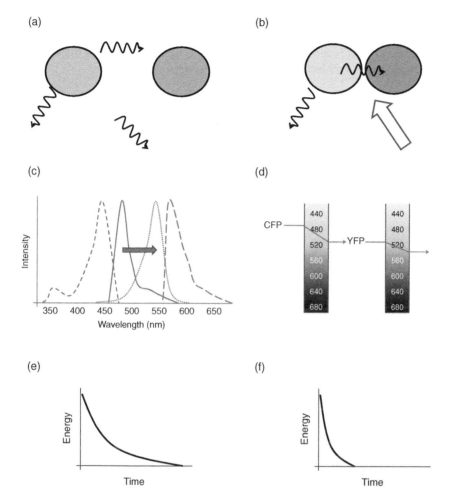

FIGURE 2.5 *Coupling of donor emission and acceptor excitation wavelengths to produce a signal in FRET and FLIM.* When two fluorophores are far away, typically 10 nm (or the diameter of a protein molecule) their fluorescent properties are independent of each other, as in (a). When they are in close proximity (b), the emission of one protein, the donor (in light shading) is captured by a second acceptor protein (in darker shading), highlighted by the arrow. This transfer is diagrammed schematically in (c), where the emission spectra of the donor fluorophore overlaps with the excitation spectra of the acceptor fluorophore. Referring to the previous figure on the change in wavelengths through the Stokes' shift for each fluorophore, the relationship between the excitation of the donor fluorophore and emission of the acceptor fluorophore are diagramed for the CFP/YFP FRET pair is shown in (d). FLIM is similar to FRET in many aspects, except that instead of measuring the luminosity of the acceptor fluorophore in FRET, the effect of energy transfer on the donor is measured (i.e., the loss of energy from the donor is accelerated because of the transfer to the acceptor). In this case, it is not luminosity or intensity, but lifetime of the donor fluorophore. In (e), the normal decay lifetime is shown schematically. In (f), the transfer in energy from the donor reduces the lifetime decay pattern of the donor emission. Thus, interaction is measured as the reduction in emission lifetime of the donor fluorophore.

the excitation spectrum (absorbance) of the acceptor fluorophore. Some examples of typically used fluorophores are BFP-YFP, CFP-YFP, GFP-DsRed, GFP-Cy3, GFP-m-Orange, YFP-RFP, and Cy3-Cy5, but new molecules are introduced frequently, so the optimal pairs of molecules is always being revised.

2.7.2 Fluorescence Lifetime Imaging/Förster Resonance Energy Transfer

Fluorescence lifetime the critical measurement for FLIM, images cells according to the decay rate for the fluorophore rather than its emission wavelength. Measuring the lifetime of a dye that has been affected by an energy transfer leads to the hybrid FLIM/FRET microscopy. The key advantage of FLIM/FRET over intensity-based FRET is that it is calibrated internally which allows for measurements to be made independently of excitation intensity or fluorophore concentration within the sample. It has been demonstrated that FLIM can provide substantially higher spatial and temporal resolution when compared to the more conventional steady state FRET imaging. Finally, the problems associated with spectral bleed-through are avoided in FLIM imaging as only the donor fluorophore lifetime is measured [21]. If an interaction has occurred that results in energy transfer, the effect is measured as a change in lifetime of the donor molecule, the acceptor functions as a quencher that is specific for the donor molecule (diagrammed in Figure 2.5 E and F).

2.8 CONCLUSIONS

The examples covered in this chapter highlight the wide array of options for labeling cells and their components. These options allow for labeling of both fixed and live cells, as well as different cell types in the same well. The breadth of options in labeling cells carries significant implications for the options in analyzing the data generated in an HCS assay. This includes integrating several events as a single assay through multiplexing, and capturing multiple measurements of the cellular architecture in a phenotypic profiling study. These topics will be covered in depth in later chapters, but the foundation for all of these studies lie in the ability to identify the components of a cell specifically and robustly.

Much of this information is essential for the development of robust cell-based assays, and its appreciation can be important, even in cases where commercially available reagents have already considered the points described here. Not all reagents will perform as advertised, and an understanding of the options for labeling cells and their properties can help in diagnosing problems and developing alternatives. In addition, commercial reagents tend to be developed and characterized in relatively narrow experimental designs (such as plating a single cell line at a low cell density, making many of the imaging and analytical challenges easier). Conversely, expertise in HCS opens up many experimental options, such as primary tumor cells in cancer studies or primary cells that can be differentiated from stem cells in developmental and metabolic studies. The ability to construct and deconstruct a fluorescence-based imaging assay is critical to these expanded options.

KEY POINTS

1. The choice of fluorescent markers available for assay development relates to the HCS instrument configuration: illumination source, filter sets. Since transfer of an assay across instruments is common, planning for mutual compatibility can save time reformatting the labeling conditions later.
2. While it is common for HCS assays to be purchased or adapted from other scientists, a firm understanding of the properties of fluorescence and of commonly used fluorophores can increase the robustness of the assay and focus troubleshooting efforts.

FURTHER READING

The Optical Microscopy Primer website is an excellent resource for concepts around microscopy, digital imaging, and fluorescence. The www site can be found at http://micro.magnet.fsu.edu/primer/index.html, accessed August 13, 2014.

Invitrogen has a very useful applet which allows one to graphically view the spectral properties of Invitrogen-offered fluorophores. You can also overlay illumination peaks of common laser sources as well as bandpass filter properties. The www site can be found at http://www.invitrogen.com/site/us/en/home/support/Research-Tools/Fluorescence-SpectraViewer.html, accessed August 13, 2014.

REFERENCES

1. Inglese, J., ed. *Measuring Biological Responses with Automated Microscopy*. Methods in Enzymology. Vol. 414. 2006, Academic Press: NY, NY. 348–363.
2. Tsien, R.Y., Ernst, L., and Waggoner, A., *Fluorophores for Confocal Microscopy: Photophysics and Photochemistry Handbook of Biological Confocal Microscopy*, J.B. Pawley, Editor 2006, Springer US. p. 338–352.
3. Howell, B.J., Lee, S., and SeppLorenzino, L. Development and Implementation of Multiplexed CellBased Imaging Assays. *Methods in Enzymology*, 2006, **414**: 284–300.
4. Lakowicz, J.R. and Masters, B.R. Principles of fluorescence spectroscopy. *Journal of Biomedical Optics*, 2008, **13**: 029901.
5. Noomnarm, U. and Clegg, R.M. Fluorescence lifetimes: Fundamentals and interpretations. *Photosynthesis Research*, 2009, **101**: 181–194.
6. Pawley, J.B., *Handbook of biological confocal microscopy* 2006: Springer Verlag.
7. Kong, X., et al. Photobleaching pathways in single-molecule FRET experiments. *Journal of the American Chemical Society*, 2007, **129**(15): 4643–4654.
8. Bogush, T., et al. Direct evaluation of intracellular accumulation of free and polymer-bound anthracyclines. *Cancer chemotherapy and pharmacology*, 1995, **35**(6): 501–505.
9. Prasher, D.C., et al. Primary structure of the Aequorea victoria green-fluorescent protein. *Gene*, 1992, **111**(2): 229–233.

10. Day, R.N. and Davidson, M.W. The fluorescent protein palette: tools for cellular imagingw. *Chemical Society Reviews*, 2009, **38**: 2887–2921.

11. Shaner, N.C., Steinbach, P.A., and Tsien, R.Y. A guide to choosing fluorescent proteins. *Nature Methods*, 2005, **2**(12): 905.

12. Cooper, J.A. Effects of cytochalasin and phalloidin on actin. *Journal of Cell Biology*, 1987, **105**: 1473–1478.

13. Kuder, K. and Kieć-Kononowicz, K. Fluorescent GPCR ligands as tools in pharmacology. *Current Medicinal Chemistry*, 2009, **15**: 2132–2143.

14. Harlow, E. and Lane, D., *Using antibodies: A laboratory manual* 1998, Cold Spring Harbor, NY: CSHL press. 495.

15. Wagner, M., Horn, M., and Daims, H. Fluorescence in situ hybridisation for the identification and characterisation of prokaryotes. *Current opinion in Microbiology*, 2003, **6**(3): 302–309.

16. NCCLS, *In Situ Hybridization (FISH) Methods for Medical Genetics; Approved Guideline.* (NCCLS documentMM7-A (ISBN 1-56238-524-0)).

17. Oh, D.J., et al. Phototoxicity of the fluorescent membrane dyes PKH2 and PKH26 on the human hematopoietic KG1a progenitor cell line. *Cytometry*, 1999, **36**(4): 312–318.

18. Pietraszewska-Bogiel, A. and T.W. Gadella, *FRET microscopy: From principle to routine technology in cell biology*, 2011.

19. Becker, W., Fluorescence lifetime imaging–techniques and applications. *Journal of Microscopy*, 2012, **247**: 119–136.

20. Dunn, K.W., Kamocka, M.M., and McDonald, J.H. A practical guide to evaluating colocalization in biological microscopy. *American Journal of Physiology and Cellular Physiology*, 2011, **300**: C723–742.

21. Chen, Y., Mills, J.D., and Periasamy, A. Protein localization in living cells and tissues using FRET and FLIM. *Differentiation*, 2003, **71**(9–10): 528–541.

3

MICROSCOPY FUNDAMENTALS

STEVEN A. HANEY, ANTHONY DAVIES, AND DOUGLAS BOWMAN

3.1 INTRODUCING HCS HARDWARE

3.1.1 The HCS Imager and the Microscope

The HCS imager is an automated microscope that includes mechanical devices to control precise movements of the objectives, the samples (typically arrayed in a microplate), filter cubes, light sources, and cameras as needed. Adjustments to camera exposure times are typically done electronically rather than mechanically. While there are many moving parts to an HCS imager, they all move around an optical microscope, both figuratively and literally. Since the operation of a microscope is so central to the concept and operation of an HCS imager, it is important to fully understand the operation of a modern fluorescence microscope before beginning to discuss the extended steps of image processing and analysis. This chapter will provide an introduction to the microscope, and will focus on critical aspects of microscopy that impact HCS, such the practical limits of resolution, spectral bandwidth and the ability to multiplex several wavelengths. Every commercialized design of HCS imager has many specific elements to its use and design that strike a balance between flexibility and ease of use. That said, all versions do essentially the same thing: capture images of cells across many wells, usually in several fluorescent channels, and in sufficient numbers per well to allow a rigorous, automated, quantitative analysis of small molecule or biological perturbations in an automated manner.

Historically, a discussion of microscopy fundamentals has combined the factors that influence the quality of the science with the care and maintenance of the microscope itself. This was regarded as essential for two reasons, one practical and one

An Introduction to High Content Screening: Imaging Technology, Assay Development, and Data Analysis in Biology and Drug Discovery, First Edition. Edited by Steven A. Haney, Douglas Bowman, and Arijit Chakravarty.
© 2015 John Wiley & Sons, Inc. Published 2015 by John Wiley & Sons, Inc.

philosophical. The practical reason was that a high quality fluorescence microscope was a significant investment for an academic investigator or department. A single microscope can support the research of half a dozen or more researchers simultaneously, and therefore its proper use is embedded in the shared interests of a research laboratory. New users have been trained in the use and maintenance of the group microscope in order to make sure that their actions do not negatively affect others. The philosophical reason is that an understanding of the rules of maintenance and use are linked to a principles-based understanding of microscopy itself, and the considerations need to be taken when embarking on a study of the role of a newly discovered protein or regulatory process. For proper experimental data collection over the expected lifespan of the microscope, it is important to know that mistreating the lamp by failing to start it and shut it down properly can erode its effective lifespan, that some objectives need to be calibrated when they are installed, and that scrupulous clean-up and maintenance of objectives (especially oil or water objectives) is essential. These common-sense precautions can minimize the risk that an old lamp can cause unpredictable artifacts in illumination leading to problems such as an uneven background, or that experiments need to be put on hold because improper cleaning of the objectives has ruined them.

High content screening changes the discussion of microscopy fundamentals, because many parts and settings of the microscope are either completely shielded from the user or are accessed through a computer software interface comprised of checkboxes and/or pulldown menus. That said, good microscopy practices are still essential, no HCS setup is completely fool-proof. A good grounding in microscopy fundamentals becomes increasingly important when problems occur. Although most troubleshooting and maintenance issues might be handled by a service manager from the imager vendor or the platform manager (who coordinates maintenance, informatics, and instrument time for end users), successful resolution of any technical issues ultimately depends on good communication between end user and service provider and/or platform manager. It is essential that the end user has a basic understanding of the critical factors involved in the setting up of a microscope and how the setup of the instrument may effect accuracy and precision of the system and hence the quality of the data collected.

3.1.2 Common uses of HCS that Require Specific Hardware Adaptations

Some types of HCS microscopy that entail specific instrument configurations are the following.

3.1.2.1 Fluorescence Microscopy HCS was originally developed for fluorescence-based detection of cell markers that were critical to drug discovery. In this regard, it may initially be considered unusual to describe fluorescence microscopy as an adaptation of HCS hardware. However, there are different approaches to enabling fluorescence microscopy, and these will affect the options that are readily available, how much work it takes to set up an experiment and whether there will be any limitations to multiplexing fluorescent reagents. In addition to basic fluorescence microscopy, adaptation of the HCS platform to enable extensions of fluorescence microscopy, such as FRET and FLIM (introduced in Chapter 2) has been accomplished.

3.1.2.2 Confocal Microscopy   Confocal microscopy is an adaptation of fluorescence microscopy that eliminates out of focus light from contributing to the exposure. This optical feat is achieved by directing the light through a small pinhole which eliminates background light from other optical planes. This narrows the illumination and detection plane to a relatively thin section through the cell. Two common confocal technologies are: (1) point scanning where the laser illumination beam is scanned in an XY pattern through a single pinhole to a photomultiplier tube to generate an image and (2) a Nipkow spinning disk, which contains an array of pinholes that project the confocal image to a CCD camera. Several HCS platforms are capable of confocal microscopy, but not all instruments can be adapted for it, so a decision about whether it will be needed must be made at the time the instrument is purchased. As compared to standard HCS systems, confocal HCS units are different in the way that light to and from the sample are gated (see discussion on filter cubes, below). Also, very high quality objectives are required for confocal microscopy. To make matters more complicated still, some commercial implementations of confocal microscopy are not truly confocal and may not suit all applications. If confocal microscopy is desired, it is necessary to spend enough time looking into the options available.

3.1.2.3 3D and Image Deconvolution   While related to confocal imaging, the distinction of 3D imaging is typically invoked when multicellular bodies or structures are analyzed. While 3D imaging can be achieved using normal fluorescence and objectives with long working distances, it can also employ other illumination methods, such as two-photon fluorescence, which can penetrate deep into tissues. Deconvolution is a software-based method for removing out-of-focus light. It takes advantage of the point spread function, which describes how the microscope objective blurs a point object, and reassigns light to its origin effectively removing out-of-focus light. Deconvolution can be used to increase the resolution of widefield and confocal images, and is frequently used when multiple images in a Z-stack are combined into a single 3D image. It can also increase the signal-to-noise ratio within the image, resulting in potentially better image segmentation results. These methods are not high throughput, but they are starting to appear in commercial HCS instruments.

3.1.2.4 Brightfield Microscopy   Brightfield microscopy includes general widefield microscopy, phase contrast, and differential interference contrast (DIC) capabilities. Although brightfield microscopy has been around since the time of the first microscope, it is a relatively new adaptation for HCS and presents several technical adjustments. The first adjustment is to the hardware itself. Fluorescence microscopes used in HCS are epifluorescence designs (where the excitation and emission light paths both pass from beneath the sample and are split using the dichroic mirror), whereas brightfield microscopy is a transmitted light design, and an illuminator needs to be attached to the top of the instrument. Many instruments include the option for brightfield microscopy, which includes an additional housing that contains the brightfield lamp and is fitted over the stage. This also requires appropriate clearance to allow for plate handling robots to access the stage.

Analysis of brightfield images can generally be performed by standard applications that are used for fluorescence microscopy, but it requires understanding of what a brightfield image represents, particularly with respect to intensity measurements. In fluorescence microscopy, more intense staining generally corresponds to higher quantity or concentration of the stained entity, for example, of a protein bound to a fluorescently tagged antibody. Phase contrast, by definition, is sensitive to borders between phases or materials (particularly membranes and fluid phases) and such boarders are visualized as changes in intensity. Thus, intensity measurements in brightfield images provide information about shape, orientation, and organization parameters only. There are applications where cellular matter can be quantified as dark regions in a brightfield image, such as precipitates of nonfluorescent dyes, such as X-gal. These are limited by the contribution of the cell mass and borders that will also register as darker areas, so careful image analysis is needed to specifically quantify desired regions.

3.1.2.5 *Live-Cell Imaging*

Live-cell imaging encompasses many types of studies, from transient events, such as calcium fluxes that occur on the timescale of minutes or seconds, to cell motility and cell division processes that take place over hours and days. These different applications place different demands on the imager. For transient events, the imager will capture images in rapid succession and the experiment will have to be read-out well-by-well in a sequential manner, where initiation of cell signaling through the addition of a ligand being performed on the imager itself through the use of integrated automated liquid handling. For events that occur over longer timeframes, the imager will scan the entire plate at the beginning of the assay and revisit the well locations repeatedly until the assay is complete. Each type of live-cell imaging presents its own sets of challenges. Keeping cells alive and healthy while being imaged, as well as tracking cells (that might move) over time, requires adaptations to the basic HCS instrument as well as its software. Adaptations to the instrument include onboard liquid handling for transient events (discussed above), as well as temperature, CO_2, and humidity control, such that the imager is able to manage the environment of the cells as if it were a cell culture incubator. For long-term studies, such as motility of cells or growth patterns of neurites over one or several days, revisiting wells at the exact location where a previous image was taken is possible on many instruments, but microscope stage and objective movements need to be engineered with tight tolerances for this to work well; some instruments are better engineered for this kind of study than others. Beyond the operational requirements for such studies, image analysis becomes more complex, as spatial and temporal links between images of the same cell taken at different times need to be tracked and measured.

One of the obvious limitations for live-cell imaging is the inability to fix cells and stain them for indirect immunofluorescence readouts. GFP fusion proteins, and related technologies that use ligand binding domains to label proteins of interest are wonderful for live-cell imaging experiments, but can require a fair amount of reagent production and cell engineering to implement. Also there are fewer choices for labeling structures like DNA in live cells, as the available agents may be toxic.

3.2 DECONSTRUCTING LIGHT MICROSCOPY

In the following sections, each of the components of the HCS microscope, as applied to HCS, are described. The basics of light microscopy are presented in Figure 3.1. The elements of microscopy will be discussed in the order presented in the figure, beginning with the light sources, then the filter cubes (used in fluorescence microscopy), objectives, and finally the camera. A few optional adaptations to the imager are discussed at the end.

3.2.1 The Light Source(s)

HCS instruments have been developed by several groups, each with a different philosophy of what is most important. These goals affect the design of the platforms at every step, and begin with the light source. Currently, there are multiple options for fluorescence microscopy. One is a "white light" source, typically a mercury, metal-halide lamp, or a xenon arc lamp. Multiple lasers or light emitting diodes (LEDs) are also used. These generate light with very specific excitation wavelengths.

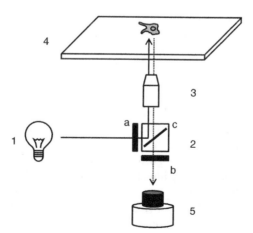

FIGURE 3.1 *Schematic of epifluorescence microscopy.* The major elements of the epifluorescence microscope are shown. (1) Light source, (2) fluorescence filter cube, (3) objective, (4) XY stage, and (5) digital camera. The filter cube (2) is comprised of three elements, the excitation filter (a), the emission filter (b), and the dichroic mirror (c). The functions of these elements are discussed in Figure 3.2.

White light sources, by being polychromatic, afford a wide flexibility for fluorescence microscopy, through the use of the filter cubes to bracket the wavelengths transmitted to the stage. Although these lamps appear as white light, they can display strong spectral patterns that result from the materials used to generate the light. The differences between the major light sources are compared in Appendix A. The differences become important when considering the most commonly used fluorescent probes, discussed in Chapter 1. Laser light sources, an alternative to white light illumination, are high intensity light sources that are limited to a narrow range of wavelengths. They provide several advantages, the most obvious of which is decreased exposure times, and therefore increased throughput. Also, confocal performance is better with narrower excitation light sources. Stronger illumination generates a brighter image, and since a brighter image is acquired faster, the time spent per well is reduced. Finally, LED illumination sources are becoming more common on HCS instruments. These LEDs are similar to lasers in that they output a narrow range of wavelengths. LEDs have many advantages, such as low power consumption, very long operating life, and compact size. There are some limitations on their output wavelengths and power, but the technology continues to evolve.

3.2.2   The Filter Cube

Precisely filtering light is the essence of fluorescence microscopy. Filtering is affected by the wavelength of light passing through each component. It will reflect the excitation light from the lamp or laser to the sample and then let the fluorescent light generated by the sample pass through the filter to the camera. The filters control the region of the light spectrum that reaches the sample, and is transmitted to the camera. These characteristics are typically diagrammed graphically, so that users can select optimized arrangements of filter sets and fluorescence dyes. The difference between the excitation and emission wavelengths is usually narrow (in fact, frequently overlapping, as shown in the example of Alexa-488 in Figure 3.2). In such cases, the switch between reflection and transmission wavelengths of the filter cube need to be carefully matched to the fluorescence properties of the dye being used.

The **excitation filter** (Figure 3.1) can be used both in lamp- and laser-based light sources. These filters reduce the polychromatic light to a narrower band of wavelengths (bandpass range) to excite the fluorescent probe. In particular, the bandpass range of the excitation filter is designed to exclude the longer wavelength light that will return from the fluorescently illuminated probe and needs to be directed to the camera, as emitted fluorescent light is lower in energy and therefore is a longer (red-shifted) wavelength. This is obviously essential for a white light lamp-based system, but is still important for laser-based systems. Similarly, the **emission filter** reduces the range of light returning from the objective before it passes to the camera. Like the excitation filter, the role of the emission filter is to remove light, but this time, it functions to specifically remove light in the range that was used to excite the dye. Working together, these filters limit the light reaching the stage to the excitation range, and the light reaching the camera to the emission range.

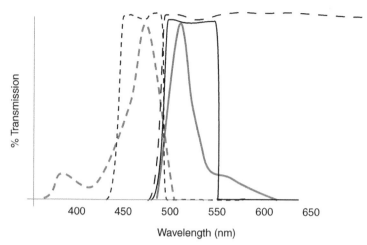

% Transmission

400   450   500   550   600   650

Wavelength (nm)

FIGURE 3.2 *How a filter cube channels fluorescent light.* The figure showing the shift in light excitation and emission from Figure 2.3 is used to diagram how filters are used to limit light transmission through the microscope. Excitation spectra, the range of light that will be absorbed by the fluorophore is on the left, while the emission spectra, the light wavelengths released by the fluorophore is to the right (lower energy and longer wavelength). The scale at the left is the percentage transmission (as it now refers to the filters instead of the light energy) of light as a function of wavelength for the three filters that comprise the filter cube. These filters will act as a mirror or a window, depending on the wavelength. The excitation filter (thin short dashed line) will only allow light in the excitation bandpass range for the fluorophore, letting this range enter the filter cube. The dichroic mirror shows very low transmission in this range, so it reflects this light toward the stage (the line is a little hard to see—it is very low at the low end of the spectrum, increases sharply at the interface between the excitation and emission spectra, and remains high at the higher wavelengths, shown as a thin long dashed line). Emission bandpass light that enters the filter is at a range where the transmission properties of the dichroic mirror are high, allowing light to pass through the filter, as also occurs for the emission filter (thin solid line).

The core of the filtering process is the **dichroic mirror**. Like the excitation and emission filters, the dichroic mirror is finely tuned to a specific bandpass range. For a filter with a bandpass range in the green wavelengths (typically referred to as a "488" or "FITC" filter), the range may run from 450 nm to 500 nm. The dichroic mirror does more than filter the light in the way that the excitation and emission filters do, it actually changes the path of the light as a function of wavelength, allowing light to pass through it (like a window), or to reflect it (like a mirror).

HCS systems will either have filters that can be set individually or can be precon-figured for common fluors as filter cubes. Filter cubes can be designed for almost any specific application. Each component is available in dozens of designs (bandpass ranges), and they can be combined to create custom sets. Several vendors provide web-based tools for comparing predesigned cubes, and for designing novel ones, and to allow fitting these cubes to most fluorescent dyes (see reference appendix). Filter sets are available for each system, and may be selected at the time that instrument is purchased, but sets appropriate for the most commonly used dyes are typically considered as part of the standard package for an instrument. Therefore, for most scientists, the issue is less about how to engineer a set of filter cubes, than it is to understand what you have and occasionally, what you may need. The system of setting filters is an important consideration. Filter cubes are easier to use, because they are standardized combinations. New users can have more trouble setting filters individually, and it happens frequently that two scientists working can set these differently for very similar assays. On the other hand, the filter cubes make nonstandard experiments more difficult (FRET, in particular), because a custom filter cube needs to be purchased.

3.2.3   The Objective

Probably the most widely reported parameter of an HCS experiment is the magnification of the objective lens. This is meant to explain the extent to which the number cells and the level of detail for each cell are captured. A 5× objective can over 2000 mammalian cells per field, and for many assays, this single field is enough to determine the effect of a treatment on the nuclear localization or abundance of a protein. Even at this low magnification, it is possible to detect changes in nuclear area (a measure of cell stress), nuclear intensity (a measure of DNA content and general stage of the cell cycle), and nuclear intensity variation (a measure of cell death, such as from apoptosis [1]. Many HCS experiments are set up using 10× and 20× objectives, which allow the cytoplasm to be captured in greater detail. Using these objectives, the cytoskeleton (actin, tubulin, and other filaments) can be visualized. Measurements of fiber length and orientation can typically be made using such objectives. Vesicle structure and patterns can also be visualized as they change (for instance, the stages of internalization of a transmembrane receptor as it moves from the cell surface to membrane pits, to internalized vesicles, and finally to endosomes). 40×, 60×, and 100× objectives represent the high end of magnification for commercial HCS platforms. These objectives allow greater resolution and can be used for colocalization studies. However, greater detail means fewer cells per field, and since statistical significance is a direct function of the number of cells analyzed, the robustness of the assay could be affected by sampling too few cells.

In addition to the magnification of the objective, its numerical aperture plays an important role in the level of detail that can be observed. Numerical aperture is commonly referred to as the "light-gathering ability" of the lens and thought of as how well the objective can "see in the dark." These are both true and useful ways to think about how two lenses of the same magnification can function differently if their numerical apertures are different. There are two additional factors that are affected by the numerical aperture of a lens. The first is logistical. A lens with a high numerical aperture has a much shorter working distance, or distance the lens needs to be from the sample. In many cases, this can be a few millimeters, and conventional plastic microplates will be too thick to allow the lens to focus on the cells in the well. There can also be problems with the outer perimeter of wells, and the edge of the plate can interfere with the lens. Offsetting this disadvantage is the increased resolution that comes with a higher numerical aperture lens. This second factor that differentiates objectives by numerical aperture can be underappreciated by scientists, who are used to thinking of magnification as specifying the detail that can be captured by a lens. However, the sensitivity of the lens, which is strongly affected by the numerical aperture of the lens, plays a dramatic role in the level of detail that can be captured [2], as shown in Figure 3.3.

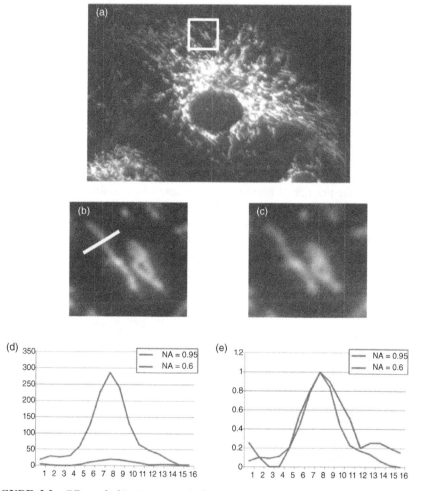

FIGURE 3.3 *Effect of objective numerical aperture on image resolution and intensity.* (a) Bovine pulmonary arterial endothelial (BPAE) cells with MitoTracker Red CMXRos (FluorCells prepared slide #1). Two images taken with objective of the same magnification, but differing numerical aperture (b) NA = 0.95 (c) NA = 0.60 are shown. (d) Linescan shows significant decrease in S/N because of lower numerical aperture. (e) Normalized linescan shows decrease in resolution of NA = 0.6 exhibited as larger line width/diameter of mitochondria.

3.2.4   The Camera

HCS platforms usually capture images through **charge-coupled device** (CCD) or **complementary metal oxide semiconductor** (CMOS) cameras, while point-scanning confocal HCS platforms use a **photomultiplier tube** (PMT). These cameras are extraordinarily sensitive, operating several orders of magnitude faster than film-based cameras. In the case of the CCD and CMOS cameras, the microscope image is projected onto an array of light-sensitive elements (pixels). This array of pixels can be transferred to the computer and presented as an image on a computer monitor. The resolution of the camera is defined by the size of the individual pixels. This grid is redrawn on a computer monitor (after scaling for the reduced dynamic range) and the raw data is used by the image analysis applications to determine cellular dimensions and organization. There is a tremendous amount of technical information concerning the design and implementation of image capture technologies, such as how the data is transferred off of the pixel array, but in all commercial HCS systems, the cameras are highly sensitive and very fast. Photons captured by each pixel or element are recorded linearly over a wide dynamic range. For a 12-bit camera, the light reaching a pixel is recorded linearly from 0 to 4095. As such, a high level of granularity is available to the image analysis software to determine the abundance of an antigen, and the shape and granularity of the nucleus or cytoskeleton. One last note about the CCD camera and its use is the practice of **binning**. Binning refers to combining individual pixel data into a square array of pixels (e.g., 2×2 or 3×3 groups of pixels). For example, by combining the pixels into 2×2 arrays, the sensitivity is increased fourfold, but the resolution (or the functional pixel size) is decreased twofold. This is diagrammed in Figure 3.4. Binning the pixels into 3×3 arrays affects these parameters by ninefold, relative to nonbinned data collection. Therefore, binning can significantly increase

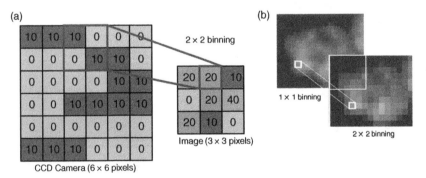

FIGURE 3.4 *Binning of pixel data to increase sensitivity.* Intensity of light registering on a 6×6 CCD camera is diagrammed. (a) The higher the number of photons reaching each pixel, the greater the intensity. When 2×2 binning is applied, groups of 4 pixels are summed to create an image with up to four times the signal but half the spatial resolution. (b) Example of fluorescent stained nuclei. Note that the image is the same size in both images, but the binned image is comprised of larger blocks of pixels, eroding the level of detail, but able to measure weaker fluorescence.

sensitivity of weak dyes, but lowers the resolution. Unlike CCD cameras, PMTs have no spatial resolution and basically collect all the photons projected onto the detection surface. These detectors are used in point-scanning confocal instruments, therefore, the image is formed by scanning the illumination across the sample field in an XY raster pattern, creating the image after scanning the entire sample area. Binning is typically accomplished either by scanning 2 "pixels" of the sample before reading out the PMT value, or by combining pixels in software after the image is formed.

3.3   USING THE IMAGER TO COLLECT DATA

The technical aspects of fluorescence microscopy have been discussed above, but it is worth walking through the process of setting up an experiment and collecting data. These steps are common to all imagers, but each system may perform these steps differently. All HCS imagers collect images after determining the focal plane for the field. How the instruments do this varies by manufacturer, but a couple of common methods are image-based and laser-based focusing. Image-based focusing is essentially a form of image analysis, but is limited to determining the focal plane, and in some cases counting objects in the field. Most applications follow a simple pattern: use a DNA-intercalating dye to label the chromatin and nucleus, and then use up to three additional fluorescent dyes to label additional proteins. Since there is so much DNA in the cell, and the dyes that fluoresce as they intercalate into the DNA are strongly fluorescent (typically Hoechst 33248 or DAPI in the UV range but occasionally DRAQ5, which fluoresces in the red range), finding the nuclei in a field is usually the first step in acquiring an image, and the very bright dyes make it easier for the instrument to find the cells and begin collecting data. This is typically done through an autofocusing routine, which steps progressively toward the sample and captures images. The images are processed immediately and are evaluated for the sharpness of the object outlines. As the objective approaches the optimum focal plane, the borders around the nuclei will sharpen dramatically, and as the objective passes the optimum plane, this sharpness will deteriorate rapidly. These transitions are measured by the image processing software, and when it is clear that the optimal plane has been found, the imager will move back to that plane, and typically a second set of fine adjustments will be made. The step size that the imager uses to identify the optimum plane is defined by the objective, and the starting position is defined by the microplate or slide used by the investigator. This is diagramed in Figure 3.5. Although a lot of effort has gone into standardizing microplate screening plates, each plate has key characteristics, such as the location of the plate bottom relative to the border, and the variance of the plate bottom for each well across the plate. These parameters need to be obtained by the plate or imager manufacturer, or need to be measured by the imager prior to the experiment. Initial values for each plate model manufacturer are loaded into the autofocusing software, but the high sensitivity of microscopy requires that the focal plane for each well be determined empirically. This is true despite the fact that microplates are required to

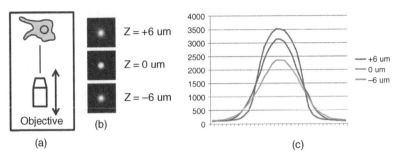

FIGURE 3.5   *Image-based focusing.* (a) Automated focusing of nuclei is based on a stepwise movement of the objective and analysis of the resulting image. (b) When the image is in focus, the boarders of the object are at their sharpest, and the slope in pixel intensities between the background and the object are steepest (see Chapter 4 for a fuller discussion of image analysis). (c) Linescan at center of object shows that when the object is in focus (top graph—0 um), the intensity is at its peak and the edge is at its highest slope.

conform to standard measurements defined by the Society of Biomolecular Screening (http://www.slas.org/resources/information/industry-standards/). These standards outline specific dimensions of microplates to allow them to be handled in automated laboratories, but there is enough variability in the design specifications to require individualized plate parameters for microscopy-based applications. If these parameters are incorrect, the imager will frequently fail to find the nuclei.

Laser-based focusing uses a laser to illuminate the surface of the microplate or slide. This laser is something of a pilot light that is only used for focusing the objective, and is independent of what is used to illuminate the sample, some imagers use a laser-based system for focusing and a lamp-based system or a set of illumination lasers. The focusing laser produces a diffracted light pattern as it reaches a phase transition, either the bottom of the plate (an air-to-solid transition) or the bottom of the well (a solid-to-liquid transition). A software routine can track the distance from the objective where these transitions occurred. In subsequent fields and wells, a time-saving advantage to a laser-based autofocusing system can begin to accrue, as it is possible to determine the offset between one of these surfaces and the nuclei during the setup phase of the process, and the skip image-based autofocusing during the image acquisition phase. The aggregate savings in time can be substantial when applied to a true HTS campaign of 500,000 to over a million compounds for which millions of images will be captured. Following the identification of the nuclei, one or more other channels are typically captured as well. Depending on the objectives used and the cell type of the experiment, other channels may require additional autofocusing, which also add to data collection times for the plate.

As with all lab instrumentation, the tolerance for extended data collection times is driven strongly by the scale of the study and the demands on the instrument. At one extreme, an HTS campaign presents significant demands on time and sample management, and the image acquisition phase is frequently limited to a single field per well, and a single fluorescence channel in addition to the nuclear marker. Such screens require a highly robust difference between the positive and negative controls; examples of this scale of assay include screens inducers of apoptosis or GPCR internalization. On the other extreme, an experiment that measures a limited number of treatments or an instrument that is used by a small group of researchers can usually accommodate a scan that incorporates cell-level measurements without any problems. In the middle are instruments that serve in core labs, where a number of groups work off of a single instrument, and time management is important. In such cases, imaging may need to be pushed to overnight scans. It is important to consider the benefits of automation in managing an imager, even for walk-in facilities that have a large number of clients performing unique experiments. Once the assay parameters have been set, the plates can be tracked by barcode and the correct algorithm will be called when the plate is loaded, allowing several projects to be run overnight.

3.4   CONCLUSIONS

Spending time presenting HCS in the context of classical microscopy helps one to consider HCS experimental design as a highly nuanced extension of microscopy, rather than as a plate reader that generates a table of data. The chapters that follow address the steps in the set up and analysis of an HCS experiment, including the detection and resolution of problems. Intrinsic to all of these discussions is that HCS is an automated microscope, and as much as the adaptations that enable screening-level systems produce a very powerful research tool, refining experimental design is driven by considering the process as microscopy. In particular, the microscope at the heart of an HCS imager is a specific brand and model of microscope, so upkeep and modification of the instrument (e.g., new filter sets) have to be made within product lines that are appropriate to the microscope itself.

KEY POINTS

1. The illumination system will define what fluorophores and filter sets are needed. A set of very robust filter sets and fluorophores define a common repertoire that is suitable for most experiments, but a much wider array of options exist for specialized applications (e.g., FRET).

2. Understand magnification and numerical aperture as it relates to what you will be looking at is an example of the type of microscopy-centric thinking that is an essential part of HCS. You do not always need to acquire images at high magnification to measure a particular biological phenotype.

3. HCS is a highly integrative technology. Proper separation of physical and spectral events are essential, but the imaging system is very powerful, particularly when applied to hundreds or thousands of cells, so very subtle changes can be detected with confidence. At the same time, very robust responses (such as strong nuclear localization) can also be detected with confidence, but with far

fewer cells, enabling HCS to be an effective method of primary drug screening. Learning to integrate microscopy, image analysis and data analysis is the core of the science and practice of HCS.

4. Consider including a platform manager on your team, who is knowledgeable about microscopy concepts, these principles are embedded in all aspects of experimental design in ways that go beyond understanding the physics of a laboratory plate reader.

FURTHER READING

The Optical Microscopy Primer website is an excellent resource for concepts around microscopy, digital imaging, and fluorescence. Microscope vendors also have useful information. Invitrogen, Chroma, and other reagent and filter vendors have very useful applets that allow one to graphically view the spectral properties of their products and how they interact with each other and with associated reagents. You can also overlay illumination peaks of common laser sources as well as bandpass filter properties. Web sites for these can be found in Appendix A.

A.J. North wrote a very perceptive commentary in the Journal of Cell Biology (see references).

In addition, the following references are in-depth compendiums of topics introduced in this chapter.

Goldman, B.D. et al. *Live Cell Imaging: A Laboratory Manual.* Cold Spring Harbor Laboratory Press, Cold Spring Harbor, NY, 2009.

Spector, D.L. and Goldman, B.D. *Basic Methods in Microscopy: Protocols and Concepts from Cells: A Laboratory Manual.* Cold Spring Harbor Laboratory Press, Cold Spring Harbor, NY, 2005.

Yuste, R. *Imaging: A Laboratory Manual.* Cold Spring Harbor Laboratory Press, Cold Spring Harbor, NY, 2010.

REFERENCES

1. Haney, S.A. *High Content Screening: Science, Techniques and Applications.* Wiley-Interscience, 2008.

2. North, A.J. Seeing is believing? A beginners' guide to practical pitfalls in image acquisition. *The Journal of Cell Biology*, 2006, **172**(1): 9–18.

4

IMAGE PROCESSING

JOHN BRADLEY, DOUGLAS BOWMAN, AND ARIJIT CHAKRAVARTY

4.1   OVERVIEW OF IMAGE PROCESSING AND IMAGE ANALYSIS IN HCS

In this chapter, we will explain the process involved in obtaining biologically relevant information from digital images. This area of HCS bridges multiple disciplines including engineering (signal processing), computer science (algorithm development) and, of course, biology. Most biologists who perform HCS usually accept much of the "math" involved with image processing at face value. In addition, the available literature for image processing is biased to the extremes of either engineering (with mathematical nomenclature difficult to understand for most biologists) or biology (with many black boxes used to describe the process). Since a clear understanding of the concepts behind extracting information from digital images is invaluable when designing high content assays, we attempt here to describe in detail image processing and analysis but without complex mathematical descriptions. For more formal descriptions of this field, the reader is directed to a number of excellent references [1–5].

In the following text, we will first explain the storage of information in digital images and how this information is processed by computer software algorithms. We will then describe the steps involved in a typical image analysis workflow, starting from preprocessing steps and ending with a set of values related to the identified objects within the image.

An Introduction to High Content Screening: Imaging Technology, Assay Development, and Data Analysis in Biology and Drug Discovery, First Edition. Edited by Steven A. Haney, Douglas Bowman, and Arijit Chakravarty.
© 2015 John Wiley & Sons, Inc. Published 2015 by John Wiley & Sons, Inc.

4.2   WHAT IS A DIGITAL IMAGE?

A digital image is a matrix of numbers. Each "box" in the matrix is called a picture element, or pixel. Images can be either color or grayscale (monochromatic). However, since almost all HCS systems collect and analyze monochromatic images, we restrict our discussion here to grayscale images. In a grayscale image, each pixel is assigned a number proportional to the amount of light (the number of photons) that was collected by the detector, typically a CCD camera. Multiple fluorescence markers are captured in separate monochromatic images by using different filter combinations as described in earlier chapters. Usually the array of numbers that makes an image is rectangular with typical dimensions of 1280×1024 pixels. Images of this type therefore contain 1.2 million numbers and so can store a lot of information.

The amount of grayscale or intensity information captured per pixel is termed the bit depth of the image. Many CCD cameras used in HCS collect 12-bit images, and therefore light intensity is encoded in the integer range of $0–2^{12}$, that is, 0–4095 gray values. This high dynamic range enables the quantification of a wide range of both dim and bright objects in a single image. It is important to note that many imaging systems can detect and store light information over a greater dynamic range than the human eye can perceive and the monitor of desktop computers can display. As such, a procedure is needed to allow humans to access (view) this information— this procedure is often called image or display scaling. In addition to the captured image, other intermediate images are used in the image processing routines, and these images can have a different bit depth. These include binary images, often referred to as "masks," in which each pixel is either "0" or "1". Such images are used for defining objects (specific cellular compartments such as nucleus or cytoplasm) for the image analysis algorithm. For example, a nucleus mask can be calculated based on a fluorescent nuclear marker and then "overlaid" on a second wavelength so that only the nuclear and nonnuclear portion of the image is identified. Image analysis algorithms perform the mathematical functions using greater dynamic ranges (16 bit, e.g., providing a range of 0–65,536) to preserve accuracy but then convert them back to the original bit depth when any intermediate images are needed (such as for the visualization of segmentation overlays).

4.3   "ADDRESSING" PIXEL VALUES IN IMAGE ANALYSIS ALGORITHMS

All image analysis algorithms analyze images, or specific subregions of images, using the same basic principle—each and every pixel is analyzed sequentially. Computer algorithms raster scan the array of numbers that represent an image (Figure 4.1). The analysis of each pixel requires the application of a function, the output of which will change the value of that pixel. The input parameters for the function applied to each pixel are derived either from the single pixel itself, as occurs when the contrast of an image is changed (Figure 4.1a), or from multiple numbers derived from surrounding pixels (Figure 4.1b). Obviously, the latter approach is essential for considering groups

(a) **Single pixel input for image analysis,** for example, contrast adjustment

(b) **Multiple pixel input for image analysis,** for example, image convolution

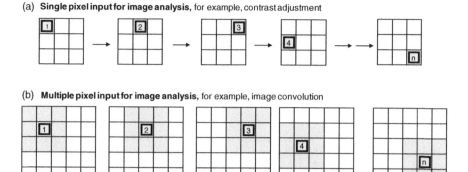

FIGURE 4.1 *Schematic diagram to demonstrate the methods of pixel processing in image analysis.* (a) Single pixel processes modify each pixel separately. An example is background subtraction where a value is subtracted from each pixel. (b) Multiple pixel processes modify each pixel based on the surrounding pixel values. Examples include sharpen or blur techniques.

of pixels that make up objects. When considering neighboring pixels, each and every pixel is still "addressed" but the calculation performed on each pixel is dependent on the values of specific neighboring pixels. With the exception of boundary pixels, all pixels within an image have eight neighbors. As discussed in detail below many different types of functions are used in image analysis. Each of these functions, either used alone or sequentially with other functions, produces a unique change to the original image that is useful for information extraction.

4.4 IMAGE ANALYSIS WORKFLOW

The image analysis process can be divided into five stages as illustrated in Figure 4.2.

1. Raw images are processed to remove noise, reduce background fluorescence, or correct for nonuniform illumination. This process is often referred to as preprocessing.
2. Images are thresholded to first identify signal above background, and then the images are segmented to "identify" biologically meaningful areas, such as the nucleus and cytoplasm of cells. This is the most challenging step of image analysis and often requires careful optimization of key parameters in order to avoid false positive and false negative "identification" of objects. Object-based processing is performed to further refine the object identification in the previous step. One example is object splitting or merging in the case of mis-segmented nuclei where the shape and size of the objects of interest are known a priori. For instance, if one is analyzing cell nuclei and an object is identified as something shaped like a figure eight, the algorithm can split this object into two nuclei.

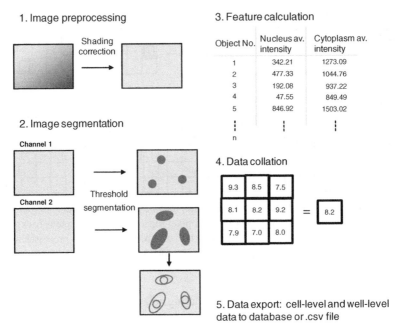

FIGURE 4.2 *Five stages of the image analysis process in HCS.* There are five basic steps for HCS image analysis. 1—Images are preprocessed to remove background or correct for illumination artifacts. 2—Image segmentation algorithms process the image to identify objects of interest. 3—A series of features are collected for each identified object. These objects could be nuclei, cells, or subcellular structures. 4—Features are summarized for each image and well. 5—Finally, cell-level and well-level data is exported for further analysis.

3. Once objects have been identified, the algorithm calculates a variety of features related to each object (size, shape, intensity, texture, etc.). The user then identifies the features which are important for the biological process under study, such as the average intensity within the cytoplasm, or the number of puncta identified per cell.

4. Features are collated and compared for the endpoint assessment of the experiment, such as averaging of triplicates within different conditions to generate concentration–response curves. This is discussed in more detail in Chapter 9.

5. Finally, the cell-level and well-level data is exported to either a database or file (.csv) depending on the HCS platform.

4.4.1   Step 1: Image Preprocessing

The pixel intensity of a digital image represents the sum of the "signal," fluorescence from the biological process of interest, and "noise," fluorescence from nonspecific staining, background fluorescence, or instrument noise. The difference between the signal and the background is referred to as the signal-to-noise ratio. The signal to noise ratio of a fluorescence-based assay is determined by a multitude of factors: reagents, wet-lab protocol, instrument characteristics (filters, illumination system), and acquisition settings (exposure time, magnification, binning) as described in Chapter 2 and Chapter 3. Since it is impossible to eliminate all the noise from the imaging system, it is important to understand what is responsible for the noise in order to minimize it. Chapter 11, discusses some strategies to increase signal to noise for both the wet-lab protocol and the image acquisition process. Although numerous approaches are used to minimize noise within the imaging system, such as cooling the CCD chip of a digital camera, various image processing methods can be applied to reduce noise and maximize the detection of the desired cellular objects. A few functions that are often used to preprocess images to reduce noise and increase signal-to-noise ratios are discussed below.

4.4.1.1 Illumination Correction   This is a specific form of processing used to correct the uneven illumination described above, that is, the center of an image having greater illumination than the corners. However, with illumination correction, the values used to modify the image of interest are not derived from that image. Basically, an image, or set of images which are then averaged, is acquired from a control sample that has uniform fluorescence, a blank coverslip, for example. A correction image must be acquired with the same acquisition settings as those used for the experiment acquisition because variation in illumination is specific to the hardware used for acquisition, such as the objective and filter set. The correction image reveals the illumination intensity "pattern" for a particular system configuration. The pixel values of all images of interest are then divided by the corresponding pixel values of the illumination correction image (Figure 4.3a). Illumination correction is typically performed automatically by the acquisition instrument using a correction image collected by the platform manager. Users should be aware that the illumination variation may vary over time and may require recalibration.

4.4.1.2 Background Subtraction   In its simplest form, background subtraction corrects for nonspecific fluorescence not associated with the signal of interest. One can empirically determine the average background intensity from an image, using a blank region in the image, and subtract this integer value from all pixels within the image. This should reduce nonsignal pixels to approximately zero. However, it is rare for all background pixel values to have similar intensities. This is because it is almost impossible to obtain uniform illumination of the specimen even after the illumination correction procedure. As a result, there may still be intensity variations across the image. Obviously, simple subtraction of a single light intensity value will not correct for this varying background intensity. As such, local rather than global background subtraction methods can be employed. Various algorithms have been deployed to locally subtract background but the concept for all is similar. Within a small defined region of the image, 20 pixel square for example, all the intensity values are convolved effectively blurring this region. The blurring of the 20 pixel square region is carried out across the entire image such that local blurring occurs for the whole image. The pixel values of the original image are then divided by the corresponding pixel values of the new blurred image (Figure 4.3b).

(a) **Illumination correction**

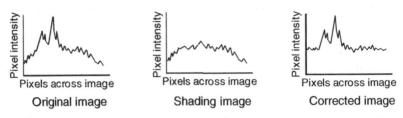

Original image   Shading image   Corrected image

(b) **Local or adaptive background subtraction**

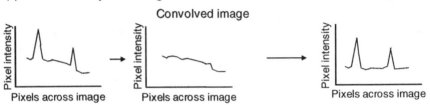

FIGURE 4.3 (a) *Image processing to correct for uneven illumination.* A representative line scan is show which exhibits the uneven intensity across an image. A separate, shading image, is collected from an evenly fluorescent sample. Finally, image processing is performed to correct the original image. (b) Adaptive background subtraction is attained by estimating the local background by smoothing the original image and then processing the original image, creating an image with a flat background across the image.

4.4.2   Step 2: Image Thresholding and Segmentation

4.4.2.1   Image Thresholding   The next step is to identify what pixels within an image are brighter than background and represent objects of interest. This step invariably requires thresholding, or selecting the pixel intensities that represent the signal of interest above background and then performing segmentation of the resulting groups of pixels, or objects, such that the identified groups of pixels are a true representation of the biological specimen of interest.

Global Image Thresholding. Thresholding is the process of classifying pixels into particular groups depending on their intensity value. In the simplest form, an image can be thresholded based on a single intensity value, all pixels with a value above the threshold value are selected while those below are not. In more complex analysis routines multiple threshold values can be applied to the same image to define multiple groups of pixels.

Many algorithms are available for calculating a threshold intensity value based on the intensity values of the image. However, the simplest thresholding procedure does not use an algorithm at all but rather the threshold value is empirically determined by the user (Figure 4.4a). This approach works well when the signal-to-noise ratio is high and the range of absolute intensity values across all images of the experiment is similar. However, if the absolute intensity of pixel values in images of an experiment

(a) **Global empirically determined threshold value**

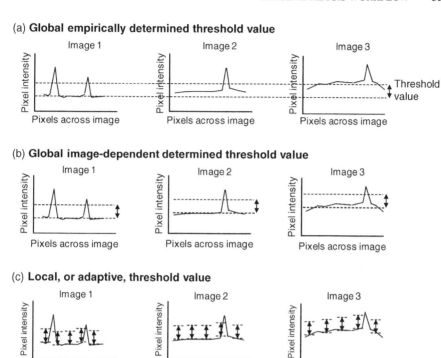

(b) **Global image-dependent determined threshold value**

(c) **Local, or adaptive, threshold value**

FIGURE 4.4 *Thresholding techniques.* (a) An empirically determined threshold is a value determined by eye using multiple images that best captures the regions of interest within an image set. Once established this absolute threshold value is applied to all images. (b) An image-dependent threshold is an empirically determined value that is dependent upon a metric of each individual image, such as average, or minimum, value. As such, the image-dependent threshold may be different for each image and may take into account different background intensities across images. (c) An adaptive threshold is a value that is based on relative intensity and can change across a single image. This enables the detection of objects across a varying background.

differs over a range that is similar to the threshold value itself then this approach will fail. It is often unavoidable that the absolute intensity value of all images in an experiment changes over a specific range. One approach for overcoming this problem is to calculate the threshold value based on a property of each individual image. The simple way to do this, for example, would be an intensity value that is set to be two or three standard deviations above the average intensity value of the image (Figure 4.4b). Again, this approach works only with images that, despite having a relatively large range of absolute intensity values, do not exhibit large fluctuations in pixel values as may occur with cell debris. For this reason, robust algorithms are used to determine the single threshold value and these are usually based on the pixel intensity distribution (the frequency histogram) of all the pixels in the image.

Local, or Adaptive, Image Thresholding. Even with a robust algorithm, if the background intensity of an image varies across the image, a single threshold value will not effectively identify all of the objects in an image. Therefore, local, or adaptive threshold algorithms have been developed (Figure 4.4c). These algorithms are similar to the above process but basically apply an ever-changing (adaptive) threshold across an image based on the local neighborhood of pixels. An analogy, graphically described in Figure 4.4c, is finding bushes on rolling hills. Obviously, a simple elevation measurement (single threshold) would fail, whereas a "height-above-ground" (adaptive threshold) would work well.

4.4.2.2 Image Segmentation and Filtering

Once pixels have been selected using a thresholding algorithm, they are often referred to as objects. These objects may be single pixels but for the most part pixels over a threshold value are often adjacent to other pixels that have been thresholded—hence the formation of groups of pixels, or objects. In an ideal assay, each of these groups of pixels, or objects, would represent the biological specimen of interest such as a nucleus or cell. However, very often threshold algorithms produce pixel groups that represent background noise in addition to objects that represent the biological specimen. Furthermore, even if images have a high signal-to-noise ratio (allowing the use of a fixed threshold value), artifacts may occur. Two such artifacts are cell debris that appears in the fluorescence image, and the grouping of multiple closely spaced objects into one larger object. To distinguish artifact objects from the true biological specimen, additional segmentation algorithms can be applied to the thresholded image.

Segmentation algorithms consist of multiple image processing steps that either remove pixels or separate large objects into the underlying, closely spaced, objects that represent the biological specimen. As mentioned above, algorithms can modify each pixel value within an image using functions that have as their input the values of neighborhood pixels (Figure 4.1b). At the heart of segmentation algorithms are a number of simple functions that modify pixels in a nonlinear manner—meaning that the central pixel value is not modified by an equation of the type $x = x \times 2$. Rather, the central pixel value is modified dependent on a set of specific conditions established by the values of the neighborhood pixels. The set of neighborhood pixels is termed the structuring element and for the simplest functions, the structuring element is the surrounding eight neighborhood pixels. Two such simple functions are termed the "dilate" and "erode" functions. As the names suggest, these functions when applied to a binary image, an image in which each pixel is either "0" or "1," will strip away or add a layer of pixels to the outside of all positive pixels respectively. The conditional functions for each of these processes are simple: to dilate the outer layer of an object the function will make all pixels of the structuring element positive ("1") only if the center pixel has a value of "1." Conversely, to erode the outer layer of an object the function will make the center pixel "0 "if any of the pixels of the structuring element are "0."

These simple functions may not, at first pass, appear particularly useful. However, they can have unique effects that can be used for particular processing functions. For example, if two objects, that represent cells, are just touching then the erode function can effectively split, or separate, the two objects. Now two objects instead of one will be counted. Conversely, if we wish to quantify groups of cells that are very close together but the thresholding of the image separated all cells, we could apply the dilate function. Only cells that are very close together will merge into one larger object and these larger objects would represent groups of cells. Furthermore, when the erode and dilate functions are combined in a specific order very useful effects can be obtained. For example, a simple procedure for removing noise is to perform an erosion operation on the objects in the image. In its simplest form the erosion procedure removes (converts from "1" to "0") pixels that are on the periphery of an object. For large objects this simply strips the object of its outer "layer" of pixels. However, smaller objects that represent background noise will disappear with one or two rounds of erosion. Once the "noise" objects have been removed an equal number of dilation procedures can be applied to the image. Dilation performs the opposite of erosion, it adds (converts from "0" to "1") pixels on the outer layer of an object. As such, larger objects are returned to their original size. The "noise" objects cannot be returned to their original size because all pixels in these objects became "0" after the erosion and therefore they have no outer layer on which the dilation can work. Overall, these two simple pixel manipulation procedures when used sequentially effectively deleted all the small "noise" objects from the image.

A further level of complexity in segmentation and filtering algorithms can be achieved if the original grayscale 12-bit image is analyzed together with the binary mask derived by thresholding. For example, thresholding of an image may result in two or more cells being "joined" together. In many cases, the simple erosion procedure described above may not achieve separation of the cells, or may separate the cells but in so doing may degrade the mask too much. By analyzing the profile of intensity values in the grayscale image defined by the mask, one can determine the boundaries of the cells and modify the mask accordingly. For many objects the intensity is highest in the middle and lowest at the edge of the object. The boundary between the two objects can be derived by analyzing the intensity profile using a function called the watershed function. In essence, the watershed function analyzes two points, one starting at the peak intensity value of each cell. Each of the points moves down the intensity gradient until the points colocalize—this is the boundary.

If the intensity of the cells is not at its lowest value at the edge of the cells then the shape of the mask in the binary image can be analyzed. All objects that appear joined after thresholding are never round but usually have an hour-glass shape. We can determine the center of each of the two cells by calculating the least distance to the edge of the mask for each pixel within the mask—a so-called distance transform. This is achieved very simply for each pixel by successively analyzing each "layer" of surrounding pixels. If the first "layer" of pixels are all "1"s then the value of the pixel of interest is incremented and the next layer analyzed. Only when a layer is encountered that contains a "0" is the process stopped. As a result, pixels farthest from the boundary of the mask will have the highest numbers. The boundary of the two cells can then be determined by applying the watershed function to the image of the distance transform.

We have described some of the basic steps used to go from a digital image to "objects of interest." There are obviously many advanced techniques other than those described here but our intent was to review the fundamental process of image segmentation. Additional resources can be found at the end of the chapter.

4.4.3 Step 3: Calculation of Image Features

In moving from a digital image to a high content assay, the goal is typically to focus on one or more aspects of the biological effects that one is interested in, and reduce them to measurements that can be extracted via image processing. These measurements are referred to as features.

The image analysis algorithm will ultimately identify individual objects that are related to cellular or subcellular structures, or localization of a fluorescent protein. There are a large number of measurements, or features, which are derived from these objects. These features may or may not be relevant to the biological phenotype, so identifying the appropriate feature or set of features is important. Some examples are shown in Table 4.1. Chapter 8 will describe the transition from these object features to a metric, a value derived from the object features that reports on the relevant biological event.

TABLE 4.1   Examples of How Different Features from Digital Images are Associated with Biological Processes

Type of assay	Marker	Underlying biological process	Feature
Intensity	Concentration of, for example, DNA	Cell division	Intensity within nucleus Intensity within cytoplasm
Spot counting	Cells	Proliferation	Segmented nucleus
	Cleaved caspase 3	Apoptosis	Positive nucleus
	γ-H2Ax	DNA damage	Spots within nucleus
	LC-III puncta	Autophagy induction	Spots within cytoplasm
Translocation	Forkhead (FOXO)	Signal transduction	Ratio of intensity within nucleus vs. intensity within cytoplasm
	p53	Signal transduction	Intensity within nucleus and/or intensity within cytoplasm
Morphology	Neurons	Neurite outgrowth	Segmented neurons
	Mitotic spindle	Cell division	Positive nuclei
	Cells	Cell migration or differentiation	Location or shape of cell

Some of the more common features that are output from the image analysis algorithms to describe objects are as follows.

4.4.3.1   Count   A simple count of identified objects. The most common use is the number of cells as a measure of cell proliferation or the number of phospho-histone H3 positive cells as a mitotic cell count. In addition, the count of objects within a defined subregion of an image can provide an index of cell migration.

4.4.3.2   Morphological Measurements   Once individual objects are identified, there are a large number of morphology-related measurements that can be derived from each image such as the following.

- Size: area of object.
- Shape: measurement to categorize the shape of an object, whether it is circular or elongated. These are often used to help identify "real" objects versus artifacts. The shape feature can also provide information on the differentiated state of the cell and indeed when the cell has died.
- Diameter, fiber length: dimensional measurement of an object. In a heterogeneous population of cells this feature can provide an index of the relative numbers of each cells type, for example, neurons versus glia cells.
- Orientation: measurement of the angle of the longest chord of an object relative to the image. This is often used to determine if all objects are oriented in the same direction within an image. Again, this feature could provide information as to the migration of cells.

4.4.3.3   Intensity Measurements   The intensity is a measurement of the relative amount of fluorescent protein in an object. These are typically output as follows.

- Average intensity of the object, relating the average amount of protein per unit area.
- Total, or integrated, intensity of an object, relating the total amount of protein in the object.

For most HCS fluorescent antibodies are used to make specific proteins. Since the antibodies used should always be used at saturating concentrations, all of the available targeted protein should be marked. As such, the average or total measured fluorescence should be proportional to the concentration of the targeted protein.

4.4.3.4   Texture   Measurement of the relative contrast, or graininess, of an object that numerically describes the smoothness of the intensity distribution of an object.

4.4.3.5   Interpretative Morphology   Some applications require more advanced or complicated measurements that relate objects to other objects within the same image, or relate complex relationships or morphological features within the image. This is essentially a dynamic scoring process, looking at an object in relation to other similar objects to group the objects and measurements. A common example would be the neurite outgrowth assay where complex measurements help describe the complex morphology. In this assay, individual neurons are identified, and separated into primary branches that originate from a cell body, and minor branches that form off the primary branch. Measurements include:

- *Outgrowth.* Measurement of the length of each neuronal process, or the total length of all outgrowth.
- *Straightness/tortuosity.* Measurement of the relative straightness of each of the segments, where segments are defined by junctions, either the cell body or a secondary branch point.
- *Number of major and minor branches.* Measurement of the number of branches that begin at a cell body. The major or primary branches are typically the longest extensions that originate from the cell body. This is a highly flexible parameter that needs to be worked out carefully for each experiment. This could help describe the number of primary neurons. Minor branches are those that begin from a primary branch rather than the cell body.

4.4.3.6   Intracellular Object Relationships   Since most high content assays consist of multiple biological markers, there are also a variety of measurements used to describe the relationship between these markers, such as the following.

- *Colocalization/percentage positive cells.* Measurement of the percentage of objects in one image that overlap with an object in a second image. This is one of the most common metrics used in assays, for example, the number of cells positive for one marker such as pHH3 that also overlap with a second marker such as DAPI.
- *Correlation coefficient.* The correlation coefficient provides a quantitative index of the colocalization of two proteins of interest.

4.4.3.7   Intercellular Relationships   Lastly, objects within a field can be measured by their relationships to each other. For example, colony morphology, the clustering of cells across a field, is a measurement made by measuring distance between cells (or nuclei) to determine how evenly spaced they are, or whether they are grouped into patches of cells.

4.4.4   Step 4: Collation and Summary of Features

When considering how to summarize the data in a field or a well, and how to reduce these numbers to make the analysis of the experiment more manageable, it is common to summarize the individual measurements. A particular parameter is calculated for all cells in the image, for example, the average intensity of stain in the cytoplasm. If the image contains 50 cells then 50 average intensity values will be calculated. However, a summary statistic is often calculated from these 50 values, the most common of which is the average. Calculating the summary statistic will therefore give a single average intensity value for each image. If nine images are acquired per well we would therefore have nine average intensity summary statistics. These again can be averaged to provide a single average intensity value for the well. If triplicates are used for each condition of the experiment then the three average intensity values from the individual wells can be averaged to give a single value for the condition (although this would typically be done in the data analysis stage rather than when reporting feature data from the image analysis). These concepts are discussed in Chapters 9 and 10. Images can be legitimately excluded from an analysis for various reasons.

For example, debris in the image may preclude image analysis, or an image may be out of focus. In an HCS experiment, data sets can quickly grow to a large size even if only one number is calculated per image. For example, if nine images are collected per well for the center 60 wells of a 96-well plate, 540 summary statistic numbers are generated per plate. These numbers are usually exported from image analysis software and organized in a spreadsheet. Great care must be taken in the sorting and summary of the data. For experiments of a specific type that are often repeated, it is very useful to use a template into which the raw numbers can be placed. Embedded equations in the spreadsheet save considerable time in the summary of the data and also prevent user error in calculating summary statistics.

For most HCS experiments, summary statistics, such as average intensity, are invaluable. However, if the effect being measured is small and/or occurs in only a subpopulation of the cells being sampled then the effect will be lost in the averaging used to generate the summary statistic. For applications of this type, one can use powerful population analysis techniques, together with nonparametric statistical tests. To perform such analyses one must collect every measured parameter from all images for each condition. For example, if each image contained 50 cells and if nine images per well were collected in triplicate, then $50 \times 9 \times 3 = 1350$ numbers would be collected for each condition. For 60 wells of a 96-well plate, 32,400 numbers would be calculated. Such number crunching is particularly difficult to perform in a standard spreadsheet—even if custom scripting is used. Suitable commercial software is best for viewing and analyzing such large data sets. This topic is addressed in Chapter 10.

4.4.5   Step 5: Data Export and Feature Data

The product of image analysis is the generation of one or more tables of measurements for the cellular features. These can be measurements for each cell or summary measures for each well (see Chapter 8 for examples). The cell-level data can result in a very large table and lacks the data processing benefit of the well-level analyses, so well-level summaries are more commonly used in postimage analysis data processing, but there are cases where analyzing experiments at the cell level is called for. We will cover cases where each method is appropriate in Section 3.

One thing to understand at the moment is that all image analysis processes images in similar, but not identical, approaches. All platforms can measure intensities and shapes, and can reliably capture changes in these and other properties. One thing that can be confusing is that image analysis programs each have their own lexicon or jargon when it comes to naming features. For the most part, the features recorded are similar, but the naming conventions are unique to each image analysis application. When we start looking at the data generated by image analysis platforms, there will be a learning curve associated with how the measurements are reported. For the moment, it is enough to understand that the measurements made during image analysis are recorded and stored. We will spend more time in the earlier steps in HCS, producing robust imaging conditions, before we tackle data analysis.

4.5   CONCLUSIONS

Now that you have a good understanding of the image analysis process and the different features generated by the image analysis algorithms, the next challenge is to identify which metric or combination of metrics is the best readout of the biological phenotype of interest. This will be detailed in Chapter 8.

KEY POINTS

1. Understand general process of image segmentation and object identification.
2. Understand there are a number of different algorithms to identify similar cellular events, and there are a number of parameters for each algorithm. The user may need to try multiple algorithms or iteratively adjust parameters of each algorithm to get an optimal segmentation result.
3. Viewing the image segmentation overlay is critical in evaluating the accuracy of an image analysis algorithm. This overlay is typically color coded to delineate specific compartments that were identified *by the algorithm*. The user should iteratively adjust algorithm parameters while viewing the segmentation image to assess algorithm performance.

FURTHER READING

Instrument vendors typically publish application notes for each of the image analysis algorithms offered with the instrument. Application notes often explain the segmentation algorithm and provide a description of the numerous object features output from the algorithm.

Ljosa and Carpenter wrote a nice introduction of the image analysis process utilized in high content screening (see References). A more technical discussion of image analysis was written by Zhou and Wong, A Primer of Informatics for High Content Screening, in High Content Screening: Science, Techniques and Applications, edited by Haney.

REFERENCES

1. Ljosa, V. and Carpenter, A.E. Introduction to the quantitative analysis of two-dimensional fluorescence microscopy images for cell-based screening. *PLoS Computational Biology*, 2009, **5**(12): e1000603.

2. Wolf, D.E. *Digital Microscopy in Methods in Cell Biology*, S. Greenfield and E.W. David, (eds.). Academic Press, 2007, p. II.

3. Burger, W. and M. Burge, *Digital Image Processing: An Algorithmic Introduction Using Java*. Springer-Verlag, 2008.

4. Russ, J.C. *The Image Processing Handbook*. CRC Press, 2007.

5. Gonzalez, R. and R. Woods, *Digital Image Processing*. Addison Wesley Publishing Company, 2008.

SECTION II

GETTING STARTED

Now we start to roll up our sleeves. There is still a lot to cover before the first experiment can be set up, but we are getting there. As described in previous chapters, HCS workflows involve multiple hardware and software technologies, requiring expertise in areas from microscopy and automation to database management and information technologies. Chapter 5 will describe the system components (Figure II.1) as well as the many activities around installation, configuration, and maintenance of the system. Chapter 6 will describe the underlying components. Figure II.1 shows how the HCS imager is placed within the context of the network, data management, data flow, storage systems, which are the "infrastructure" of the HCS workflow. When digging into these two chapters, it is important to understand how many people can be involved in each step, setting up the platform and the information management system, particularly for larger institutions. Regardless of the size of the institution, there are many parts to each process.

Chapter 7, the last chapter in this section discusses setting up a first experiment. Setting up an experiment and capturing image data is a constantly evolving process (until the transition to a screen where focus is on locking down the process and reagents so that they do not change during the course of the screen, see Chapter 12). For setting up an initial experiment, we discuss the validation of reagents, which is critical, but also recognizes the importance of learning through doing (and revising). For well-characterized proteins, many high quality and rigorously tested reagents exist (making such a protein a good starting place for learning HCS). In these cases, a fairly good experiment can be anticipated with minimal refinement, so we present a basic protocol that works for the majority of commonly studied proteins and protein modifications (phosphorylation, proteolytic cleavage, among others). From this simple experiment, you will have images that can be analyzed in the subsequent chapters, as well as refined through basic protocols discussed in Chapter 7, and for assays that require additional work, as described in Chapter 11.

An Introduction to High Content Screening: Imaging Technology, Assay Development, and Data Analysis in Biology and Drug Discovery, First Edition. Edited by Steven A. Haney, Douglas Bowman, and Arijit Chakravarty.
© 2015 John Wiley & Sons, Inc. Published 2015 by John Wiley & Sons, Inc.

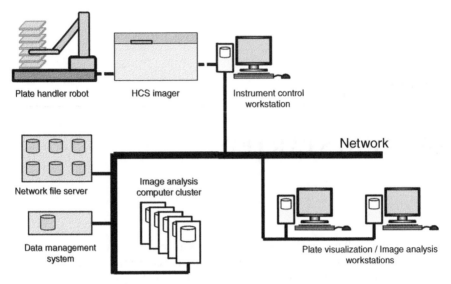

Plate handler robot

HCS imager

Instrument control workstation

Network

Network file server

Image analysis computer cluster

Data management system

Plate visualization / Image analysis workstations

FIGURE II.1 A typical high content screening platform configuration.

5

A GENERAL GUIDE TO SELECTING AND SETTING UP A HIGH CONTENT IMAGING PLATFORM

Craig Furman, Douglas Bowman, Anthony Davies, Caroline Shamu, and Steven A. Haney

5.1 DETERMINING EXPECTATIONS OF THE HCS SYSTEM

The decision to establish an HCS platform commits significant money and resources. The first steps in creating a system that delivers the expected results are to have a clear idea of who will be using the equipment, and what kinds of experiments they will be performing.

- Will the system be used exclusively by a single research group or will it be shared between multiple departments?
- Will the system be used for large-scale screening projects (e.g., using compound or RNAi libraries), or primarily for assay development?
- What kind of endpoints, such as apoptosis, or nuclear translocation, will be measured?
- Will it be operated as a core facility by a dedicated team or will it be available on a free access basis with only baseline technical support?
- For specialized equipment, such as confocal imaging or fluidics for compound additions, are these real needs or nice-to-haves that might be used in the future?

The vision of how the platform will function triggers important decisions on what instrumentation and software to purchase, and how it should be configured.

An Introduction to High Content Screening: Imaging Technology, Assay Development, and Data Analysis in Biology and Drug Discovery, First Edition. Edited by Steven A. Haney, Douglas Bowman, and Arijit Chakravarty.
© 2015 John Wiley & Sons, Inc. Published 2015 by John Wiley & Sons, Inc.

5.2   ESTABLISHING AN HC PLATFORM ACQUISITION TEAM

Setting up and managing an HCS platform requires work from several key individuals. Contacting them early in the process (such as during the vendor evaluation and purchasing process) can go a long way toward ensuring that the instrument selection process, installation, and operation proceed smoothly. This idea of delineating the necessary job functions (or skill sets) and then finding support for them was used in Chapter 1 of this book to identify the necessary functions for running HCS experiments. Here we describe several specific functions necessary to set up a functional HCS platform from system evaluation, system installation, system maintenance, and use and training of the image acquisition and analysis software. These functions include the following.

5.2.1   The Platform Manager

The platform manager will be centrally linked to all other members of the platform team, as well as to outside users. Although other scientists will contribute to the development and execution of assays run on the instrument, the dynamic nature of how the essential roles are filled will make the platform manager something of a Jack-of-all-trades, as the roles that fall to the manager will change for different projects or in different settings. For example, this person might be required to provide microscopy advice for researchers developing HCS assays. In addition, the platform manager may be responsible for integrating HCS instrumentation with lab automation and data management tools. This work entails controlling the instrument through the computer interface, initializing the related hardware (plate handling, pipetting, etc.) and managing data storage, that is, deciding where experimental information is to be retained after it has been collected.

If a new instrument is being purchased, the platform manager will play a significant role in the evaluation process, as each instrument will have different methods for setting up an experiment, including the selection of objectives and filters, autofocusing and exposure control, and may need to integrate with a number of other platform technologies in the department. For some instruments, these methods can be fairly easy to establish but leave little room for control once set up, while for other instruments it is possible to set up a complex pipeline for collecting images, but a sophisticated user interface is required. Obviously, the former situation might be better for a facility with a lot of independent users, whereas the latter might be better for instruments used for more complex assays but that have access to ongoing support from an expert manager/microscopist. Furthermore, modifications to the HCS system will be required from time to time, such as installing new objectives or filter sets, as will periodic maintenance. Many of these functions can be performed by the platform manager while others might require a vendor field service engineer visit. For a successful purchase, the needs of the team and users need to be clear in advance, and the manager should use this information to evaluate both the hardware and the software.

5.2.2   The Department or Research Head

The person releasing the money for the HC platform has the direct authority (or can meaningfully ask others) to redirect resources needed to complete the project. This person may also be responsible for defining the project and throughput needs, which will drive many of the decisions on specific platform selection. If this person is not leading the effort, they should be kept informed about the purchase and installation process, what support is needed, and the projected costs. This is essential, as many weeks or months can be lost waiting for building upgrades or required funding to be released.

5.2.3   Facilities Management/Lab Operations

Even with limited accessories, installing a high content imager requires space planning. Additional work for facilities may include installing an instrument-grade vibration isolation table, blinds on the windows or light curtains, data ports, electrical outlets, and CO_2 lines. Vendors will have installation guides, but you should review them with the vendor and a facilities/operations contact to assure their relevance to the system (including all accessories) that will be purchased.

5.2.4   Local and Institutional IT Personnel

Information technologies (IT) are essential to HCS, and they fall into micro- and macro-level functions. In large companies, they may be handled by different groups, whereas for small companies and academic settings, they may be played by a single group. Successful setup, use, and maintenance of an HCS system requires that local and institutional IT personnel are consulted before a system is purchased to ensure that server and data management software are installed and integrated properly, and that data management, including database administration and data archiving, is addressed. The IT expertise and considerations important for running HCS systems are covered in detail in Chapter 6. Note that in some settings, institutional IT services have a responsibility for enforcing institutional policies (e.g., data security, archiving, and access policies) in addition to supporting research. Thus, on occasion institutional mandates may conflict with the expectations of the researchers.

5.3   BASIC HARDWARE DECISIONS

5.3.1   Consider the Needs of the Users and the Lab Setting

Before beginning the process of selecting an instrument, it is important to assess the lab setting and anticipated usage. An instrument requiring sophisticated user control may be a problem for a lab with many independent users, and can place an undue burden on the platform manager. An instrument with some basic automation can greatly ease congestion when multiple users are running studies in parallel.

5.3.1.1   Existing Instrumentation   What is the current state of the lab? Is this a facility that is being constructed de novo, or is the instrument an extension of a functional lab? If assays are prepared through automated plate handling, is it desirable to integrate the imager directly into an existing automated system? A survey of the needs of the lab, and the capabilities that the imaging platform is meant to provide, is a good place to start when choosing an HCS system.

5.3.1.2   Number of Independent Users   The number of users that a platform will need to accommodate plays a major role in many decisions. While the user workflows continue to improve in commercial HCS systems, user controls can be quite complex and some instruments may have nonintuitive user interfaces that are challenging to interact with. Inexperienced users will require hands-on training before they can run the system independently. In selecting an HCS system, the ease of use of the instrument and the need of investigators for sophisticated user controls must be weighed, while keeping in mind the time available from the platform manager for training and supporting microscope users.

5.3.1.3   Throughput Requirements   Will the instrument be used primarily for assay development, for screening, or both? True high throughput screens (>500,000 compounds) can only really be run efficiently on a few commercial systems. However, almost all platforms can be configured for smaller screening projects, including genome-wide RNAi screens and chemical genomics screens that encompass thousands to tens of thousands of samples. Still, these smaller projects require explicit configurations, including automated plate handling, in order to run smoothly and in a way that can accommodate other users. This will require expertise across multiple functional areas including automation, imaging, informatics, and IT to manage assay plates and the large volumes of image and numerical data.

5.3.1.4   Biology Considerations   Many complex assays can be handled by any imager. There are some cases, however, where a biological problem will require a specific instrument or accessory. More sophisticated assays such as migration assays, spheroid cultures, some colocalization studies, and live-cell assays might place specific demands on the instrument because of the 3D nature or resolution requirements of the assay, or needs for environmental chambers to incubate cells during imaging. If multiple assay plate or slide formats are required to accommodate different assays, it is worth paying attention to the ability of the imager to accommodate new plastic or glass assay plates. While most instruments have tools to train the instrument on setup and configuration of the various parameters for new plate types, for some instruments, it can take several hours to "teach" the instrument to properly focus on and image new plate types.

5.3.2   Instrumentation Options

Now that you have defined the types of experiments that you intend to perform on your HC system and have established a team to help in the decision process, it is time to consider what system to purchase. The system and its features need to be evaluated in context with the experiments that will be performed, as well as their installation requirements and impact on the manager who will maintain the system. While many options sound great on paper, some options turn out to be rarely used, and some are mutually exclusive. It should be noted that a microscope-based system may not be required. If subcellular localization is not required, then a laser scanning cytometer (instruments are available from Molecular Devices and TTP Labtech), which collect data at low resolution may be sufficient. These systems use laser scanning rather than a microscope to capture images. They are relatively inexpensive, robust, and easy to use, but have important trade-offs. Such systems offer excellent throughput and data storage requirements are generally substantially less than more conventional HCS systems. More advanced applications, those requiring subcellular resolution, will require a microscope-based HCS platform. The following is a brief overview of key hardware choices that need to be considered when choosing a microscope-based system. These functions are described in Chapter 3. Here they are placed in a context of equipment choices, with their commitment responsibilities and an eye toward their costs.

5.3.2.1 *Wide-field Epifluorescence versus Confocal*

One basic choice is whether to select a traditional wide-field epifluorescence or confocal microscope. Described in more detail in Chapter 3 confocal systems restrict out of focus light and therefore allow the imaging of thick specimens or better resolution of internal structures within 3D objects. Confocal systems can generate high quality images gathered from a particular plane in the Z-axis with a reduced background signal.

While confocal microscopy is a powerful tool, there are trade-offs associated with choosing a confocal system that need to be carefully considered. Microscopes with confocal capability are typically more expensive than standard epifluorescent HCS microscopes. For many assays, confocal imaging does not add significant benefits and therefore does not warrant the extra expenditure. Furthermore, by restricting the light to a particular Z-plane, only a small amount of the total light is captured by a confocal system, which translates into a decrease in signal intensity and potentially longer acquisition times compared to epifluorescence systems. Also, confocal systems are extremely sensitive to focusing, because small focus errors can result in a large decrease in signal intensity.

Before choosing epifluorescence or confocal options for an HCS purchase, it is recommended to image your samples on both types of systems to determine which is best for your application(s), with particular attention given to exposure times and image quality. Keep in mind that not all confocal systems offer the same degree of confocality, so it is advisable to collect a Z-stack (by moving through the Z-axis of your sample at a set interval) on several different systems. This will allow a comparison to determine which system does the best job at restricting out of focus light. Some systems offer both confocal and epifluorescence microscopy on the same platform.

5.3.2.2 *Illumination Light Source*

There are typically three types of light sources for fluorescent microscopes: arc lamps (mercury, xenon, metal-halide), lasers, and

LED systems. Each illumination source has its advantages and disadvantages. Arc lamps typically have a short lifespan (in hours: Xenon = 1000, Mercury = 200, Metalhalide = 2000) as compared with lasers and LED light sources, though have a more uniform emission across the visual spectrum. Arc lamps operated beyond their life expectancy have significantly decreased light output *and can explode*, so their usage should be routinely tracked and they should be replaced as required. Not surprisingly, lamp-based systems are declining, but many such instruments are still operational. Lamp replacement often involves a "break in" period and optical realignment, so the longer-lived lamps have an obvious advantage. Some manufactures have turned to supplying lamp sources as a drop-in module, eliminating the alignment process. Though relatively expensive, lasers are also used by some systems, particularly confocal-based systems where the intense light helps to compensate for the loss of light endemic to these systems. There are often limitations on illumination options available on specific HCS instruments, and therefore it is important to understand which fluorophores will be used to help define specific laser or LED options.

5.3.2.3 *Wavelength Requirements*

To discern signal from different fluorescent dyes, the microscope passes light through excitation and emission filters. Some systems have these filters paired together in a cube for a given fluorescence signal, while others have filters installed in separate wheels, where the excitation, emission, and dichroic filters are set separately. The cube format is simpler and makes it easier to standardize protocols, but it does not give as much flexibility and usually provides for only the most common fluors. If users wish to go beyond this limited selection, filter cubes must be swapped out. While not difficult to accomplish, this activity should be restricted to personnel familiar with the process. Also, after swapping cubes, the system will need readjustment by resetting the background correction files (and potentially other alignment functions), a time-consuming operation. If work will be limited to standard assays or the instrument will have many users, filter cubes make more sense. If there will be work on novel approaches (such as FRET), setting filters individually provides some necessary flexibility. For the purchasing process, it is important to gauge these scenarios accurately. If the FRET experiments never materialize, users will unnecessarily struggle with a more complex system.

5.3.2.4 *Automatic Focusing Systems*

There are two general types of automatic focusing systems on the market, image based and laser based. Image-based focusing acquires an image at different focal planes, and determines the optimal focal plane based on the assumption that the best focus gives the strongest signal and highest contrast. Image-based focusing is important for assays that involve small subcellular objects which require imaging at high magnification. There are a number of scenarios where image-based focusing strategies can fail—when images are overexposed resulting in saturated images, or when there is highly fluorescent debris. Image-based focusing also requires exposing cells to large doses of energy as the system repeatedly captures images to determine the optimal focal plane, leading to photobleaching and, in the case of live-cell imaging, increased phototoxicity. On the other hand, laser-based focusing employs a small laser system to detect the air/plastic interface at the bottom of the plate and/or the plastic/fluid interface at the bottom of the well. The system then takes this information and moves a set distance, which the operator has predetermined, to arrive at the correct focal plane. The better systems allow a different offset to be applied for each wavelength, a feature that is handy when imaging objects that reside in different focal planes and to correct for chromatic aberration. Laser-based focusing systems avoid photobleaching and toxicity problems associated with image-based focusing, and are significantly faster. Laser focusing systems are highly dependent on the accurate entry of plate measurements into the database to allow the system to accurately detect the bottom of the well. This is a time-consuming procedure that involves a bit of trial and error. The process of optimizing a plate type should be restricted to highly trained operators and users should be encouraged to limit their choice to preoptimized plates whenever possible. Since both focusing techniques have their advantages, it is advised to choose a system that offers both options.

5.3.2.5 Magnification/Resolution Requirements   Most systems house the objectives on an automated turret, eliminating the routine handling of the objectives. However, there are some systems which require the manual replacement of one objective with another when changing magnification. These configurations can result in dirty and damaged objectives due to handling error, especially in facilities with multiple users. Magnification choices will depend on specific applications. For instance, if you are performing colony formation or migration assays, the resolution requirements are minimal and you can work with a low magnification objective such as 2× or 5×. If you are performing a subcellular localization assay, you will need higher magnification objective such as 20× or 40×. In extreme applications, a water immersion objective may be required for sufficient signal to noise. Some HCS instruments support water immersion objectives.

5.3.2.6 Live-Cell Imaging Capabilities   The capability to image living cells is essential for certain assays requiring time lapse microscopy, and is frequently advantageous in other situations. For example, live-cell imaging can eliminate sample preparation work such as fixation, staining, and wash steps. However, to avoid environmental fluctuations from affecting the assay, live-cell imaging requires an environmental chamber on the microscope to provide temperature and CO_2 regulation. Thus, this feature requires a source of balanced CO_2 which must be supplied by either tank or house CO_2 line.

5.3.2.7 Robotic Automation Compatibility   For screening projects that require reading multiple plates per session, vendors offer various robotics options that can be integrated with HCS systems. These can include relatively simple "standalone" configurations with a stacker-type plate loader to allow for running multiple plates overnight, a robot arm to move plates back and forth from stage to integrated incubator, or they can include fluidic handling capabilities for live-cell assays such as those measuring calcium flux. These options add expense and complexity to the overall system, but can be well worth it depending on the needs of users. The addition of a liquid handling system generally precludes light microscopy such as phase contrast since the space normally occupied by the condenser must be kept open for access by the robotics.

5.3.2.8   Stage Configuration   Another consideration for microscope placement, in addition to having the physical space to accommodate the microscope and robotics, is whether the system needs to be shielded from room light. Most systems image the plate in an enclosed chamber which allows them to be placed in standard lab space, but certain manufacturers keep the plate unshielded, for example, the Olympus ScanR. Furthermore, light microscopy typically necessitates an unshielded configuration. For these situations, light pollution becomes a potential concern. If the equipment is situated in a room used for multiple purposes, people may turn on and off lights while the microscope is imaging, resulting in varying background signal caused by light pollution. With unshielded systems, it is advisable to house the equipment in a designated room or space protected by a blackout curtain where ambient light levels can be controlled.

5.4   DATA GENERATION, ANALYSIS, AND RETENTION

All high content systems come with a basic computer and software to both operate the system and perform image analysis. Because image analysis and visualization software may require more computing power and larger display monitors, you may need to supplement "standard-issue" computers with additional computer power and, potentially, software. For example, a dual-monitor workstation is very useful when visualizing a 3-channel, 96-well data set. Some vendors also offer "image analysis engines" which are server-based applications that significantly increase computing capability. The applications can be hosted on vendor-provided or user-provided hardware.

These supplements represent a substantial investment which is frequently overlooked in the planning stages.

5.4.1   Image Acquisition Software

The image acquisition software is critical as it is the primary interface to set up and acquire the appropriate images for each experiment. It controls all the experimental parameters that include: which wells to acquire, number of wavelengths, number of images per well, and laser- versus image-based focusing. The software needs to provide the flexibility to accommodate numerous types of acquisitions, yet provide an intuitive interface for rapid setup. During the system evaluation, it is a good idea to try setting up as many different types of acquisitions as you can think of to make sure that the chosen system is acceptable.

5.4.2   Data Storage

As previously discussed, HCS systems generate an enormous demand on IT support, which is why it is critical to involve the IT department from the beginning. In addition to data storage (raw images and image analysis data), there are other support needs that include networking, databases, and installing client software. These are covered in more detail in Chapter 6.

5.4.3   Image Analysis Software

HCS systems usually come with their own dedicated software for image analysis and include both preprogrammed modules and development platforms. Pre-programmed modules are designed for specific types of assays (e.g., counting cells, calculating mitotic index, monitoring nuclear translocation) and are easy to set up but generally not very flexible. "Development" platforms have large number of tools (image correction, thresholding, measurement, etc.) that a user can "script" together to form an analysis workflow. These platforms are more flexible and can support more sophisticated analyses, but generally require more programming experience then possessed by the typical bench scientist. Third party, standalone software, not integrated with HCS instruments, is also used for image analysis but is obtained separately and may require the use of a command-line interface (typing commands and writing scripts) or significant programming expertise (e.g., MATLAB, Definiens, CellProfiler, ImageJ, Pipeline Pilot). Image analysis packages vary widely in their capabilities, and an assessment of user needs (flexibility, ease of use) is critical before purchasing a system. For commercial packages, access to the software is usually limited by the vendor to a given number of licensed seats. While it can be challenging to correctly predict how many seats will be desired, it is usually easy to add additional seats later.

5.4.4   System Configuration

Another IT consideration is how to setup the data analysis system. This is covered in detail in Chapter 6 but important points are summarized here. After images are captured and stored, they need to be accessed for analysis. Analysis can be run on desktop computers with access to the image files, or run remotely on computational clusters. As analysis requires computer power beyond the average desktop computer, higher end work stations are necessary if analyses are run locally. It should be noted that some analyses require several hours of processing time depending on size and complexity and can tie up a computer during this period. For this reason, many people opt to run analyses using a computer separate from the computer that is used to acquire the images.

5.5   INSTALLATION

5.5.1   Overview

The installation and initial training process represents a window of opportunity to work out any issues and to get the operation off to a smooth start. Sometimes this process is structured around a Site Acceptance Test (SAT)—these are tests the vendor and user can perform to verify whether the system is operating to specific operational specifications. A user may also require a specific "key" assay to be validated on the system. For instance, a well-characterized assay could be acquired on the system and compared with historical data.

It is to your advantage to have everything in place before the installation team arrives. This includes physical facilities such as power, instrument table, and computer workstation table. It is important to have your database and analysis computers ready to go, since the installation team will be onsite and able to make sure the data is acquired appropriately. During installation, the vendor will establish background correction files and help set up autofocusing parameters. Thus, it is optimal to have on hand all anticipated objectives, filter sets and, if possible, representative experiments on the correct plate type to make the most of this opportunity. It is also useful to have a simple assay plate available—something the user can verify by manual observation on separate microscope. The time of installation is a great opportunity to set up projects, user groups, administrative rights and passwords. An understanding of the projects that will be involved, which users will be assigned to each project, and how these will be organized will help get the most out of this process. During the installation, users should be grouped by project or application to meet with the installation and setup crew to get an overview of the system as installed.

Part of the purchase contract usually includes initial user training. This can either be done on site, or offsite at the vendor's location. Onsite training is advised, since it provides another opportunity to fix problems that may have gone unaddressed during installation. Generally speaking, bringing too many people into a training session dilutes the experience for everyone. This probably is not a concern for a core facility, but for a more loosely structured walk-up situation there may be more casual users than can be accommodated. While the optimal choice is different in each situation, it may be preferable to appoint a small group of end users to attend the vendor-led training sessions and then have them provide training to others at the site. Individuals attending training should be encouraged to take thorough notes and to have actual experiments ready for imaging and analysis during the session. It is also beneficial to schedule a follow-up training after a period of time to provide additional training and answer any questions that arose during initial experiments.

5.5.2 Ownership of Technical Issues

5.5.2.1 Service Contracts The importance of Service and Maintenance agreements cannot be overemphasized. Although functionality and price of the instrument are key considerations when purchasing equipment, it is very common for the customer to overlook the finer details of the maintenance or service agreement. In short, after the equipment specifications, the maintenance/service contract is the most important factor to be considered. Indeed for the successful running of your HCS platform and ancillary technologies, it is essential that the details of the maintenance/service agreement are clearly understood and appropriate for the needs of your institution and users. Maintenance contracts are a high recurring expense, with an annual cost of approximately one-tenth of the purchase price of the instrument. As a general rule, it is best to negotiate the terms and conditions of the M/S agreement before final purchasing decision is made. Important considerations are the geographical location of the vendor service base (you may be charged for technician travel costs, including airfare and hotel) and the response time after a service call is placed. As part of the due diligence process, it is also recommended that you seek feedback from existing customers in your geographical area on service provided by vendors under consideration.

5.5.2.2   Vendor–vendor Relationships   Some thought should be applied to whether you purchase all of the required platform components (automation, image analysis, data analysis, data management) from single or multiple vendor/supplier(s). Both options carry inherent risks and benefits. The single vendor approach offers the most simple solution if technical problems occur. If you encounter any technical difficulties, then the single vendor must accept responsibility and ownership of your problem. In addition, service contacts may be less expensive due to cost bundling. Buying from a single vendor may however not be feasible as their product portfolio may be limited and they may not be able to provide all HCS capabilities that you require. For example, a vendor may not offer the components that are best matched to your needs, such as computer hardware (servers, workstations), automation (plate handlers), image analysis software, integration software, etc. Therefore, if you require customized solutions which cannot be offered "off the shelf" by a single vendor, it will be necessary to purchase components from multiple sources. The main drawbacks of this approach are that it may require time to get the various component systems configured due to compatibility issues and, in the event of a system failure or breakdown, it may be unclear which component has malfunctioned or indeed which vendor/supplier is responsible. In a worst case scenario vendors could end up blaming the problem on each other rather than attempting to solve the problem.

5.6   MANAGING THE SYSTEM

After the system is installed, routine management is the next practical concern. If the HCS system is run by staff in a core facility, these tasks are easily centralized and all users are highly trained and accountable. In a more loosely organized, walk-up facility, oversight becomes more problematic and there is a need for a dedicated platform manager or small group of superusers to be responsible for managing the system. Tasks break down into two categories: keeping the equipment operational and helping to coach and train users.

5.6.1   System Maintenance

Routine maintenance includes ensuring that the system is properly shut down at the end of each day, keeping track of lamp hours, and, when required, changing bulbs and recalibrating the microscope. The installation of new hardware, such as different filters and objectives represents another significant source of effort which will be necessary from time to time as users' needs change. The installation of new software versions will also be necessary as vendors release software updates which may contain bug fixes or new features. Some groups may choose to adopt a standardized Quality Control process to regularly monitor system performance, often running a standard assay and confirming historical results or running a Standard Acceptance Test to confirm the system is running properly. Finally, having a designated person accountable for organizing routine service and repair will help ensure minimize down time and maintain a smooth relationship with the vendor.

5.6.2   New User Training

In all facilities there is some degree of turnover in personnel using an HCS system. Depending on their experience, new users may need advice on experimental design, and may need training that would include running the system and performing image analysis. This training represent a significant amount of work, especially in a walk-up facility where many users are only transiently users of a particular instrument. As with any shared equipment, there are typically two approaches to train new users.

5.6.2.1   Platform Manager   If a single person is responsible for the system, this person will also train other users. This, naturally, standardizes the use of the system, as the platform manager could create standard protocols or preconfigured equipment settings for the different uses of the instrument. As the number of applications grows, the platform manager would update the protocols, which would simplify training.

5.6.2.2   Experienced Lab Members   In a lab environment where there is no dedicated platform manager, there are typically other experienced users who would help train based on their individual experience. One user may train on the acquisition instrument while another may train on image analysis algorithms.

5.6.3   Scheduling Time on the System

If the system is heavily used, it will be necessary to create a method for scheduling time on the system. Solutions range from the simple paper calendar to electronic calendars. Electronic calendars can be created using many websites (e.g., Google Calendar) or integrated with institution calendars (Outlook). Electronic calendars have the advantage in that they can be accessed remotely from user desktops and make it easy to review historical usage.

5.6.4   Billing

For many facilities, cost recovery is mandatory, to both offset depreciation of the equipment and to cover maintenance costs. Some common methods of cost recovery are listed below.

5.6.4.1   Advance Prepayment   Individuals or groups pay an upfront fee for the use of the equipment and software. This approach is often not favored, as it is difficult to predict usage in advance and may lead to inequalities in usage and payment between users.

5.6.4.2   Pro rata Payment   A simple payment for machine time at a given fiscal rate. Equipment usage can either be recorded by the user themselves, quite often this can be as simple as filling in a time or booking sheet next to the instrument, but enforcement and accounting by this method at times can be difficult. Another approach would be to set up the equipment with individual user accounts. In this case the logon times and logoff times can be recorded and then charges made accordingly. This is not always possible and depends on the IT environment and the vendor software. Alternatively there are commercially available software solutions, which monitor and record the activity of preregistered users. These types of systems allow the administrator to preallocate specific amounts of time on a prepay basis to a given user. In this case the allocated time counts down only when the user is logged on.

5.7   SETTING UP WORKFLOWS FOR RESEARCHERS

Running the HCS platform as an ongoing activity is a challenge. The operation of the system as part of a core facility will typically be done by personnel with additional scientific responsibilities.

5.7.1   Introducing Scientists to HCS and the Imager

New users to an HCS platform will typically arrive in waves. There is often a lot of interest as the imager first comes online, and scientists will want to participate in introductory seminars and training sessions that are commonly provided by the platform manager and by the instrument vendor. These may be attended by persons with an immediate problem that can be addressed by HCS, and by persons interested in a working knowledge of HCS, but no imminent plans.

5.7.2   Superusers

Superuser arrangements are common in HCS. A superuser is someone who has a good working knowledge of the instrument and its use in several projects, but has no direct responsibilities in maintaining the system. Following from the axiom that nothing succeeds like success, scientists who develop an important HCS assay frequently feel encouraged to develop more assays, and to work harder to understand the minutiae of the instrument. These people can play an invaluable role in developing assays for new users and in enforcing expectations of fair use of the facility. It may be helpful to informally designate a group of superusers and to ask them to work with new users from designated departments.

Superusers may also have had more formal training than other scientists. Most manufacturers provide their own customized training, as well as regional users group meetings. In addition, vendor-agnostic training is available from several sources (some options are detailed in Appendix A). Superusers typically agree to assist new users in exchange for preferential access by their departments for such opportunities. Superusers are a valuable bridge between the platform manager and the greater HCS community at a research site.

5.7.3   Initial Experiments and Assay Development

When a scientist approaches the platform manager or superuser about a potential experiment, they may have invested a lot of work characterizing the biology of their problem, or they may have read about an assay that they would like to try. Initial discussions will focus on what is needed to validate the reagents and cell system prior to shifting to the imaging platform and developing a quantitative assay. Technical considerations for reagents were discussed in Chapter 2. In general, it should be expected that a researcher has validated reagents and defined strong positive and negative controls to show that a biological event is a candidate HCS assay (see Chapter 7).

In the initial stages of developing an assay, a fair amount of instrument time is necessary. This is particularly important for new users, although much of their time is needed in learning quantitative image analysis, and the tools of the tutorial may provide some opportunity to advance their understanding without demanding time on the instrument itself. A mixture of independent learning and specific attention to the specific question is most helpful in developing an effective cell-based assay in a busy work environment.

When time comes for larger screens, more instrument time and time in larger blocks are inevitable. All commercial instruments have the capacity for automated plate loading, and if screens of multiple plates are at least moderately frequent, this is an invaluable investment. Setting up screening plates to begin in the late afternoon and have them run through the night can take a lot of pressure off of other users and enables better utilization of the instrument.

5.8   CONCLUSIONS

Selecting and installing an imaging platform raises numerous issues for a research group and an institution. It is a major expense and requires the efforts of many parties. Focusing too much on maximizing the options available for each of the instruments under consideration can lead to problems bringing the instrument online, including spending too much on options that will never be used, physically integrating the instrument into the research site, and training users. We have stressed the importance of considering the work environment, the user base, and dedicated support level that will be committed to management of the instrument and the experimental needs first, and then selecting the system that fits best.

KEY POINTS

1. **Research, research, research.** Understand the needs of the group before evaluating and purchasing an HCS instrument: from what assays will be run to who will be using and maintaining the system.

2. **Quality control procedures are critical.** Develop a schedule to replace consumable items (arc lamp, filters). Develop a process to routinely confirm operation of the instrument. This can be a well-characterized assay which is run routinely and results compared to historical values.

3. **Count on colleagues.** Running a multiuser facility, even one with a dedicated platform manager, depends on responsible behavior. Set high expectations for system usage and feedback.

4. **Be critical of anticipated work.** Many purchases include supplemental hardware or a complex system because there is no real response to "what if" questions about potential work. This can result in a complex system being purchased for a large number of users or an inflexible system that makes some studies more difficult. Neither situation is completely unavoidable but can be minimized, and the instrument manager will play a big role in either situation.

FURTHER READING

While we have spoken at length about how HCS is an integration of diverse technologies, this chapter is the first occasion where we have brought in the importance of team management. Such "soft skills" are an important area of laboratory science—difficult to teach but easy to spot when such skills are absent. As such, a recommendation for further reading is less focused on technical aspects of HCS platforms, and offers a chance to consider how to develop strong interactions between colleagues in a scientific environment.

Cohen, C.M. and Cohen, S.L. *Lab Dynamics: Management and Leadership Skills for Scientists.* Cold Spring Harbor Laboratory Press, Cold Spring Harbor, NY, 2012.

6

INFORMATICS CONSIDERATIONS

Jay Copeland and Caroline Shamu

6.1 INFORMATICS INFRASTRUCTURE FOR HIGH CONTENT SCREENING

6.1.1 The Scope of the Data Management Challenge

High content screening (HCS) systems produce huge numbers of large image files. For example, consider the case of the following experiment: the researcher images all the wells of a single 96-well plate in three wavelengths, four fields per well. This experiment will result in over 1000 images! If each image is 1280×1024 pixels then this single plate will produce over 2 GB (GB = 10^9 bytes or 8×10^9 bits) of image data! If a full 384-well plate is used there will be over 4000 images comprising over 9 GB of image data. A project might involve imaging dozens or hundreds of plates. Based on numbers such as these it is easy to see how a single research project can generate hundreds of thousands of images and terabytes of data. Consider further the challenge of moving data on this scale. Moving a 1 TB dataset at an actual data transfer rate of 100 Mbps over Gb Ethernet will take over 22 hours. Even simply reading this amount of data on a disk can be cumbersome. The data transfer rate from disk to computer for SATA (serial ATA) disk drives is 300 MB/s. Thus to sequentially scan an entire 1 TB SATA drive takes about an hour! Nonsequential-data read operations like random reads or writing will take considerably longer.

Keeping track of and safely storing data on this scale is a significant challenge for all HCS facilities and researchers. Commercial HCS instruments are usually sold with software for data management, typically consisting of data acquisition, analysis, and viewing modules. Such software would also usually include a database for

An Introduction to High Content Screening: Imaging Technology, Assay Development, and Data Analysis in Biology and Drug Discovery, First Edition. Edited by Steven A. Haney, Douglas Bowman, and Arijit Chakravarty. © 2015 John Wiley & Sons, Inc. Published 2015 by John Wiley & Sons, Inc.

managing images, image analysis results, and experiment metadata. The metadata includes data about the image, such as the plate barcode, well location, optical channel or wavelength, imaging device characteristics such as degree of magnification or type of objective used, and image field location within the well. Experimental metadata would also include protocol information such as cell line, descriptions of cell treatments, control wells, time points recorded, etc. For many researchers, the vendor-supplied data management solution is adequate and the researcher only needs to be concerned with learning how to use the software. However, many researchers find that the vendor solution does not meet their needs. Common problems with vendor solutions are inadequate image analysis tools, inconvenient or expensive user software licensing, and inadequate image storage capacity. These problems will lead many HCS users to build custom image storage and analysis infrastructure, often working with the support of local IT groups. This chapter is intended to provide an overview of possible data storage and management solutions necessary to support HCS.

6.1.2   Do-It-Yourself Data Storage Solutions

One of the first decisions that you will have to make is whether to develop your own informatics infrastructure or to work with a central IT department to develop your solution, or some combination of the two. For many labs the initial choice is to come up with a quick and cheap solution that relies on lab research staff to assume some level of responsibility for doing their own IT management. This approach is very appealing especially in light of the extremely low costs of consumer-grade high capacity disk drives. One-terabyte disk drives are now available inexpensively. Moreover, the consumer storage industry offers many high capacity storage devices, such as external disk drive arrays, in response to the growing amount of music, movie, and image data that consumers have on their home computer systems. Such devices offer redundancy and management features that rival enterprise IT storage systems of a short time ago, and are flexible and convenient to use. The researcher who uses a portable external disk drive array to store project data can easily move it from one computer to another in response to their needs and workflow. This kind of data portability and convenience is sometimes difficult to achieve in centrally managed IT environments, where there may be restrictions on which computers can access which data due to security policies or limitations in file access protocols.

However, the cost of raw disk drives is only one component of a complete data storage solution. What drives up data storage costs and makes so-called enterprise storage solutions appear to be very expensive are the various layers of redundancy and management capabilities built into complete storage products. Because disk drives are devices prone to wear and tear and, ultimately, failure, they are actually treated as replaceable consumables in a complete storage solution. What the HCS user actually needs is not simply a disk drive for image storage, but instead reliable data storage capacity that is accessible to the relevant HCS devices and analysis systems. The $/GB metric represented in the pricing of consumer-grade disk drives does not come close to capturing all the costs associated with managing HCS data. The true costs of an HCS informatics solution must also include the following.

1. Redundancy in the hardware to prevent data loss due to inevitable hardware failures.
2. The cost of integrating the data storage with the entire HCS data life cycle. The HCS data life cycle includes: data acquisition, initial data processing, data browsing and curation, information extraction and data analysis, data back-up, data archiving, and data sharing.
3. Costs associated with managing storage and analysis hardware and software. IT management is now a mature profession and, as HCS informatics pipelines are developed, it is generally useful to involve dedicated database administrators. Having research personnel (graduate students or postdocs) perform IT management takes away time for doing science and often presents a challenge to providing continuity in data management as lab members leave.

6.1.3   Working with Central IT Departments

If your lab or HCS facility has access to a centrally managed IT department there are a number of services that you may be able to obtain to help with managing your HCS data. The most important services are storage, directory services, workstation management, networking, and application support.

6.1.3.1   Storage   Being able to store HCS images and data on a centrally managed storage server offers a number of benefits. Network fileservers offer a level of performance, reliability, and manageability that are difficult to achieve outside of a managed IT facility. Modern network fileservers are designed such that all hardware components have some level of redundancy to protect them from hardware failures. Disks, disk controllers, network adapters, power supplies, fans, even entire fileservers are fully redundant. Most such systems are able to continue operating even in the event of significant hardware failures, and to provide self-monitoring and notification of hardware problems. For example, if a disk fails, notice can be sent automatically to the fileserver vendor requesting a next-day shipment of a replacement disk.

Replacement of failed components can often be performed without taking the system offline or shutting it down. Power supply, fan modules, disks, etc. can be replaced without disrupting user access to data. Such capabilities are of obvious importance to a large, centrally managed datacenter. Likewise, these systems also have the ability to have their capacity expanded without disrupting services. Increasing the size of disk volumes that are in use on standard desktop fileservers is usually impossible without taking the storage offline. Online capacity expansion capabilities are a great advantage for labs and facilities that have instruments that receive a high level of use. Screens can proceed according to their own schedule and the managers of the fileservers do not need to ask screeners to stop their work when routine maintenance needs to be done. This is a nontrivial benefit particularly considering the potential cost of disrupting a running screen.

6.1.3.2   Directory Services and User Account Management   One of the most important services that a central IT department can provide is user account management, or more broadly, directory services. The primary function of user accounts managed in a central directory is to provide appropriate access of users to the resources they need and to restrict access to resources that they do not need. For example, a user of an HCS instrument needs to login to the computer controlling the instrument and save image files and data to a folder either on the local disk of the instrument controller computer or to a folder on a network fileserver. In principal, the user's login credentials should not allow them to overwrite, modify, or, in many cases, even view the files of other users unless there are explicit policies allowing these actions. Furthermore, the user's credentials should not allow the user to modify the software or settings on the instrument controller computer beyond what is necessary to run a normal experiment. Some HCS systems allow active user role management to control which users have access to view or modify specific datasets. Within some HCS applications, roles can be specified for specific types of users, such as project managers, who might have access to files of multiple different users. More commonly, however, users may be assigned to groups so they can share data within a lab or project.

In an environment with access to a central directory service, such as Microsoft Active Directory, all of this can be accomplished based on the credentials provided when the user first logs into the instrument controller computer. In many labs and facilities, however, instruments are set up to run continuously under a single privileged account. This is often the case because scientific instrument manufacturers typically do not develop instrument control software that can work correctly with an ordinary unprivileged user account. Another common cause of this situation is that vendors prefer to install new instruments using local machine accounts with default account names and passwords. This simplifies the installation and maintenance tasks for the vendor since they do not have access to user accounts in the customer's central directory. It also makes it simpler for service engineers and technicians to gain access to the machines and devices they maintain. Unfortunately a common side effect of this practice is that it can take a considerable effort to integrate a device into the lab or facility network environment. Sometimes it is necessary for a special service account to be set up in the central directory services system for the device to operate correctly. Or special groups with administrative access to specific computers need to be created so that researchers and laboratory technical staff have sufficient access to maintain and operate the devices. Making all these adjustments can usually be accomplished with cooperation between IT, laboratory staff, and the vendor.

6.1.3.3   Workstation Management   Workstation management is a service commonly offered by central IT departments. Such services typically begin with the initial purchase of the hardware and software based on standard configurations and vendors, initial installation of the operating system and application software including volume-licensed software, and installation of mandatory security and antivirus software. Initial setup of the computer will also include configuring its network settings and perhaps setting up printers, and automatic software update settings. After a computer goes into use, there are additional management tasks including installing software updates, security patches, and bug fixes that may be performed by either the vendor or the IT department, depending on the nature of the vendor support agreement. Most HCS vendors typically include a PC as part of their HCS system. Therefore the IT department might not have as much control over the initial setup and maintenance of this HCS instrument PC. Occasionally there may be compatibility issues between the HCS software and the host computer operating system or shared software libraries. It is recommended that the person managing the HCS instrument coordinate major software updates with the HCS software supplier. Likewise, if the HCS software supplier needs to update the operating system or shared software libraries, the IT department should be informed. Major software changes by the vendor might require the IT department to update settings for networking, security, backup, or other aspects of PC maintenance.

6.1.3.4 *Networking*

HCS instruments require several networked computers to function effectively. Because of the large number and size of the images being produced and analyzed it is essential that the network to which the component computers are connected is reliable and of high quality. Ideally all computers should have at least Gigabit per second connections to the network. The network itself should provide sufficient bandwidth and reliability so that automated data acquisition runs and data management-scheduled tasks that rely on saving files to network fileservers are not interrupted midrun. Network "glitches" or congestion that occurs in the middle of a run have the potential to ruin an entire data acquisition run. If such problems appear to be happening, it is vital to have the network management staff examine the logs of the network hardware that might be involved. It may be determined that some network hardware should be upgraded in order to accommodate the bit rate requirements of the HCS system. Temporary network issues are less likely to affect data acquisition runs if all data is initially written to local disk. This approach requires that data, especially image data, must be moved off the local HCS PC on a regular basis to avoid filling up the disk drive. Utilities for moving the data to a network fileserver may be available from the HCS vendor. If not, the manager of the instrument should consult with their IT department to develop a strategy to keep the local disk from filling up.

6.1.3.5 *Application Support*

The HCS vendor usually provides HCS application support. However, it is a good idea to have, in addition to the HCS instrument manager, central IT staff involved in the maintenance of the HCS software. Troubleshooting HCS software issues may require good knowledge of the computer's general configuration and its network configuration. A member of the central IT staff may be in a better position to address any changes that might need to be made in the base system configuration in order for a problem with the HCS software to be resolved. IT managers in many large organizations deploy software updates on workstations remotely by pushing updates to workstations over the network using technologies like Microsoft Active Directory or specialized scripts. In this way many computers can be efficiently managed and kept up to date.

6.1.3.6 Integration of the HCS System into the Existing IT Environment Successful integration of any HCS system into an existing IT environment requires engagement of IT managers and technical staff. Unfortunately, purchasing decisions and even shipment, delivery, and initial setup of scientific instruments often proceed before IT staff is brought into the process. There are understandable scientific and organizational cultural reasons for this but it can lead to delays, frustration, and misunderstandings down the road when the laboratory or facility finds itself needing the assistance of IT staff to make their HCS system fully functional. Therefore it is always a good idea to consult with and engage IT personnel as early as possible in the process of acquiring and deploying research instruments with significant IT requirements. Ideally, device vendors would require their customers to engage with their IT departments in the process of developing the specifications of the device and its physical requirements.

6.2 USING DATABASES TO STORE HCS DATA

6.2.1 Introduction

In this section we will discuss the role that databases play in the storage, analysis, and management of HCS data. Databases are a vital component of HCS environments and are essential for managing data created by screening instruments. Databases are used in the creation and maintenance of the HCS infrastructure necessary for preparing screens. They can be used for tracking plate layouts, reagent libraries, and experimental and analysis protocols. When screens are run and data are collected, databases are used for the storage of images, metadata, and analysis results. A screener may encounter a number of databases in the course of their work ranging from simple spreadsheets to vendor-supplied HCS applications, to large-scale customized HCS database applications. Such databases, which are involved in various separate, but related, parts of the overall HCS workflow, are sometimes not well integrated with one another. For the purposes of this discussion the most important database for the screener will be the HCS database containing the images, image metadata, and image analysis data. Image metadata are typically collected separately from the image files. Since the image files and their associated metadata are stored separately, the link between the two forms of data is inherently fragile and must be carefully maintained. If the metadata files are moved, lost, or corrupted, it is possible that the relevant image and image analysis files will be essentially useless. Image analysis data is also typically stored separately from the image files and image metadata.

6.2.2 Types of Data

6.2.2.1 Image Data Image data starts with the raw image files produced by the detector hardware in the instrument. Currently available detectors for epifluorescence and brightfield microscopy instruments include 12- and 14-bit CCD cameras, 14- and 16-bit Scientific CMOS cameras. Confocal laser-scanning-based instruments typically use photomultiplier tubes as detectors that produce either 10-bit or 12-bit images. In most cases the images are stored in 16-bit grayscale TIFFs. Acquisition software assembles the data produced by the detectors into an image data file that is written to a hard disk drive. Image files can be compressed at this point or later to reduce the amount of data stored on disk. Lossless data compression algorithms that allow the original file to be regenerated, such as bzip2 or gzip, should always be used. Lossy compression algorithms, such as those used in generating JPEG image files should only be used for generating human-viewable galleries, web pages, presentations, and other media intended only for direct viewing. Lossy compression destroys the information content of a scientific image and makes it useless for quantitative analysis.

6.2.2.2 *Metadata*

Metadata is the nonimage data about the image, experiment, detector, processing of the image data, etc. that is stored in association with the image data. For example, the metadata can include pixel size, type of detector, the name of the experimenter, the name of the project or dataset, and information about experimental conditions (e.g., objective used, filters, exposure, time points, etc.). Metadata is of critical importance in HCS applications because without it interpretation of the experimental results is impossible.

At the time of acquisition, image metadata may be written to one or more text files in various formats and associated with the image files through standardized naming conventions, or written to special image headers that accept text-based information. Some vendors use the TIFF header "tags" for recording metadata, such as cell line, treatment and dose, time point, etc. In some cases, limited, basic metadata (e.g., plate ID, well #, filter) might be encoded into the name of the images files. Following initial acquisition, the metadata may then be imported into a database to make it accessible to the sort, search, and query functions of the database or any applications that can access the records in the database.

Because metadata and image data are fundamentally different data types, they are often stored separately. Metadata is stored in text files or databases with pointers to the images files. Image files are usually stored in a file system dedicated to image storage. It is possible to store both types of data together by encoding the metadata and image data into a common file format. The advantage of this approach is that it keeps the metadata and image data together. Some HCS vendors take this approach using their own proprietary file formats. Unfortunately, such proprietary file formats are often only readable by the vendor's application software. Exporting the data to a format that is readable by third party analysis software often entails a loss of much of the image metadata. The OME-TIFF and OME-XML file formats are attempts at common file formats for pixel and metadata that are open and readable by a wide variety of third party software tools [1].

6.2.2.3 *Derived Data*

Derived data is tabular, numerical, or textual data resulting from the processing, analysis, and interpretation of the image data. Whether derived data is stored with the image and the metadata depends upon the degree to which image analysis and interpretation occurs within the same software environment used for collecting the data. Some HCS environments, such as Thermo-Fisher's Cellomics system, can perform image analysis at the time of acquisition and encode derived data, such as cell counts, in the image file. This approach offers speed and convenience but it comes at the price of data portability since it is impossible to export the data out of the original file format without losing the associated derived data

Another approach is to conduct basic or initial data analysis within vendor-supplied software packages and then to move the derived data into separate data analysis environments, such as ImageJ, MATLAB, or CellProfiler, for additional or more customized analyses. The implication of this kind of workflow from a data management perspective is that copies of the original image files tend to proliferate and meaningful connections between the final analysis results and the original raw images are harder to maintain. Unless the researcher is very disciplined and organized there will be a tendency for intermediate results, duplicate, and redundant data, to accumulate on disk drives. This can be expensive in terms of data storage costs and problematic in terms of keeping track of all the steps used in processing a large collection of datasets. Documenting all the steps in a complex, multistage, data analysis workflow is just as important as documenting wet-bench protocols.

6.2.3 Databases

"A database consists of some collection of persistent data that is used by the application systems of some given enterprise." [2] Databases have also been described as collections of self-describing data, usually in table format. These definitions mainly cover the data store itself and do not address how the data is added to the store, modified, updated, analyzed, or queried. These other functions are often thought of as being distinct from the data store itself and are rather part of the "database management system" or "client applications" that access the database. The most commonly used database management systems today are Relational Database Management Systems (RDBMS) that use some form of the database query language SQL (Structured Query Language) for accessing and querying the data. RDBMS databases have been developed primarily to meet the needs of managing transaction data in business applications. Unfortunately, RDBMS databases are not optimal for the storage and management of large collections of images because reading, writing, and processing image files cannot be performed efficiently within relational databases. Storing image files on standard file systems is still the preferred method for large image file collections. File systems are optimized for fast input/output of large files, while databases are optimized for querying, indexing, and sorting information and not for fast file input/output.

The standard approach that is used in nearly all HCS image management systems is to store images on a file system and to use a database to store image metadata, including pointers to the image file locations in the file system. This is the approach used by well-known consumer applications such as Apple's iPhoto as well as major HCS instrument vendors including Perkin Elmer's Opera system and Molecular Devices Corporation's ImageXpress Micro system. While this architecture is nearly ubiquitous, it does have some limitations. First, there is always the risk that either records in the database or in the image file system will be lost or separated, leaving records that are missing their associated image files or orphan images that do not have corresponding records in the database. Most applications attempt to prevent these problems by concealing the image files from ordinary user access through the file system. The image file names and the directory structure is not intended to be "read by humans." The disadvantage of this, of course, is that the data is not self-describing outside of the HCS application.

6.2.4   Basic Features of an HCS Database

When an automated scanning device collects images from one or more multiwell plates the images must be stored in a manner that keeps track of the well location, field, wavelength or channel, plate number, etc. of each image. Standard image file formats, such as the TIFF image format, are able to store some of this image metadata through the use of extended headers. However, most image file formats are not well suited for storing the various types of metadata associated with complex HCS datasets.

The most common HCS database options range, in the order, more simple to more complex, from self-maintained spreadsheets and image collections, simple electronic lab notebook applications, vendor-supplied screening databases and applications, to custom built databases and open source databases. Databases may be stored on instrument control computers, data analysis workstations, or on share network fileservers or database servers. Figures 6.1–6.3 illustrate where each of these systems typically reside in an HCS workflow. Each of these options have tradeoffs and advantages in terms of simplicity, complexity, cost, portability, and features.

One of the most important factors driving database choice is the structure of the data—the file and metadata formats—automatically generated by the imaging system that the screener is using. In most cases the database is part of an integrated hardware/software platform. This will be the screener's starting point and will dictate what options the screener should consider for downstream data analysis and management. Many screeners will choose to rely upon the vendor-supplied solution for simplicity and practicality. Vendor solutions, for all of their drawbacks and shortfalls, usually come with technical support and a community of users with shared knowledge including screening facility users and staff.

6.3   MECHANICS OF AN INFORMATICS SOLUTION

6.3.1   Introduction

In this section we will describe the basic mechanics of how data is created, viewed, processed, analyzed, and managed in an HCS environment. The steps involved can be thought of as a "pipeline" or workflow through which the data proceed from the start to the completion of the project. What is presented here is idealized and the emphasis will be on describing data management challenges that might arise. Figures 6.1–6.3 illustrate data paths in several possible data acquisition/analysis pipelines.

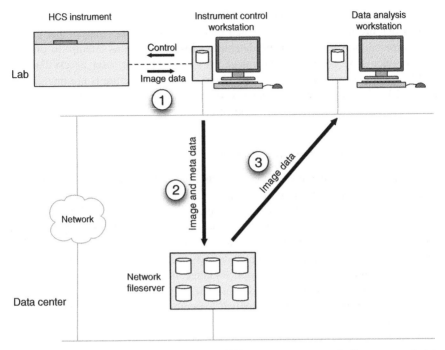

FIGURE 6.1 *Basic data flow between an HCS instrument, network fileserver, and data analysis workstations.* Bold arrows represent logical data paths between components of the system. The IT network topology is indicated by gray lines. Connections between the HCS instrument and the instrument control workstation can be either direct or via a network, as indicated by a dashed line. In this basic configuration metadata is encoded in the image file. After acquisition, image data and metadata would flow first to the instrument control workstation (step 1), then is transferred to a network fileserver (step 2). For data analysis, the image data is transferred to a data analysis work station (step 3).

6.3.2 Data Life Cycle Management

6.3.2.1 Acquisition   Acquisition starts with a plate containing samples being imaged. Images of labeled cells are captured and then written as image files to a file system. Given the large number of images that the screener will be collecting it is important for images to be collected and stored quickly. Therefore the rate at which the detector can read out data to be written to the file system is critical. Other factors influencing the rate at which a plate is imaged include the speed of the robotics positioning the plate in the microscope, autofocus speed, time needed to switch between imaging wavelengths or channels, and the actual exposure times for each channel. Modern research-grade CCD cameras can read out data at more than 500 Mbps. It is extremely important that the rate at which the image data can be written to disk is not a bottleneck in this process.

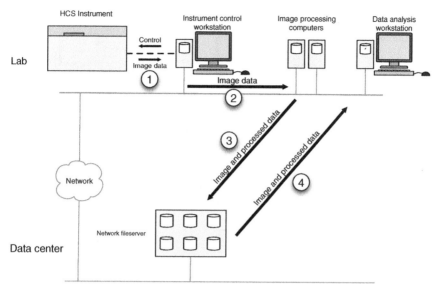

FIGURE 6.2 *HCS data flow with initial image analysis carried out as part of the image acquisition process.* Bold arrows represent logical data paths. The IT network topology is indicated by gray lines. Connections between the HCS instrument and the instrument control workstation can be either direct or via a network, as indicated by a dashed line. In this configuration, after acquisition, image data flows first to the instrument control workstation (step 1), then is transferred to a specialized data processing computer (step 2) for an initial round of automated image processing. Images and processed data are then transferred to a network fileserver (step 3). For further data analysis, image and processed data are transferred to a data analysis workstation (step 4).

In most cases the detector is connected to a special adapter card installed on the instrument PC's PCI bus. The data is then passed across the PCI bus and written to a local disk also on the computer. Modern PCs with SATA disk drives can easily handle this readout rate even when the camera is reading out almost continuously. Writing data over the network to a remote networked disk is more prone to failure. Most laboratories are equipped with gigabit per second Ethernet (GigE). The actual data transfer rate of GigE is actually much less than a gigabit per second because of network protocol overhead, thus the bandwidth of the network connection is often barely more than the camera readout rate. While it may be possible to write directly to network file systems when relatively small images are being collected at a low rate, it is usually safer to write to a local disk. In fact, many HCS instruments will halt operation completely if they encounter a significant delay or error in writing data to disk. Thus, it is often necessary that image data be initially collected on the instrument PC and then moved to a network fileserver in a second step. Separate data analysis PCs are then able to access the data from the fileserver without interfering with the operation of the instrument control PC. (Figure 6.1, steps 1–3). It cannot be

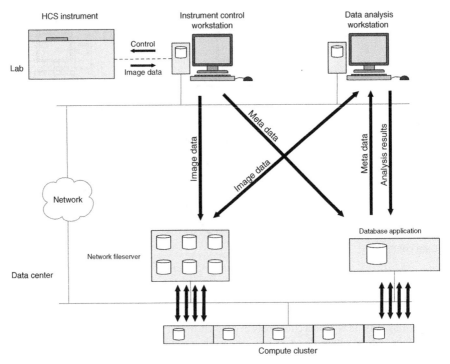

FIGURE 6.3 *HCS data flow in a distributed computing environment.* Bold arrows represent logical data paths. The IT network topology is indicated by gray lines. Connections between the HCS instrument and the instrument control workstation can be either direct or via a network, as indicated by a dashed line. In this configuration, after acquisition, image data flows first to the instrument control workstation. Subsequently, image files are transferred to a network fileserver and the metadata is transferred to a database. For image analysis, image files and corresponding metadata must be read back to a data analysis workstation. For very large datasets, image analysis can be accelerated by running the analysis algorithms on a distributed cluster. Image analysis results are stored in the database. The ImageXpress Micro HCS system by Molecular Devices uses a configuration for storing image files and metadata similar to the one shown here.

stressed enough that thinking this step through carefully is key to developing a robust infrastructure.

6.3.2.2 Initial Review and Processing   In some HCS environments the image data is first written to one or more specialized data analysis computers where an initial round of image processing is performed as part of the data acquisition process. (Figure 6.2, steps 1–3) Instrument control and initial image processing are carried out on separate computers so that neither process interferes with the other and work is distributed. The Opera (Perkin Elmer) and ImageXpress (Molecular Devices) systems take this approach. The screener can specify a set of analysis scripts to be performed immediately upon acquisition. The scripts are run on the image processing computers before any subsequent data transfers.

6.3.2.3   *Image Analysis*

After initial image processing the image data and processing results can then be transferred to a network fileserver. (Figure 6.2, step 3) This allows for additional processing and analysis using separate image analysis computers. (Figure 6.2, step 4) Separating the initial and subsequent image processing allows additional processing to occur concurrently with running screens without interfering with the acquisition process. This method, while simple and straightforward can have some drawbacks. It requires that the researcher copy the data back to an analysis workstation from the fileserver (Figure 6.2 step 4). With large datasets this step can take a tremendous amount of time. Such data copying can lead to redundant data and creates data management difficulties, such as key data and results being stored on individual PCs or disk drives that are subject to failure and data loss.

Another approach is to store the image data in a high performance fileserver and to store the metadata and analysis results in a database server. In addition, image processing and analysis can be carried out in parallel on nodes of a computer cluster. Systems of this type can be challenging to design and deploy because they require integration of the storage, databases, and image processing subsystems. An example of such a system is illustrated in Figure 6.3. Some vendors now offer cluster computing options for running their own image analysis software, for example Molecular Devices Corporation's MetaXpress PowerCore technology. MetaXpress PowerCore can significantly speed up the analysis of very large high throughput datasets by distributing the work of processing the image analysis algorithms across a group of PowerCore client computers. Installing and maintaining such clustered computational systems can be expensive and complex and requires significant IT support and expertise.

6.3.2.4   *Documenting Protocols and Workflows*

How HCS data is processed and analyzed should be considered part of the "protocol" of the experiment and therefore should be documented as carefully as the reagents used and how the experiment is run. As a practical matter it can be very difficult to keep track of complex, multistage image analysis workflows. One problem is that the experimentalist is often not a computer scientist and therefore is likely to use image analysis software they did not develop and that they do not understand fully. The use of software packages, algorithm libraries, and code that the researcher does not understand is analogous to using a lab instrument like a centrifuge without training. It is sufficient to document the name and model of the instrument, or software package and algorithm that was used, so that it is possible to repeat the experiment in principle.

Complex workflows are also difficult to manage because they often contain multiple, essentially manual, operations that do not lend themselves to easy documentation. Unless the researcher has created entirely automated routines for transforming data at each step of the analysis, key steps in the process may remain hidden from the experimental record or become difficult to repeat. This is especially true when researchers use programs like Microsoft Excel for manipulating tabular data. It is possible to encode all the steps involved in using a program like Excel with macros. Even when a researcher uses automated scripts and macros for data analysis it will be essential to document how they were used and fit together in a larger workflow. Ideally, all data analysis steps should be automated and documented and all source code maintained for reuse.

6.3.2.5   Exporting Images and Analyzed Results   In later stages of a project using HCS, the researcher will need to export images as well as numerical data resulting from image analysis. Some, but not all, commercial HCS software is capable of producing publication-quality graphs from quantitated image data. More commonly, the researcher will export the analysis data to software specifically designed for producing publication-ready figures. This is most commonly accomplished by exporting to one of the common tabular file formats such csv or xls and then reading that file into another application for producing figures, such as Microsoft Excel.

6.3.2.6   Data Sharing and Transfer   Data sharing may take place at almost any stage of the data analysis pipeline. Collaborators may require access to raw data or there may be a need to share raw and derived data postpublication as required by journals or funding agencies. If the collaborators are affiliated with the same institution then the simplest solution is to provide the collaborators with read-only access to the data on a network fileserver. If the people needing access to the data are from outside the institution, this kind of sharing can be satisfied by providing remote access to the data (via VPN connections or remote desktop), by transferring a copy of the data over the Internet via ftp or scp or, if the dataset is large, copying the data to external USB disk drives and shipping the drive.

6.3.2.7   Archiving   In later stages of a project, there will come a time when archiving of the original data must be considered, usually because of the high cost of storing data live on disk-based file systems for immediate access. Archiving in this context means moving the data to a lower cost medium. Retrieval of the data needs to be assured but it does not necessarily have to be instantaneous. Tape is a common archive format. Tape has the advantage of being relatively inexpensive compared to disk drives. A tape cartridge does not consume any energy, other than what is necessary to transport and store it in a safe location. The disadvantages of tape include the difficulties inherent in keeping track of what is on the tape, the impracticality of browsing the contents of the tape, the possibility of an entire archive being lost as the result of a relatively small failure in the tape medium, and the lack of good mechanisms to detect and correct errors in the data once it is on tape. Many of these technical problems have solutions but they can be extremely costly and difficult to implement. Some laboratories may choose to make multiple copies of data on external disk drives simply to avoid potential pitfalls of storage on tape, despite the fact that external disk drives have their own problems. It is recommended to consult your local IT group to find an archiving solution appropriate to your circumstances.

6.3.2.8   Curation   Data curation is an oft-neglected step in the data life cycle process. Few researchers have the time or inclination to carefully document, cull, and manage their data in the course of carrying out a significant research project involving HCS. However, what that means as a practical matter is that the key data that provides scientific insight can easily be lost amidst a vast collection of unimportant data. Data generated by test screens, failed protocols, and misbehaving instruments are often intermixed with good data.

The best way for these curatorial problems be avoided is for there to be some sort of protocol for the researcher to organize, annotate, or classify their data as it is being collected, perhaps to sort and classify datasets as important or trivial as the data are being produced. Also, there are technologies coming into use that make curation easier for large digital datasets. One such technology is the tagging and annotation capabilities of the Open Microscopy Environment (OME), which is discussed in more detail below. Data stored in the OME can be tagged with an arbitrary set of user-defined tags much in the same way that browsers allow for tagging of web site bookmarks and blogging software allows for tagging of blog posts. Such tagging can help create a locally defined taxonomy of datasets based on various classification criteria, such as descriptors of result quality (tagging out of focus wells or technically compromised wells) or descriptors for various control or experimental conditions.

6.4   DEVELOPING IMAGE ANALYSIS PIPELINES: DATA MANAGEMENT CONSIDERATIONS

6.4.1   Using Commercial Image Analysis Software

Many screeners will choose to analyze their data with the off-the-shelf software integrated with their HCS instrument. For screens using common cell types and assay protocols (e.g., those counting cells, or monitoring neurite outgrowth, or cell cycle state) this might be the easiest and fastest solution. The software will come with support, documentation, and possibly training. Another advantage of commercial, off-the-shelf software is that the data and metadata tend to be easier to manage when it stays within the confines of a single data analysis platform. Since the data is not being moved from one platform to another, either via data copies or exports, there is less likelihood of data duplication. This reduces the complexity of managing the data generated over the lifetime of a project since the bulk of the data, metadata, and derived data will be maintained within the commercial software environment.

Full HCS data analysis software and storage systems can be quite expensive. They are generally sold as complete hardware and software packages, including database and storage servers, and user data analysis workstations. Some data analysis suites are also sold as software-only products that can be installed on any suitable computer. While these systems can be quite costly to purchase and maintain under service contract, many labs will find that this approach is less expensive than developing custom software solutions. For many applications, off-the-shelf analysis packages will be the quickest approach for getting a new instrument into production.

6.4.2   Using Custom Image Analysis Pipelines

Despite the advantages of using standard software tools provided by the HCS instrument vendor, many screeners will choose to develop custom data analysis solutions. As described elsewhere in this book, there are a host of reasons that a screener may choose to develop customized image analysis tools and techniques. The off-the-shelf software may not be capable of extracting the necessary information, perhaps because assay protocols are used to monitor cell phenotypes that were not anticipated by the software vendor. The screener may also prefer to use software tools and algorithms, for example, MATLAB or CellProfiler that are more open and amenable to sharing with the research community, so that peer review of both the data and data analysis techniques may be carried out. Commercial, closed source image analysis software is problematic in that it does not allow either the researcher or the peer reviewer to scrutinize basic steps in the processing and analysis of the research data because the analysis algorithms are concealed within precompiled binaries. This section describes some issues to keep in mind when working with a custom data analysis pipeline.

6.4.3   Data Duplication and Uncontrolled Data Growth

One of the pitfalls of using a custom image analysis pipeline is uncontrolled data duplication and growth of derived data. A common practice is to export image files from the commercial, off-the-shelf data repository to another location prior to analyzing the data with custom software. In fact, it is not unusual to create a new copy of the data at each step in the analysis chain in order to preserve all the steps in the workflow. At some point, however, the screener will have to decide what to do with all the copies of the data and the derived data. Data storage costs and size limitations may force the screener to delete most of the derived data generated in the data analysis pipeline. One approach to managing this kind of data growth is to write all intermediate data that will not be saved to a temporary or "scratch" file system that is automatically purged on a regular schedule. How to manage raw data and derived data should be evaluated carefully in relation to the cost or feasibility of recreating the original and derived data. In next generation sequencing, for example, it is common practice to delete raw data after it has been analyzed because the cost of recreating the data is less than the cost of storing it.

6.4.4   Metadata Loss

One of the most common problems encountered when moving data to a custom analysis pipeline is the loss of metadata. Frequently, metadata is encoded in a proprietary file format or database application that is unreadable by any software other than that supplied by the vendor. Usually it is possible to export the image data to a common file format, such as TIFF, with basic screen–plate–well–field–channel data encoded into the TIFF file name. Any metadata that cannot be encoded into the file name is generally lost. Metadata lost in this fashion is a major impediment to researchers considering analyzing their data outside of the software environment of the instrument vendor. Solutions to this problem are needed in the field. The common data format being developed by OME is one solution that is being developed currently.

6.4.5   Data Movement, Network Bandwidth Limitations, and the Challenges of Moving Large Datasets

When using a custom image analysis pipeline it will likely be necessary to move large amounts of data across the network. For large datasets this can be a rate-limiting step, especially if the files need to move between institutions, as may be the case for collaborative projects. One approach to moving vast multi-terabyte datasets has been described as "TeraScale SneakerNet." [3]. This method of moving large amounts of data consists of shipping complete data storage servers or disk arrays using common freight carriers.

6.4.6   Problems with Handling Very Large Numbers of Files

Custom image analysis workflows involving large HCS datasets can easily result in the creation of many thousands of files or directories. Opening, reading, listing, and copying such a large number of files or directories can cause problems with even the most capable modern network fileservers. One commonly used approach to avoiding these problems is to store files in hierarchical directory trees in such a way that no one directory contains an excessive number of files. For example, when storing images from a collection of plates one could have a separate directory for each plate. Each plate directory would then contain separate directories for each well. The well directories could then contain the actual image files or be further subdivided into field or wavelength subdirectories, and so on.

Even when using a hierarchical directory approach it can be a significant challenge to perform file system operations encompassing large datasets containing thousands of files. A relatively new technological approach is to use solid-state disk drives (SSDs) to improve file system performance. SSDs use solid-state memory for storing data. Lacking any moving parts they are much faster than conventional electromechanical disk drives, particularly at performing seek and random read operations. This makes them especially useful in operations accessing very large numbers of files. Unfortunately, the current high cost of SSDs makes them impractical for large-scale storage applications. However, disk drive and storage manufacturers are beginning to employ SSDs in hybrid configurations by using them to store portions of frequently accessed data to speed up overall file system performance. This speeds up certain kinds of file access operations dramatically when there are very large numbers of files. Disk firmware automatically moves frequently accessed data to the solid-state memory. In theory this improves disk performance significantly. It remains to be determined whether such technologies will prove useful in storing large HCS data collections.

6.4.7   Parallel Data Processing

Image analysis algorithms can be computationally intensive, sometimes requiring hours to process and analyze images from even a single 384-well plate. To speed up analysis, parallel processing techniques can be used. Parallel processing is when a computational problem is divided into a number of separate calculations that are carried out independently of one another. For example, one could analyze the images collected from multiwell plates by spreading the computational load across a group of servers in a compute cluster. This approach is fairly straightforward, yet it requires that the programmer manage the distributions of the data and processing jobs across the cluster and then assemble the results.

There are several vendors that offer parallel processing solutions for their analysis software, such as Molecular Devices' MetaXpress PowerCore technology, which runs on a cluster of several server computers and a number of client image processing worker computers. Other image analysis software, such as Definiens and CellProfiler, can be run on general purpose computer clusters. Note that licensing terms for some software packages might constrain the total number of processing cores on which the software can be used. If a screener has access to a computer cluster and is able to employ parallel processing techniques, special consideration should be given to how to handle data access. In order to avoid performance issues, one should consult with the compute cluster administrators and determine the best approach for accessing the data on multiple compute nodes. In some cases it may be best to copy frequently accessed data to each compute node. In other cases it is better for the data to reside in a single network location and to have all compute nodes access the data concurrently. Which solution is chosen will depend on the performance characteristics and design of the compute cluster and its data storage environment. (Figure 6.3)

6.4.8   Workflow Documentation and Automation

One of the challenges of working with custom image analysis pipelines is documenting all the steps in the workflow, particularly as it is being developed, to ensure that the process can be repeated consistently. Steps that should be documented include the image analysis as well as data analysis steps such as replicate averaging, production of dose response curves, etc. Ultimately it will be desirable to automate as many steps as possible. The results and data derived from the workflow should be associated into the workflow used to create them. One of the benefits of using automated workflow tools (e.g., Pipeline pilot—http://accelrys.com/products/pipeline-pilot/) is that they reduce the need to store intermediate derived data and can be accessed from an easy-to-use web-based site.

6.4.9   Software Development and Maintenance: Managing Software Development Projects

While not strictly a consideration in the management of the research data itself, management of the custom software code is an important component for creating a custom analysis pipeline. The code and the algorithms and workflows that they contain are an important component of the overall research design. HCS researchers developing their own software should use a code version control system, such as Git or Subversion, and document what versions of their software was used in a particular workflow or analysis. This is standard practice in professional software development efforts and is beginning to be used in research labs as biologists become more familiar with standard software development practices. Unfortunately, many biologists who are relatively new to software development do not use versioning systems and instead manage their code in an *ad hoc* fashion. This is an undesirable practice, especially when multiple users are working with each other.

6.4.10   Software Sharing, User Training

One of the most important benefits of developing custom image analysis pipelines is the opportunity that exists for sharing new techniques with the wider screening community. Custom code developed using open source software tools can be shared more easily, either by hosting it on a webserver or by providing downloadable code both of which are common practices in the software community. One obvious advantage of this is that shared code can be improved and the improvements then shared again with the screening community. In addition to maintaining well-documented code, it may also be helpful to develop user training materials such as online tutorials or video screencasts.

6.4.11   Image Repositories

Custom image analysis pipelines offer greater flexibility, but this may come at the cost of increased complexity, particularly when it comes to managing large collections of image data. This is especially true if you attempt to manage your files manually as just a collection of files. The number of files and the aggregate size of the datasets will make it extremely difficult to keep track of where the data are.

A possible solution to this problem is to implement some sort of image repository. A commonly used architecture for image repositories is to store and manage the image files using a front end, database-driven application. The image files still exist as individual files stored on a file system. However, the naming of the files and their location in the overall file system hierarchy of folders and subfolders is managed exclusively by a database-driven client application. This is also how most HCS data analysis applications are structured. Actual pixel data—that is raw image data derived from images imported into the system—are stored in a file system hierarchy that is only intelligible to the system and is not intended to be directly accessed by people. All access to the images is mediated by the server and client applications. For large, complex collections of image data this approach is safer and more efficient because it discourages direct user access, which can be prone to errors. Of course this means that the software controlling access to the data needs to be robust and contain safeguards against data loss and corruption.

6.5   COMPLIANCE WITH EMERGING DATA STANDARDS

Data standards are being developed in a wide range of biological disciplines for a variety of purposes. One purpose of these cooperative standards development efforts is to establish a consensus for how metadata should be recorded and reported in experiments. Many such data standards are in the form of minimum information checklists, such as the Minimum Information About Microarray Experiments or MIAME standard [4] or Minimum Information About Cellular Assay (MIACA) (http://miaca.sourceforge.net). These were developed so that experimental results can be reported, compared, and interpreted unambiguously, and so that sufficient information is recorded to allow for experiments to be reproduced. There are many Minimum Information checklists for many research subdisciplines, all organized by the MIBBI Project [5]

Some data standards focus on facilitating data portability between different hardware and software environments. This approach to data standards is motivated by the wide variety of instruments and analysis applications in use today. Within a single lab or facility there often exists several instruments that produce data formats that are completely incompatible with one another. Many researchers would prefer to consolidate their data into a single data analysis environment and not be constrained by the tools provided by individual instrument vendors. Moreover, as development of custom data analysis pipelines becomes more commonplace, image analysts often work with data produced by multiple imaging instruments.

It is important to emphasize that data standards generally do not apply to the internal format of the data generated by the HCS instrument or data analysis application. Each instrument and application has its own particular requirements and therefore should not be expected to adhere to a particular internal data format or standard. However, to enable both adequate description of the experimental conditions and transfer of data to other analysis environments, it is necessary that the source data contain the requisite information specified by the data standard. To ensure this it may be necessary for the instrument software to "enforce" compliance with the standards by requiring that the researcher enter and correctly specify all mandatory metadata. If this task is considered excessively tedious or burdensome the researcher may try to circumvent the data entry requirements. This can be a barrier to widespread adoption of new data standards. Therefore the design of data standards must strike a balance between being general enough to encompass all important metadata types and not so overly specific that it is difficult to comply with in day-to-day practice.

The ability to move HCS data from one instrument or analysis platform to another is largely determined by whether the platforms can interoperate with open and widely adopted data sharing file formats. Historically, many HCS device manufacturers have been reluctant to open their file and metadata formats to open standards because they believe this would conflict with the business model that seeks to maximize profits through the sale of analysis software as part of an all-in-one data acquisition and analysis system. Device vendors will argue that such all-in-one systems provide significant benefits to their customers by giving them a nearly complete HCS image acquisition and analysis environment. While it is certainly understandable that many customers will want the convenience of a complete, turnkey system, it can come at high cost, both financially and in terms of flexibility. Fortunately for researchers seeking alternatives there are a number of open data sharing standards that allow data to be moved between different acquisition and analysis platforms. Most vendors support image data export to TIFF format, a standard image file format. Where many vendors fall short, however, is in the export and sharing of image metadata.

In an effort to meet the need for standardization of metadata in microscopy file formats Laboratory for Optical and Computational Instrumentation (LOCI) at the University of Wisconsin–Madison has developed Bio-Formats. Bio-Formats is a Java library for reading and writing life-science image file formats. (http://www.loci.wisc.edu/software/bio-formats) Bio-Formats contains a collection of file importers that can read a variety of file formats and convert them to several standard image file formats including the image file format called OME-TIFF. Over 75 file formats are now supported by Bio-Formats which include several produced by HCS instruments.

OME-TIFF is based on the OME data model [6] and was created by the OME consortium (http://openmicroscopy.org) The OME also has created the OME Server and OMERO Server. The OMERO Server in its version 4 release began supporting HCS data in a screen/plate format. Perkin Elmer has adopted OMERO as basis of its Columbus Server product which is the common image server for the Perkin Elmer HCS product line. (http://www.cellularimaging.com/products/columbus/) The OMERO server is open source and available for free download and includes the most recent version of Bio-Formats built into the file importer client.

The ultimate goal of the OME consortium and collaborators like LOCI is that OME-TIFF and the OME data model will become widely adopted by vendors of life-science imaging instruments and software. Supporters of these efforts note that standards have benefited many industries and actually spurred innovation and technological development (http://www.thinkstandards.net/). Unfortunately, at this stage of the development of HCS technology, movement toward standards in the area of metadata and file formats is in its early stages. Of course in other aspects of HCS technology, standards have played a very important role in adoption and growth of HCS technology—the 96- and 384-well plate formats are perhaps the most obvious examples. In the mean time Bio-Formats will continue to be incrementally expanded and improved as support for more proprietary file formats are added and improved. To see which file formats are currently supported see http://www.loci.wisc.edu/bio-formats/formats.

6.6   CONCLUSIONS

Managing HCS data is a significant challenge. Ideally, lab managers and scientists should work closely with their IT departments early in the planning for the acquisition of new HCS instruments. It is important to consider how to manage HCS data throughout its full life cycle from acquisition to publication and management. While it is sometimes tempting to build standalone solutions for storing such data, there may be considerable benefits in taking advantage of the IT resources already available.

KEY POINTS

1. The many files generated by imaging experiments, including images of each field for each wavelength used for each experiment, quickly become a logistical challenge.
2. While not required, a formal database not only helps organize files, but is integral to some data analysis platforms, particularly for tracking experimental metadata.
3. IT personnel are becoming aware of image-based experiments, but discussions between experimenters and IT are critical for setting the data management system itself, as well as data retention rules and a growth plan.

FURTHER READING

Swedlow, J.R. et al. Bioimage informatics for experimental biology. *Annual Review of Biophysics*, 2009, **38**: 327–346.

Swedlow, J.R. et al. Open file formats for high-content analysis. *High Content Screening: Science, Technology and Applications*. Haney, S.A. (ed.) Wiley Interscience, Hoboken, NJ, 2006.

REFERENCES

1. Linkert, M. et al. Metadata matters: access to image data in the real world. *The Journal of Cell Biology*, 2010, **189**(5): 777–782.
2. Date, C.J., *An Introduction to Database Systems*, 6th edn. Addison-Wesley systems programming series, Addison-Wesley Pub. Co., Reading, MA, 1995, xxiii, 839 p.
3. Gray, J. et al. *TeraScale SneakerNet: Using Inexpensive Disks for Backup, Archiving, and Data Exchange,* 2002.
4. Brazma, A. et al. Minimum information about a microarray experiment (MIAME)-toward standards for microarray data. *Nature Genetics*, 2001, **29**(4): 365–371.
5. Taylor, C.F. et al. Promoting coherent minimum reporting guidelines for biological and biomedical investigations: the MIBBI project. *Nature Biotechnology*, 2008, **26**(8): 889–896.
6. Goldberg, I.G. et al. The open microscopy environment (OME) data model and XML file: open tools for informatics and quantitative analysis in biological imaging. *Genome Biology*, 2005, **6**(5): R47.

7

BASIC HIGH CONTENT ASSAY DEVELOPMENT

STEVEN A. HANEY AND DOUGLAS BOWMAN

7.1 INTRODUCTION

Developing a high content assay involves taking a biological observation and converting it to a quantitative readout. The factors to consider in advancing this initial observation into a high content assay are biological, statistical, and technical in nature. Biological considerations include kinetics, for instance, if the observation is present for only a short time (such as 5–30 minutes after stimulation). It might be a challenge to use it to develop a large-scale screen, but a small-scale screen may be possible. Statistical considerations are relevant for image analysis, especially if the event only occurs in a small fraction of cells, say less than 1%, and a large number of cells need to be imaged and analyzed to produce significant data. In this chapter, we will discuss the basics of setting up an HCS assay and highlight some of the technical considerations.

7.2 INITIAL TECHNICAL CONSIDERATIONS FOR DEVELOPING A HIGH CONTENT ASSAY

7.2.1 Plate Type

HCS instruments typically screen standard microtiter plates, particularly 96-well and 384-well plates that conform to SBS guidelines. These guidelines facilitate the use of a plate across different instruments, including plate-handling robots and liquid-handling instruments that can process plates before, during, and after imaging. HCS

An Introduction to High Content Screening: Imaging Technology, Assay Development, and Data Analysis in Biology and Drug Discovery, First Edition. Edited by Steven A. Haney, Douglas Bowman, and Arijit Chakravarty.
© 2015 John Wiley & Sons, Inc. Published 2015 by John Wiley & Sons, Inc.

instruments are also generally capable of imaging cells on standard microscope slides, and lower and higher format plates including 6-, 12-, 24-, and 1536-well plates can also be used. The ability to set-up multiple experimental conditions (cell line, drug concentration, antibody conditions, replicate wells) and to pipette by hand using single or multichannel pipettors makes use of 96-well plates, a virtual standard of HCS assay development. Many assays can be set up using standard objectives with relatively long focal lengths, however, experiments requiring high magnification or resolution may require thinner plates; plastic and glass plates with thinner well bottoms are commercially available.

7.2.2 Choice and Use of Staining Reagents and of Positive and Negative Controls

For a new assay, it is common to evaluate several primary antibodies (the minimal plate design discussed in 11.3.2 can be extended for several antibodies in a single plate). Since high content assays can be subtle, including counting or measuring puncta, they are very vulnerable to such effects. For proteins that respond to a treatment or culture conditions, comparing cells with and without treatment is essential. Common examples are changes in nuclear localization or phosphorylation state after treatment with interferons, growth factors, DNA damaging agents, etc. RNAi knockdown to reduce levels of the target protein is another valuable control experiment [1]. Wells with only the secondary antibody and those without primary or secondary antibodies are very important for evaluating new reagents.

The choice of assay controls is strongly affected by the nature of the biological problem. These may not be the same controls that were used to screen the antibodies. Assays where proteins, si/shRNAs, or small molecules will be tested should use control molecules of the same type as will be screened, as the magnitude or time course can be distinct for each class of molecule. Previously, we were concerned that our antibody may not be specific for the protein we thought it was. Here, we are considering how strong our assay response is. Treating cells with a powerful reagent (e.g., staurosporine) is fine in the former case, but if we will be performing an siRNA screen, siRNAs that induce apoptosis or cell death are the relevant positive controls. While all experimental scientists are familiar with the concept of controls, some specific aspects of control setup are worth mentioning here. At this point, we are looking for a change in one or more cellular responses. The extent of this change will be the basis of our assay. How the visual feedback from the controls compares with the actual data once converted to numbers is an important part of the process, but for the moment, the key thing is that some definitive change has occurred that can be recognized as different. This latter point is important, and it is always a good idea to pull in a colleague and compare images (doing this blinded is even better).

7.2.3 Plate Layout

The distribution of samples across the plate plays an important role in the assay, such as the occurrence of **edge effects**, changes in assay data that result from samples being locate in the outer wells of a plate. As such, plating all of the positive or negative controls in the outer wells can distort the data. In setting up an assay, plating replicates in both outer and inner wells can alert one to the presence of edge effects, although many investigators bypass the problem entirely by only using the interior wells of a microtiter plate. There are other layout-related factors that can affect an assay. Setting up a dose–response assay in the same direction across a plate can mask a time-dependent effect, such as precipitation of a compound, magnifying or blunting the true effect of the treatment. Things like this are unusual, but they do happen, so changing the orientation of treatments during the assay validation stage is a good idea. Sometimes, if a spatial effect on a plate is suspected, it is desirable to run a "uniformity plate" which consists of negative (or positive) control wells across an entire plate. Visualizing a specific metric (number of cells, fluorescence intensity, etc.) of a plate as a heat map is a useful tool to identify trends within a plate. This can often be visualized with the vendor software or by exporting data to Excel.

7.2.4 Replicates

As mentioned above, replicates are essential to determining assay stability and overall response. How replicates are set up can affect how much they vary, and therefore whether they are a true measure of how robust the assay is. For smaller assays carried out in a single plate, replicate wells across a dose response or time course is necessary, as critical concentrations or times can give sharp shifts in signaling changes, which can be difficult to identify accurately if one is not appropriately accounting for background noise. Given the scale of image acquisition available in a high content screening, running identical samples over several plates is valuable for determining true variability and is not very difficult to set up. Doing so allows for absolute differences between plates to be measured, and is useful for determining the impact of various plate normalization methods, such as normalizing to control wells or taking a plate average of all of the wells. A summary of the different types of replicates is shown in Table 7.1.

7.2.5 Cell Plating Density

The density at which cells are plated affect multiple aspects of an experiment. From a biological point of view, plating density can affect certain cellular properties, such as growth rate, which can be attenuated as the cells reach confluence, and so in many screening experiments, cells are typically plated at low densities. On the other hand, primary cells and differentiating cells (such as stem cells) may require high cell densities or even feeder layers in culture. Perturbations, particularly siRNA delivery, are also dependent on cell plating densities. In general, higher densities have reduced knockdown efficiency, but if a phenotypic measure such as cell death following transfection with siRNAs against PLK1 is used, the extent of cell death is also reduced because the cells are proliferating less. Image analysis algorithms are very sensitive to cell density. Some image segmentation methods may explicitly sample the background before measuring the labeled regions of the cells and this can be difficult or impossible for experiments which are plated at high densities. It also

TABLE 7.1 Replicate Types

Replicate type	Example	Uses	Limitations or caveats
Instrumentation (imaging) replicates	Identical samples on the same plate, can also be on different plates	• Consistency of image capture and analysis (e.g., frequency of out-of-focus images) • Determining response magnitude and temporal range during assay development • Important for determining plate artifacts (such as edge effects)	When a single master mix is aliquoted into multiple wells, replicates do not adequately measure assay variability
Automation replicates	Identical samples on different plates (when applied using some form of automated sample preparation or delivery)	• Assesses consistency of automated sample preparation and dispensation • Indication of stability of events over time (when plates are prepared or processed over a length of time) • Allows testing of plate normalization methods • May take the form of a pilot screen	May under-represent biological variability when automation merely aliquots a single set of samples (as above)
Biological replicates	Independently prepared samples, either on the same or on different plates	• Best measure of assay variability	The more valuable the measurement, the more the effort to set up (such as setting up assays on different days) Such experiments may also use technical replicates
Laboratory replicates	Same assay run in different laboratories and/or with different personnel	• Important when an assay is being transferred to another group, such as for high throughput screening, or as a prerequisite for publication of a novel or unusual result	Unrecognized ambiguities in a protocol or reagent specification can introduce discordances which can be difficult to resolve

can be difficult to segment individual nuclei and therefore will result in an inaccurate cell count. As such, it is impossible to optimize an assay around one criterion, and all aspects of assay development that are affected need to be checked.

7.3 A SIMPLE PROTOCOL TO FIX AND STAIN CELLS

The mechanical process of staining is a highly empirical optimization of primary and secondary antibodies, although a good general protocol will account for the majority of HCS assays, and allow for the introduction of protocol optimization steps. As such a protocol is offered and diagramed in Figure 7.1. The protocol comprises all of the basic steps required to process samples. It is tailored for an indirect immunofluorescence study where the secondary antibodies are labeled and detect the primary antibodies. This protocol assumes that at least one unlabeled primary antibody will be used, requiring a labeled secondary antibody for imaging. The grid outlined in Figure 7.1 can be replicated several times on a single plate, either to compare different antibodies, antibody validation controls against the same antibody, or the potential assay controls, as differentiated above. Different reagents, such as directly labeled primary antibodies (generally well validated and used in flow cytometry) or phalloidin, and cell lines that

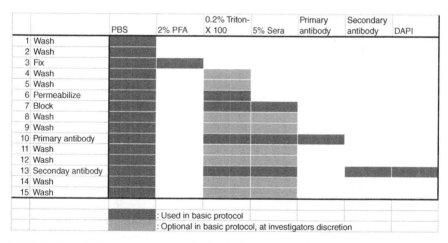

		PBS	2% PFA	0.2% Triton-X 100	5% Sera	Primary antibody	Secondary antibody	DAPI
1	Wash	■						
2	Wash	■						
3	Fix	■	■					
4	Wash	■		░				
5	Wash	■		░				
6	Permeabilize	■		■				
7	Block	■		░	■			
8	Wash	■		░	░			
9	Wash	■		░	░			
10	Primary antibody	■		░	░	■		
11	Wash	■		░	░			
12	Wash	■		░	░			
13	Seconday antibody	■		░	░		■	■
14	Wash	■		░	░			
15	Wash	■		░	░			

■ : Used in basic protocol
░ : Optional in basic protocol, at investigators discretion

FIGURE 7.1 *A diagrammatic view of a basic fixed-cell labeling protocol.* The protocol is frequently successful, and so represents a good starting point for increasing sensitivity, cell structure, or characterizing primary antibodies. The sequence of steps runs from top to bottom, with the composition of reagents for each step indicated by the gray boxes in each column. Details of such adjustments are described in the text. The shading scheme is intended to provide some flexibility, even in this basic protocol. PBS is used in all steps, PFA is used only in the fixation steps. Many labs will add Triton X-100 to all postfixation steps, but it is not necessary. Some labs will also maintain serum levels in all steps subsequent to blocking, but frequently it is not necessary.

use GFP-fusion proteins will have a different protocol (elimination of the secondary antibody incubation and washes for these examples). A short discussion of each stage follows, and more extensive advice is presented in Chapter 11.

7.3.1   Washing Cells

It is typical, but not essential, to wash the media from the wells prior to fixation. In cases where the cells are loosely attached, this step can be eliminated.

7.3.2   Fixing Cells

Two to four percent paraformaldehyde (PFA) is most common for fixing cells for HCS. Formaldehyde, formalin (formaldehyde/methanol), methanol, and ethanol are also used. Alcohol fixatives are more likely to alter morphology. For PFA, a 2% solution is a good place to start. Solutions containing PFA or formaldehyde are toxic, so fixation steps should be carried out in a hood. If an automated system is performing this process, the system should also be housed under a containment hood. PFA oxidizes in air so purchasing sealed ampoules is recommended. Nonvolatile alternatives (e.g., Mirsky's Fixative) are also available for screening labs and other circumstances where vapor containment is a problem.

7.3.3   Permeabilization

Cells may need to have the lipid bilayer extracted with a low concentration (e.g., 0.1%) of a nonionic detergent in order for antibodies and many stains to access intracellular proteins. A few reagents do not require permeabilization, including some DNA intercalating dyes. Receptor uptake assays will sometimes avoid a permeabilization step, so that cell surface-localized proteins are preferentially stained.

7.3.4   Blocking

Blocking with sera reduces the nonspecific interactions between the antibodies and the cells. Blocking with bovine serum albumin (BSA) is common, as BSA is readily available to cell culture labs. Blocking with serum of the species that the secondary antibodies were raised in is generally recommended to reduce nonspecific binding by the labeled secondary antibodies (frequently donkey or goat). Some investigators prefer a single specific blocking protein (such as BSA fraction V). Five percent is considered common for protocols where blocking is done for an hour, 2% is more common when blocking is performed overnight. It is common to use this solution ("blocking buffer") for all antibody-based steps.

7.3.5   Postblocking Washes

It is possible to skip these steps and go straight to the primary antibodies, as the primary antibody solution frequently is blocking buffer. Reasons to skip these washes include when cells are lightly adhered to the plate and can be washed off in excessive treatments.

7.3.6   Primary Antibody Application

When beginning a study, typically a high concentration of antibody (1:100) can be used. Most binding occurs within an hour at room temperature. Altering antibody concentrations, longer incubations, and reduced temperature (4°C) can help improve the signal or reduce background. This step is one of the most frequently adjusted, although if staining is not sufficient in 2 hours, longer incubations rarely improve it significantly, and this is the most common factor in selecting primary antibodies. The minimal effective dilution will be affected by the illumination source of the instrument, with stronger light sources (lasers) capable of producing satisfactory signals with lower antibody levels (1:1000 or lower). Although lower dilutions conserve antibodies, note that this is because the illumination can compensate for the lower signal through increasing excitation light intensity, and that as the lower limit is reached, the loss of signal intensity will increase variability in the assay, discussed later.

7.3.7   Postprimary Antibody Washes

This is an important step, and can often be extended to three washes. These are performed with either PBS/detergent or blocking buffer.

7.3.8   Secondary Antibodies

Commercially available labeled secondary antibodies are very robust and have low background. As such, there is typically less optimization required than for primary antibodies, but coordinately optimizing the primary and secondary antibodies (see Chapter 11 for examples). It is common for protocols to have the nuclear dye (DAPI or Hoechst) used in a separate step shortly before imaging, but it can be combined with the secondary antibodies without a problem. A dilution of 1:100 or 1:200 is typically a fine starting point, with a 1 hour incubation at room temperature. If the primary antibody is labeled, this step and the following washes are skipped. In such cases, the nuclear dye could be combined with the labeled primary antibody.

7.3.9   Postsecondary Washes

These are obviously necessary to remove the majority of the labeled antibodies and nuclear dye. Two washes are usually sufficient, but three or even four can be performed if nonspecific staining is observed.

7.4   IMAGE CAPTURE AND EXAMINING IMAGES

An initial image capture may be useful to check for reagent performance and significant processing difficulties. Visualizing images early in the process will confirm many experimental parameters, such as cell plating densities and correct/expected cellular localization of specific antibodies. This step also provides the images needed to develop image analysis algorithms.

7.4.1   Resolution, Magnification, and Image Exposure

Low power objectives (5× or 10×) are usually sufficient for image capture for many assays. Higher power objectives, are more difficult to focus (due to the shorter working distance), which increase errors in image capture. Use of high power objectives also increases the time it takes to read each plate, because fewer cells are captured in each field of view, and more images must be acquired per well for statistically significant results (see Section 7.4.2). If the event is very subtle, such as the localization of a protein to an organelle, it may require a higher power objective, but very robust data (even for morphological changes) can be captured with lower magnification objectives. Typically, a pilot experiment that is analyzed by one or more image analysis routines will help to establish the extent of magnification necessary and will provide an indication of the number of cells that need to be analyzed per treatment.

In order to acquire images with a higher signal-to-noise ratio, one must adjust instrument parameters related to camera exposure to maximize the acquired signal without saturating (overexposing) the detector. HCS instruments typically have an automated exposure adjustment feature to help determine an optimal exposure. For dim signals, this may result in a long exposure time, which may not be desirable as it can result in photobleaching or phototoxicity (especially for live-cell imaging where images are acquired over a period of time.) Confocal imaging, as described in earlier chapters, can also damage a sample, even if the exposure is adjusted by the instrument, because confocal imaging captures a small fraction of the light used to illuminate the sample. In either case, one should avoid saturation/overexposure as this will prevent one from analyzing the intensity within your sample, and it may result in errors in the image analysis algorithms.

In cases where the object of interest is very dim, a couple of options exist. First, using the strongest fluorescent dye available and making sure that the filter is optimized can help (see the discussion above on primary and secondary antibodies). There are also configurable parameters within the instrument software that can also help increase signal intensities. The first option is binning pixels, as discussed in Chapter 4. A second option is to increase the gain. This maintains a single pixel resolution, but pixel measurements are multiplied by common factor, raising the overall image intensity after exposure. This raises both signal and background, so its use is limited to raising a faint image to the point where it can be analyzed, but cannot resolve a flat image with a high background.

7.4.2   Number of Cells to Acquire for the Image Analysis Phase

This an empirically driven issue, and is ultimately addressed by whatever statistical test is relevant to the assay. In an assay development phase, several fields can be captured, and the assay quantified using different numbers of cells. How this can be done will vary significantly from instrument to instrument, but in general, reanalysis can be performed *post hoc* and the number of cells to be quantified can be set. Alternatively, all the cell-level measurements can be exported and studied in a data analysis application such as Spotfire™.

There are a few general guidelines that can help. For highly robust assays, such as for transcription factor translocation to the nucleus, few cells (in the range of 100–200) may be sufficient for an assay, and these kinds of assays perform well in HTS, where capturing a single field per well is all that might be necessary. Rare (less than 1% of cells) or infrequent events (say, 5% of cells), such as mitosis in unsynchronized cells, will usually require 1000 or more cells for an assay to be sensitive enough. Morphology measurements, particularly in machine learning or phenotypic clustering of many features, may require several thousand cells. All of these estimates would rise if the assay examines a continuous response rather than a discrete threshold for a "positive" versus a "negative" response. Two examples would be a complete measure of the cell cycle or a rate of movement over time.

7.4.3   Performance of Positive and Negative Controls

Once you have captured images, an important reality check is to look at the cells themselves, before completely moving over to working with the numerical data. Does the fluorescent antibody exhibit the expected cellular localization and/or intensity change? Do they report the observation as you expect (cell death, translocation, etc.) across all of the replicates? For comparisons of different antibodies or treated/untreated comparisons, images from the same dilution conditions should be compared side-by-side. Ultimately, each of these conditions should be quantified, as signal-to-noise may be better for one of the conditions, or it can turn out that it is fairly flat across all conditions, and that the effect of increasing antibody concentrations is to create brighter staining.

7.5   CONCLUSIONS

At this stage, you have plated, treated, stained, and imaged your cells and have essentially completed an experiment. The next stage can be to try to optimize these steps, particularly if the results are wildly unexpected, or try to set up an image analysis algorithm and extract quantitative data from your images. As you can imagine, this is a highly iterative process, working around the wet-lab, image acquisition, and image analysis phases.

Developing an HCS assay requires a lot of time spent refining an initial observation, but getting started may be intimidating to persons very new to HCS. The approach outlined here is to jump into the process with a very general protocol to see how things stand, and then use these initial results to begin the assay development process.

Although the focus of this chapter is on the wet-bench work, the data generated during the early phase of assay development also allows the exploration of image analysis approaches. The standard for evaluating potential improvements to the assay must be data driven, meaning based on the performance of the image analysis algorithm, but the algorithm itself will be under development. Image analysis algorithm development works in conjunction with optimization of the cellular treatments.

KEY POINTS

1. HCS assay development is a highly iterative process that ultimately optimizes a variety of components that include the wet-lab protocol, reagents, image acquisition, and image analysis algorithms.
2. Define appropriate positive and negative controls early. These are important in defining the dynamic range of the assay and are used to optimize image analysis algorithms.
3. Define plate layout early. This will enable the development of a standardized method and potential assay automation.
4. Understand sources of noise. Optimize staining protocol, then instrument, then analysis algorithm. Strategies are outlined in Chapter 11.
5. Understand the right metric (out of many image analysis features) which will measure the biological phenotype.

FURTHER READING

There are many sources for immunofluorescence assay development. Several popular examples are:

Current Protocols (many titles available, including Cell Biology, Chemical Biology and Cytometry) www.currentprotocols.com

Harlow, E. and Lane, D. *Using Antibodies: A Laboratory Manual.* Cold Spring Harbor Laboratory Press, Cold Spring Harbor, NY, 1998.

Spector, D.L. and Goldman, B.D. *Basic Methods in Microscopy: Protocols and Concepts from Cells: A Laboratory Manual.* Cold Spring Harbor Laboratory Press, Cold Spring Harbor, NY, 2005.

Yuste, R. *Imaging: A Laboratory Manual.* Cold Spring Harbor Laboratory Press, Cold Spring Harbor, NY, 2010.

REFERENCE

1. Appasani, K. (ed.). *RNA Interference Technology: From Basic Science to Drug Development*, 1st edn, Cambridge University Press, Cambridge, UK, 2005, 510p.

SECTION III

ANALYZING DATA

The following chapters bring a shift to our introduction of HCS with a dramatic move to working with the data generated through image analysis. From the measurements made during image processing and analysis, outlined in technical terms in Chapter 3 and presented pragmatically through the Tutorial, image analysis generates the measurements that will be used to define which siRNAs regulate a pathway or which compounds cause cells to arrest in G1 or S phase. A connection between conducting cell-based experiments and reviewing the tables of data generated is the essence of "getting" HCS. However, many experimenters skip this process. This is because many HCS assays performed on commercial instruments do this through canned algorithms that are developed by the manufacturers to preformat the image capture and data analysis. As we are focusing on the principles of HCS, we will cover the data thoroughly. Even if a scientist ultimately gathers most of the their data through such preconfigured algorithms, it is important to appreciate the breadth of the data available, and the properties of such data, including the options for quantifying a given object, and the intrinsic heterogeneity of cellular imaging data, which can have important consequences on the appropriate method of analysis.

Chapter 8 introduces the data tables themselves, and makes some distinctions in both terminology and concepts. Most importantly, we spend time on the notion of how these many measurements relate to the biology or pharmacology of the experiment and leads to the notion of the assay metric—the feature or features that are specifically relevant to the experiment. Chapter 9 starts the process of analyzing data and discusses the common method of analysis for HCS data—the traditional methods of analyzing data at the whole-well level, where the average measurement for a feature for all cells in a well is used for analysis. Replicate wells are used to determine variability for the assay and standard screening statistics, such as the Z' measure of assay robustness are calculated. In this regard, HCS assays are similar to other screening assays, looking

An Introduction to High Content Screening: Imaging Technology, Assay Development, and Data Analysis in Biology and Drug Discovery, First Edition. Edited by Steven A. Haney, Douglas Bowman, and Arijit Chakravarty.
© 2015 John Wiley & Sons, Inc. Published 2015 by John Wiley & Sons, Inc.

at cell states at the whole-well level, but substituting enzyme activity with cytological events. Almost all cell-based assays are evaluated with such methods, and therefore, so are most HCS experiments. What makes Chapter 9 a little different is that we take advantage of the cell-level measurements that are available in HCS to evaluate the quality of the data before analyzing the actual experiment. This is something that ought to be expected of all analyses, and is commonly done when an experiment is analyzed by a statistician or bioinformatician, but it is frequently omitted from "canned" or commercial algorithms. In addition to serving as a potential quality control measure, they can alert one to cases where the data should not be analyzed by standard methods and either the data should be transformed or analyzed by alternative methods such as single cell analysis. Fortunately, we cover these methods as well. Chapter 10 introduces some important concepts in how cells respond to events in culture and how they are measured. Historically, these methods have been more common in flow cytometry, but there is increasing appreciation of the atypical (or more precisely, nonnormally distributed) responses by cells in culture. These patterns indicate that cells are frequently highly heterogeneous and may even indicate the presence of subpopulations that respond very differently from a single average response. Chapter 10 presents alternative assay methods that account for intrinsic heterogeneity, such as when a treatment induces a gradient of responses, including when more complex assay systems are used, such as differentiation of cells in culture or in coculture systems. Finally, they form the basis of complex analytical approaches such as phenotypic profiling, clustering and machine learning, concepts covered in the last section of this book.

8

DESIGNING METRICS FOR HIGH CONTENT ASSAYS

ARIJIT CHAKRAVARTY, STEVEN A. HANEY, AND
DOUGLAS BOWMAN

8.1   INTRODUCTION: FEATURES, METRICS, RESULTS

At this point in the book, you understand the conceptual aspects of image processing and microscopy, and have a clear picture of what it takes to set up and run a simple high content assay out of the box. Congratulations, you are (almost) ready to start designing your own high content assays! In graduating from running traditional whole-well assays (such as well-based enzyme activity or viability assays) to designing your own HCS assay, you will face the new challenges of starting from a biological effect, using image processing to extract features, converting the features to metrics, and mapping the metrics to a final result. While this can be more difficult and time-consuming, the effort pays off in carefully considering how the assay reports on the biological process of interest.

Before we leap right into all this, it is worth defining a few terms explicitly. **Features** are the cellular structures that are measured by the image analysis programs; they are quantitative properties derived from each image acquired in a high content assay. Some features are recorded for each cell and may be summarized for each well (e.g., the length of individual neurites), while others are made on a per-well basis (e.g., number of nuclei per well), or are aggregates of other features (e.g., the ratio of nuclear and cytoplasm abundance levels or the ratio of length to width of the nucleus). In these last examples, it is clear that some features can be redundant or highly correlated. **Metrics** are what is measured to determine whether the experiment has produced a result. Only features that report on the relevant biological event are

*An Introduction to High Content Screening: Imaging Technology, Assay Development, and Data Analysis
in Biology and Drug Discovery*, First Edition. Edited by Steven A. Haney, Douglas Bowman, and Arijit Chakravarty.
© 2015 John Wiley & Sons, Inc. Published 2015 by John Wiley & Sons, Inc.

metrics. Some experiments will only use one or two features as metrics, others will integrate dozens or even hundreds. **Feature data type** is the kind of measurement that is made. Some feature types are direct measurements, such as a length, an area, or a perimeter; some are intensity measurements (brightness). Different data types may require different methods for analysis. These distinctions are discussed in this and subsequent chapters. This chapter will help differentiate features and metrics, and tie them into the final result of an assay. Some of the points here may be obvious to a practicing biologist, while others are a little more unique to high content assays.

8.2   LOOKING AT FEATURES

We begin the process of analyzing experiments by looking at features that are calculated and tabulated from image analysis. We will work through an example that is used in the tutorial, the inhibition of nuclear translocation by a series of compounds. A graphic of the experimental design is shown in Figure 8.1a. Here several compounds are used to treat cells in decreasing concentrations, moving from left to right in columns 2–10. Each row is a different compound. Column 11 is used for negative and positive controls. The design is replicated in two additional plates, so replicate treatments for each compound and dose are found in the corresponding location on the other plates. Figure 8.1b shows a graphic of one way to measure the results, a ratio of nuclear to cytoplasmic abundance. Compounds that are more potent inhibit nuclear localization to a greater extent at a given concentration. Increased nuclear/cytoplasmic ratio is depicted as a larger circle. For many platforms, the ability to generate data views such as this is part of the instrumentation software or a supplementary data analysis package. As a primer, and in an effort to remain platform-agnostic, we will export the data and review them directly. This is actually quite important, as it may not be obvious what options exist or how much data is recorded, particularly if analytical streams are borrowed from others or are canned approaches.

Measurements made by image analysis programs are recorded in files, typically text or comma-separated values files, (".csv files"). The image analysis programs available for quantifying high content images all have their own lexicon or jargon when it comes to naming features. For the most part, the features recorded are similar measurements with different naming conventions. Thus, getting started in data analysis is partly a matter of getting familiar with the naming conventions of the analysis platform you are using. Each platform will assign default names for each feature, and if specific names for the labels are entered when the algorithm is set up, these will be used to amend the default headings. Typically, an image analysis algorithm will record data in two files. One file will be the individual measurements for each cell, or the cell-level data. The other file will be the well-level summations of these cell-level measurements, or the well-level data. Depending on the application, the files may be retained for analysis *post hoc* (by applications such as Spotfire^{TM}), or used by the image analysis application itself. Most data analyses are done on the well-level data, particularly screening and routine assay work, whereas highly multivariate such as Clustering (Chapter 14) use cell-level measurements. If data are

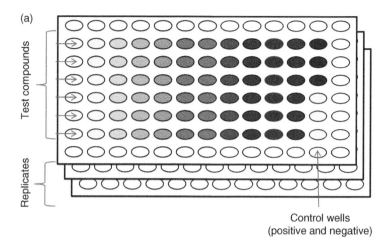

Control wells
(positive and negative)

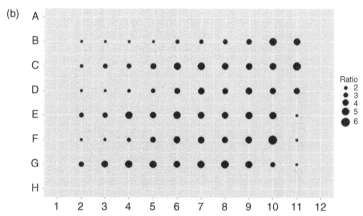

FIGURE 8.1 *Platemap of the transcription factor translocation experiment.* (a) A general schematic of the experiment showing the seeding of cells in the inner 60 wells, the addition of compounds by row, and the use of replicate plates with identical treatments. Compound dose is shown by degree of shading, each row is a different compound. (b) Nuclear/cytoplasmic ratio for each well, scale is shown at right.

to be analyzed by non-parametric statistical tests (even for screening or assay work), then the cell-level measurements are required. This will be discussed in detail in Chapter 10. The data analysis process begins with opening these files.

Examples of the data files for well and cell-level features are shown in Tables 8.1 and 8.2, respectively. These are just portions from the data. The naming of these features change according to the image analysis application and how the information about the assay conditions was entered. *Cell:mean compartment area* is how Imag-eXpress names one of the features in this assay. Using a Cellomics imager would yield essentially the same measurement, but it might be called *CH2TotalArea* in

TABLE 8.1 Well-Level Data from the Translocation Assay in the Tutorial

Stage label	Z position	Compartments	Mean compartment area	Mean compartment average intensity	Integrated inner intensity	Integrated outer intensity
B02	11195.86	135	102.4099467	591.7543031	15,508,400	13,125,311
B03	11189.96	99	110.5533909	599.4955527	14,265,166	11,286,382
B04	11184.28	119	116.7666807	937.3800299	19,665,785	14,715,857
B05	11179.94	105	98.32453714	1062.057664	15,489,827	12,101,019
B06	11175.96	57	117.1146868	971.0335286	12,422,553	6,309,241
B07	11171.42	94	100.1203995	1120.591062	21,149,918	10,134,985
B08	11172.8	151	104.2734978	1229.905726	44,816,224	18,009,883
B09	11151.9	164	109.4424791	1205.162529	57,469,870	19,594,161
B10	11153.58	123	102.1561226	1111.97295	45,244,258	12,876,231
B11	11156.86	151	100.5843755	1191.070176	51,294,208	15,721,034
C02	11190.6	109	100.2238576	610.9705625	20,729,123	12,561,276
C03	11181.22	107	114.9278783	611.7466423	28,374,525	11,745,803

The table is a selection of the data in the complete table.

TABLE 8.2 Cell-Level Data from the Translocation Assay in the Tutorial

Stage label	Z position	Cell: assigned label no.	Cell: compartment area	Cell: compartment integrated intensity	Cell: mean compartment intensity	Cell: inner area	Cell: outer area	Cell: integrated inner intensity
B02	1195.86	1	31.201875	33,785	450.4666667	19.9692	32.033925	17,317
B02	1195.86	2	52.835175	59,143	465.6929134	38.690325	31.6179	40,974
B02	1195.86	3	58.659525	98,015	695.141844	38.690325	34.11405	29,561
B02	1195.86	4	56.995425	96,779	706.4160584	33.698025	34.11405	38,963
B02	1195.86	5	45.346725	62,886	576.9357798	32.865975	40.77045	40,824
B02	1195.86	6	65.315925	106,728	679.7961783	43.682625	50.339025	54,610
B02	1195.86	7	64.06785	82,975	538.7987013	42.43455	34.9461	57,015
B02	1195.86	8	59.07555	72,910	513.4507042	38.2743	56.163375	49,924
B02	1195.86	9	26.209575	35,928	570.2857143	13.728825	32.033925	20,668
B02	1195.86	10	69.476175	81,502	488.0359281	43.682625	60.323625	39,937
B02	1195.86	11	64.483875	56,833	366.6645161	48.674925	40.77045	30,621
B02	1195.86	12	78.628725	76,631	405.4550265	56.5794	53.667225	41,748
B02	1195.86	13	32.865975	34,692	439.1392405	22.049325	34.9461	16,811
B02	1195.86	14	73.636425	68,079	384.6271186	57.41145	45.346725	32,473

The table is a selection of the data in the complete table.

the default. Image analysis applications can incorporate **metadata**, data about the experiment, into the output files. Examples of metadata include information about the treatments for each well, either the compound and dose or the RNAi reagent. The application will add the metadata to both of the cell and well-level data tables in one or more columns. These data can then be used to combine replicates and construct the dose-response curve or to identify hits.

Many experiments use the well-level measurements only. Table 8.1 is a portion of such a table. The complete table measures 60 rows (one for each well) by 18 columns. This file has been reduced from the original table of features captured during image analysis in order to make it more manageable during the tutorial. Typically, additional data are also collected. These include measurements that are frequently highly correlated, but may capture important differences for some assay types (as an example, mean intensity, median intensity, and total intensity may all be reported for cells per well). Some data are linked to technical parameters, such as column 2, which reports the Z position, or the extent to which the objective needed to be moved in order to capture the image in sharpest focus. Typically, such measurements have little impact on the true metrics, but are important for some applications, or can be used to identify problematic wells or fields (i.e., if the value is far out of range of the other values) [1]. Table 8.2 presents the same data, but at the cell level. The complete table reports feature measurements for each cell, one cell per row, which in this case number more than 6500 rows, and 23 columns of data for each cell (including the well location and a number for the cell once it is identified in a field; only the first 14 rows (representing the first 14 cells in well B02) and 9 columns are shown in the table. Reporting events at the cell level makes for a very redundant table, but the table enables some very powerful analytical methods, discussed in subsequent chapters.

8.3   METRICS AND RESULTS: THE METRIC IS THE MESSAGE

A lot hinges on the way in which the basic quantitative measurements (the features) are boiled down into a single number that encapsulate the biology of interest for the assay (the metric). The metric is the take-away message from the experiment, just as it would be for a whole-well assay that measures ATP or enzyme activity levels. The complication of an HCS assay is that there are many features to consider as the assay metric, and the choice can influence this take-away message in ways that do not exist for other assay formats. As a first example, let us consider a simple (non-high content) multiwell viability assay. When cells die *in vitro*, they stop utilizing energy, lose their membrane integrity, and (often) fragment their DNA and proteins. There are a variety of viability assays on the market that will focus on one of these effects, such as energy utilization (WST, ATP-lite), protein content (sulforhodamine B (SRB)), membrane integrity (neutral red/trypan blue), and quantify it as a metric of cellular viability. Screening based on the selected viability assay will be successful only to the extent to which it measures the biological events in question directly. Any disconnect between the biology of interest and the selected metric will pose a threat to the eventual success of the screen. For instance, screening for compounds that kill cells by altering cellular metabolism would be better off using a viability readout that is unrelated to energy utilization, since it is already known that the compounds can affect metabolism, but the question is which ones actually kill cells.

This is all very good, you are wondering, but what does this have to do with high content assays? In a high content assay, the problem of metrics amplifies greatly because even for a simple effect, there are multiple ways to quantify the changes. For example, in designing a high content assay for cellular senescence, there are changes in cell volume, nuclear volume, nuclear morphology, and cell shape that can be used as a measure of the desired biological effect. However, even when the measure of desired biological effect is obvious, there may be a number of ways to boil it down to a single metric. All HCS assays will measure the total intensity and the average intensity for an object (a nucleus, cytoplasmic region, or spot). Either will suffice as a metric if the area of the object does not change, but if area changes as a function of treatment, then total and average intensities are no longer equivalent. These changes actually occur frequently in HCS, particularly in toxicology and proliferation inhibition assays.

There is great value in thinking through the options at this stage of assay design. Often it comes down to tradeoffs in terms of complexity, implementation, or in terms of runtime for the algorithm. For example, simply implementing a cytoplasm as a "donut" around a nucleus has the definite advantage of requiring one less fluorescent stain to optimize for the identification of the cytoplasm. It is faster in terms of implementation and saves on image and data storage. However, this choice can lead to poor assay behavior if the distribution of the protein being measured is uneven in the cytoplasm, or if the size and extent of the selected "donut" is too broad for the cell type. Examples are discussed in the Tutorial.

In other cases, different choices of metric may behave differently in terms of the statistical stability of the assay. For example, dividing nucleus by cytoplasm (which is what we think of intuitively based on the term "nucleo-cytoplasmic ratio") has the undesirable property of being extremely noisy when the intensity in the cytoplasm is low. Subtraction of the cytoplasmic intensity from the nuclear intensity is one alternative. In the next section, we will take a look at the types of high content assays classified by metric, and some design tips on what makes a good metric.

8.4   TYPES OF HIGH CONTENT ASSAYS AND THEIR METRICS

The choice of metric in a high content assay can often determine the challenges that you will face during its setup and running. Some common categories of high content assays are described just below this paragraph (Sections 8.4.1–8.4.5). Developing a metric impinges on the statistical analysis. For example, if the sample is defined by a mean value per well, a distribution of values or a percent-of-control, the appropriate statistical test to determine significance may change (discussed in detail in Chapters 9 and 10). To begin with, even for a specific marker, such as chromatin, a fluorescently labeled protein, or a cellular feature, there are many potential measurements. Examples are shown in Figure 8.2; these are based on the mechanism by which the image-processing steps can associate intensity patterns to cellular features,

Nuclear or cell measurements:
Number per field or well, area,
average intensity, total intensity,
intensity variation, perimeter/area
ratio, length of longest axis,
length/width ratio

Puncta measurements:
Area, average intensity, total
intensity, intensity variation,
number per cell, number in
cytoplasm, number in nucleus

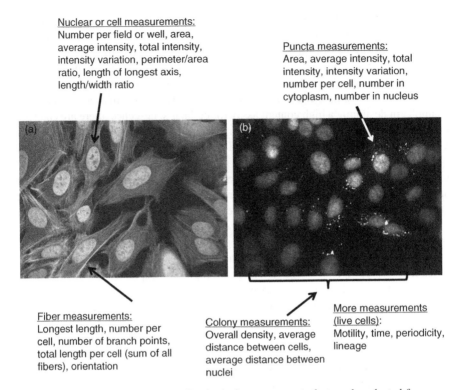

Fiber measurements:
Longest length, number per
cell, number of branch points,
total length per cell (sum of all
fibers), orientation

Colony measurements:
Overall density, average
distance between cells,
average distance between
nuclei

More measurements
(live cells):
Motility, time, periodicity,
lineage

FIGURE 8.2 *Assay metric types.* Cytological measurements that can be adapted for quantitative cellular assays. (a) A549 cells labeled for chromatin, cytoplasm, and fibrous actin (b) A549 cells labeled for nuclei, cytoplasm, and fluorescent puncta that accumulate during autophagy. Original images are composites of separate labels, but are shown in a common gray scale. Surrounding text highlights the potential methods of quantifying each label.

as was discussed in Chapter 4 and listed in Table 4.1. For any given marker, there are numerous ways of measuring it. Alternatives are detailed below.

8.4.1 Intensity

In a sense, all HCS measurements are intensity measurements, as all regions of interest, spots, fibers organelles, and so on are based on areas of fluorescence intensity in an image. However, using these measurements as metrics can affect the image and algorithm corrections that are needed, and how robust the measurements are as metrics. We discuss each of these in turn, beginning with the simplest, a measure of the brightness of a region of the cell as a metric. The use of phosphorylation-specific antibodies means that the activity of protein kinase signaling pathways can be readily measured by the fluorescence intensity of a cell labeled with such an antibody. The expression of a gene using an antibody against the protein, and the modification of a protein through proteolytic cleavage are also common intensity-based assays. Methylation, acetylation, ubiquitination, and other modification are possible. These are widely used because they are easy to format, but the challenge to intensity-based measurements is that they are prone to artifacts in background intensity levels, requiring careful image correction. Most image analysis platforms have illumination correction routines (some will have several methods), and corrections are easy during the assay development and piloting stage of a screen, and as such, are not usually a problem for individual plate assays, but can be a problem for full-size screens, as noted next.

8.4.2 Area Above Threshold

For an assay that will be run over a long period of time, an immunofluorescence assay that is quantitative on a per-plate basis may nevertheless report different absolute values over time. Changes in lab conditions (e.g., temperature and humidity), lot-to-lot variations in reagents and aging of the lamp (for some systems) can all affect the absolute measurement of fluorescence intensity.

How then can one quantify a readout that varies in direct proportion to the total amount of protein being measured, but that is less sensitive to flux over time? The simplest way to assess such a protein is to establish a threshold (minimum intensity cutoff) and count the area or number of objects in the field that are brighter than this threshold. Thresholding is an effective way of dealing with background variation, and enables the design of an assay that focuses in on the population of interest. Thresholding can also be used to deal with day-to-day fluctuations in signal intensity. These approaches are similar to the process of background correction, which was diagrammed in Figure 4.3, but here we are not differentiating between the signal and the noise *per se*, but whether the pixel intensity is sufficiently great that the region (and the corresponding cell) be classified as a positive or a hit.

However, thresholding has the undesirable property of converting continuous data "artificially" into binary (yes/no) data. Such truncation leads to a loss of information, and can make the results of the assay dependent on the selected threshold. As a general rule, thresholding is easiest to apply when the data clearly fit into two categories. Of course, in real biological settings, such truly "dichotomous" or two-cluster data are the exception rather than the rule. If you are forced to apply a threshold on continuous populations, consider setting the threshold on the basis of the control population. For example, for a marker whose abundance increases on treatment, you can select a threshold such that only 5% of the area in the control images is positive. Alternatively, once you have a threshold, you may decide to work with other parameters such as the size or shape of the thresholded regions to identify cells. Spot counting assays are covered in the next section, and they have their own peculiar challenges.

8.4.3 Spot Counting, Including Nuclei or Cell Counting

These assays are based on counting the number of objects in one field. The simplest and most robust spot counting assay is counting nuclei, which readily translates into cell counts and a measure of proliferation. Objects may be defined (segmented) using a variety of different properties, such as their size, shape, intensity above threshold, or intensity above local background. Regardless of the approach used in segmentation, once counted, a spot is a spot. Spot counting assays are most appropriate when counting objects that are all equivalent, for instance, in a nuclear counting assay for cell viability. For example, DNA damage foci (using H2Ax or 53BP1 staining) are spots, each of which localizes to a single double-strand break locus. There are at least three ways to implement spot counting using DNA damage foci; by counting either the total number of foci, the number of foci per cell, or the cells above a certain intensity/quantity threshold as positives. The last method listed has the undesirable property of truncation (covered in the previous section). Counting cells above a certain intensity threshold also has the undesirable property of being unable to differentiate between bona fide DNA damage staining (in foci) and background spots or inappropriately bright cells that lack foci. Some assays use a signal amplification step, such as proximity ligation assays and horseradish peroxidase (HRP)-mediated tyramide signal-amplification kits which use reagents to identify cellular events (colocalization, protein modification, or expression), and follow with the generation of a deposit of fluorescent material. The amount of deposition is not linearly related to the magnitude of the event—if the event occurs, then a fluorescent deposit is generated. In this case, thresholding or object counting are more appropriate ways of characterizing events, total intensity or total area of the puncta could be misleading. However, if variability in intensity is biologically relevant, such as autophagosome formation, then using a spot counting metric may lead to a loss of information.

If you decide to implement your assay using a spot counting metric, focus on the quality of the segmentation (some platforms will have explicit spot counting assays, but all platforms count spots; you can develop an algorithm that meets the same goal). It is well worth the effort to put together a pilot dataset that covers a broad set of conditions, and tests the performance of the segmentation routine. The segmentation routine *is* your assay, and its performance sets an upper bound on the quality of the assay.

8.4.4 Translocation

In a translocation assay, the movement of a protein between two subcellular compartments is quantified. Nuclear localization assays are most common, but localization to the plasma membrane and receptor trafficking through organelles are also valuable assays. In such assays, usually there are two rounds of segmentation—one for the cell as a whole and the other for the specific cellular compartment of interest. This generates additional challenges (one more segmentation to test!), but it also brings in a significant advantage in that there is a baseline that can be used for normalization.

When designing an assay to measure changes in the nucleocytoplasmic ratio of a protein, it is worth noting that there are a number of ways to implement such a measure. One way to implement this is to segment the nuclei (identify the boundaries of each nucleus precisely), segment the cytoplasm (using a third cytoplasm-specific marker) and then divide the intensity of protein in each cytoplasm by the intensity of protein in each nucleus on a cell-by-cell basis. Alternatively, once the nuclei and cytoplasm are identified, one can divide the intensity of protein in all cytoplasm by the intensity of protein in all nuclei. As an alternative to segmenting the cytoplasm with a cytoplasm-specific marker, one might simply draw a "donut" around the nucleus and use that as a substitute for the cytoplasm. As a further alternative, instead of dividing intensities, one can choose to subtract nuclear intensity from cytoplasmic intensity or vice versa.

As an alternative to dividing nuclear intensity by cytoplasmic intensity (unstable) or subtracting nuclear intensity by cytoplasmic intensity (baseline is not utilized), consider what happens if we compute the percentage of total protein for each cell (nuclear + cytoplasmic) that is nuclear. This metric renormalizes every cell to the total intensity of protein present in it, and has distinctly different properties for noise—if the edge of the image is dimmer, this method will correct for it. It will work well regardless of whether the protein is highly cytoplasmic or highly nuclear.

8.4.5 Morphology

The final, and most challenging, category of high content assay is the morphology-based assay. If your biology of interest generates distinctly different morphologies between treated and control, quantifying cells with the "control" versus the "treated" morphology may well be the way to go. This can be somewhat challenging to implement at times due to the complexity of extracting image processing features, but there are ways to do this (discussed later) that are powerful.

One big advantage of morphology-based assays is that they are more stable to intensity variations—as long as the features that are extracted from the images have been chosen carefully. For example, consider the development of an assay for chromosome alignment, measured by quantifying the DNA stain in metaphase cells staining positive for a mitotic marker. Tightly aligned chromosomes form a tight, intense band of DNA at metaphase (a "metaphase plate"), while poorly aligned chromosomes are more diffuse and dimmer (Figure 8.3). Using a simplistic approach where the image is thresholded and the area positive for the DNA stain is used to quantify the degree of chromosome alignment will work well when the DNA stain is very robust—misaligned chromosomes will have a larger area than tightly aligned metaphase plates. On the other hand, when the DNA stain is variable, this method will fall apart if the dimmer misaligned chromosomes fall below the threshold intensity. Now the chromosomal area will be *lower* for the misaligned chromosomes. The way to get around this is to use the shape factor (the ratio of the long axis to the short axis of an elliptical shape), rather than the area of the chromosomes. Regardless of the overall area of the chromosomal mass, tightly aligned chromosomes should have a much longer shape (higher shape factor) than poorly aligned chromosomes, which should appear "rounder". In this way, the assay can be made more robust by focusing on a morphological property that is not sensitive to intensity variations. Consider using a hand-scoring blinded approach for morphology-based assays, at least during the proof-of-concept stage. This can let you evaluate the feasibility of such an assay—if you cannot spot the differences in morphologies in a blinded and randomized dataset, you probably do not have a biological effect.

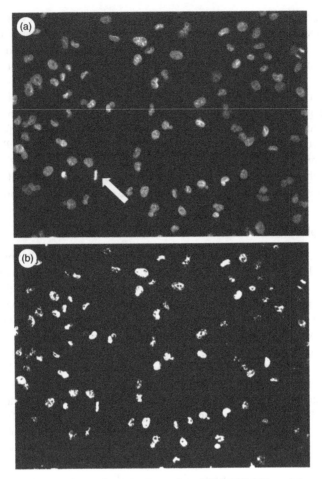

FIGURE 8.3 *Image analysis of a metaphase plate*. MDA-MB-231 nuclei are stained with DAPI. (a) A high definition image. A cell showing metaphase alignment of chromosomes is shown with an arrow. (b) The same image but where illumination and contrast have been increased to reduce texture and threshold the intensity levels.

8.5   METRICS TO RESULTS: PUTTING IT ALL TOGETHER

Converting data into results depends on both the experimental design and the type of data that is being used as a metric. Experimental design includes basic factors such as the number of replicate wells that are necessary to have confidence in a result. The data type can determine whether the experiment can be treated with standard experimental statistics, such as the *t*-test or the *z*-score (Chapter 9 and [2]), or whether cell-level statistics are needed. This process is covered in the next chapter. Here, we finish with a discussion on how the configuration of the assay could affect the definition of a metric and the analytical process.

8.5.1   Basic Assay Measurements

We will start with basic, single-metric, measurements, ones that are essentially high content versions of typical microtiter plate-based cellular assays. In a basic cellular assay, cells can be treated with a compound and an effect on the cells is observed. For instance, our earlier example of a traditional viability assay— some assay kits rely on proliferation, some rely on ATP utilization, others rely on cell count or total protein quantity. Simple high content assays transpose these events into image-based measurements. Substitute nuclei counts for ATP levels, and you have a high content version of a proliferation assay. Of course, the high content version is not identical to the traditional cellular assay. In the proliferation assays, nuclei counts and ATP levels are both indicative of cell growth, but they measure different things. A further wrinkle in comparing assay formats is that even when a very similar measurement is used, a whole-well assay and an image-based assay can appear to give different results. In cases where a lot of cell death has occurred through apoptosis, cells may be lost during the experiment, and the HCS assay could report high apoptosis per *cell*, but the whole-well assay would report low apoptosis per *well*. Both are technically correct but tell very different stories about the perturbation in question.

8.5.2   Use of Multiple Independent Measurements to Assess a Perturbation

Cell health and cell signaling assays have been developed to give comprehensive views of treatment responses. For example, an assay that measures proliferation (cell counts), cell cycle (average DNA content per cell), and apoptosis (caspase-cleaved proteins such as Poly (ADP-ribose) polymerase (PARP) or other caspases) all speak to the extent to which cells may be killed by a treatment. In the simplest analysis of such an assay, a strong concordance of proliferation rates with the effect on a signaling pathway would provide support for a specific mechanism of cell death.

This approach can be used in screening, where compounds that affect more than one pathway are valuable, or in cases where an effect on one pathway, but not the other, is observed. An example includes a multiplexed assay that looks at several transcription factors, such as NF-κB, STAT3, and c-Jun, in the same assay to measure multiple inflammation pathways at once. A basic approach would be to consider them independently, three separate assays that just happen to be run at the same time (and in the same cells). If you are interested in compounds that hit any of these three transcription factors, then you have saved time and expense by running them together as a multiplexed assay, data would be sorted for metrics of each protein independently and three hit lists would be generated. However, data for each protein could be combined in Boolean operations (e.g., a hit is defined as a strong response for STAT3 but not c-Jun). This also applies to assays that combine different feature types, for example, a threshold or localization for GADD153/CHOP, which responds to ER stress, and the abundance or morphology of LC-III-GFP fusion proteins, which form puncta during autophagy.

Integrating cellular or organelle morphology measurements can add a lot of sophistication to a multiplexed assay. Compounds that do not kill cells but put them under stress will alter their morphology. Scoring nuclei through tight specifications for what a healthy nuclei looks like means that cells showing stress through an altered nuclear morphology will be an independent measure of treatment effects, because cells under stress will start showing changes in morphology. In this way, two independent features can report on compound effects, one on a specific signaling or cell-death pathway, and another on stress to the cell.

8.5.3   Integrating Multiple Features to Measure a Phenotypic Response

Phenotypic responses that are defined by an aggregate of features are becoming common in HCS. They are used in "hypothesis-free" studies, where the effect of a perturbation is not explicitly known, but a response can be detected by monitoring multiple features simultaneously. They can also be used in classification studies, where perturbations can be grouped or stratified through the effect on a set of features. Results are frequently presented in dendrograms or plots of the principal components, analogous to transcriptional profiling experiments. These studies show that a set of perturbants are related. Such studies can be very informative, for example, if a group of inhibitors are tested and a subset show unique effects, this could be a sign of toxic or off-target effects. The features are typically morphological and are typically very general, such as actin and tubulin, each of which can be quantified in many features. The analytical process is much different, with significant data processing, and ultimately leads to machine learning methods, which are covered in later chapters.

8.6   CONCLUSIONS

This chapter begins the process of moving from cellular observations to numerically based conclusions. At this point, recognizing how a given fluorescent marker can be measured in subtly and grossly different ways is an important take-away message. The next step in quantification, recognizing how these alternatives portray the biology of the experiment differently, is equally important and occasionally less obvious. The last issue addressed in this chapter, that how the metric is chosen can affect the downstream analytical process, is still important, but something that we cover in more detail in the following chapters. All in due course.

KEY POINTS

1. Determining the metric for an assay integrates knowledge about the assay with an understanding about how measurements are made by image analysis applications, and how to select or transform feature data to develop the most appropriate metric.
2. Patterns in images form the basis of features that are reported for each cell and well in an experiment. Each fundamental pattern can be quantified in multiple ways, and each method can be recorded as a separate feature.

3. From the above point, many features are very similar, selecting which feature is most relevant draws on the biology and image analysis used in the experiment.

4. The feature (or combination of features) that best describes the biology behind the assay is the metric, the numerical value that defines the response of a cell to a perturbation.

FURTHER READING

This chapter sits at a transition point, focused on the quantification of cytological measurements. This process brings many new concepts forward, ranging from what to call these many (many) measurements to how well these measurements relate to the biological question. While the discussion up to this point in the book (and resuming shortly) has been platform-agnostic, this is a time to dig into the reference material for the particular platform you are using. What is the default name for nuclear area? Where can you change it to something connected to the experiment you are setting up? What do you need to do to access the well- and cell-level data? To get more into the spirit of this transition from cell biology to informatics, read about what each feature actually reports. How many shape or intensity measurements are made?

REFERENCES

1. Haney, S.A. High content screening approaches that minimize confounding factors in RNAi screening. *Methods in Molecular Biology*, in press.

2. Malo, N. et al. Statistical practice in high-throughput screening data analysis. *Nature Biotechnology*, 2006, **24**(2): 167–175.

9

ANALYZING WELL-LEVEL DATA

Steven A. Haney and John Ringeling

9.1 INTRODUCTION

In the previous chapter, we discussed how features can be used as metrics to generate an assay that measures a specific cellular event. In this chapter, we will process metrics to determine results and communicate findings using standard statistical tests for well-based assays and screens, including dose–response assays, RNAi screens and HTS campaigns. Since there are several excellent sources for these methods [1–3], we will cover their bases briefly here, and focus on how HCS data can be handled in an assay context.

HCS data are based on measurements of cellular events that are tracked on a per-cell basis. However, it is common for these data to be expressed as well-level aggregates of these measurements. Image-based metrics such as nuclear translocation, transit from the plasma membrane, phosphorylation of a protein by a kinase, and exit from the cell cycle (as measured by DNA content, phosphorylation of histone H3, or other event) are all amenable to being summarized at the well level. As such, much of HCS screening data analysis is a two-part process, converting the cellular measurements into well-level values, and handling the well-level values in essentially the same way that other screening technologies are used. Almost all HCS data are analyzed at the well level. However, HCS is relatively unique in that the individual cellular responses are recorded, so we will introduce a general process for evaluating data before deciding on the analytical path. We present some general data analysis concepts that help understand how well an experiment or screen has performed prior to analysis, and in cases where the data appear fine, the general well-level analysis methods are discussed in this chapter. In cases where anomalies

An Introduction to High Content Screening: Imaging Technology, Assay Development, and Data Analysis in Biology and Drug Discovery, First Edition. Edited by Steven A. Haney, Douglas Bowman, and Arijit Chakravarty.
© 2015 John Wiley & Sons, Inc. Published 2015 by John Wiley & Sons, Inc.

are detected, alternative methods of analysis are discussed in Chapter 10, which focus on cell-level analysis methods.

The data review methods described here will enable some important decisions. For screens and other assays where a high Z' has previously been established, they will provide tests for ongoing assay performance and detect problems with specific runs, with the experiment itself, image acquisition, or analysis. They can also identify alternatives when appropriate, such as when to transform data to make it amenable to the common methods of data analysis or when to use cell-level methods to analyze an experiment.

9.2   REVIEWING DATA

There are several options for reviewing data, starting with image analysis applications themselves. Most HCS platforms can readily produce graphs or heat maps within the software's user interface. These are extremely useful, providing a quick check on the experiment, and often allow several features to be compared as candidate metrics. Heat maps have additional benefits because they recall the layout of the experiment, and systematic errors at the plate level can be spotted quickly (usually edge effects, but also pipetting errors for a row or column).

We need to inspect the data, much like we needed to inspect the images during the image analysis phase. In the previous chapter on Metrics, we introduced the tables of raw data generated in an HCS experiment: tables of measurements made for each cell, tabulated by well and extending for many hundreds or thousands of rows, and measurements made for each well, the averages and variability of these features when grouped by well, with one row for each well. Image analysis platforms will provide some supplementary statistics for each feature when reviewed at the well-level. **Summary statistics** are a defined set of measures for a population that provides a good understanding of the data. Typically they are the **mean**, **median**, **range** (the **min** and the **max**), and the 1st and 3rd **quartiles** (or the 25th and 75th **percentiles**, respectively). From these data, overall trends, such as unusually large number of outliers or compressed values that could be an indication of problems in the data. Doing this at the data analysis level rather than the image level means that the summary statistics can be plotted or even analyzed algorithmically—reducing the need to review images from every site and well.

Most image analysis platforms provide the well mean and median values, as well as the %CV, standard deviation, and some will provide the minimum and maximum values for each feature as well. Therefore, calculating the formal summary statistics may not be essential to review the data, however an extended review of the data through some form of summary statistics provides a complete picture of the variance in the data. To generate the formal summary statistics, we present a method that will work for data from any image analysis platform in Appendix C using the open-source statistical analysis program R, and introduce some related analytical methods in later chapters. Note that summary statistics are generated from the cell-level data because they report on the responses of the population of cells within a well. All image

TABLE 9.1   Summary Statistics for Rows from Two Experiments

Experiment	Well	x.Min.	x.1st Qu.	x.Median	x.Mean	x.3rd Qu.	x.Max.
FKHR	B02	0.8078	1.211	1.353	1.387	1.521	2.225
translocation	B03	0.7399	1.218	1.293	1.397	1.529	3.199
	B04	0.8359	1.245	1.395	1.469	1.649	2.442
	B05	0.7767	1.306	1.461	1.55	1.747	3.17
	B06	1.458	1.961	2.417	2.503	2.894	5.545
	B07	0.8018	2.079	2.6	2.782	3.366	7.629
	B08	0.8057	2.535	3.337	3.628	4.477	16.81
	B09	0.891	2.701	4.645	4.403	5.928	12.53
	B10	0.7107	3.704	5.572	5.77	6.918	24.82
	B11	1.078	3.116	4.666	5.102	6.887	11.48
FYVE puncta	D02	0	0	0	1131	921.8	44,060
	D03	0	0	674	2284	2394	20,680
	D04	698	3921	6905	11,390	14,930	71,480
	D05	0	14,570	30,850	44,170	53,360	234,300
	D06	8037	21,900	43,200	66,490	80,220	543,600
	D07	2239	22,170	31,210	42,200	54,630	235,300
	D08	0	15,410	32,660	42,990	53,910	236,600
	D09	865	20,670	38,020	59,920	81,010	444,300
	D10	1427	19,180	27,560	40,310	49,400	222,300
	D11	377	19,880	27,620	31,900	38,500	136,700

analysis platforms have methods for exporting the cell-level data and the process is typically straightforward, but you may need to consult the documentation or technical support for assistance. For the time being, we discuss the data itself.

In Table 9.1, summary statistics are shown for one row from two experiments. The table shows one row from each experiment, which keeps this exercise as simple as possible, but presents enough data to see some of the trends in the experiment. In the first experiment, values for Row B are shown. The data report on the nuclear–cytoplasmic ratio for the FKHR (FOXO-1a) transcription factor after treatment with an inhibitor, an experiment that is highlighted in the tutorial. Looking at any of the columns (but usually the mean or the median) it is easy to see a trend towards increasing ratios. The mean and the median are fairly similar throughout the data, suggesting that there are some outliers in the experiment, but the experiment does not seem to be overwhelmed by them. Data from Row D of a FYVE-puncta formation assay are shown in the second half of the table. Puncta form when cells are under stress and trigger the autophagy response [4]. The data shown are the integrated area for all puncta per cell (an alternative metric for this assay could be the number of puncta per cell). Here the data look a little different. For the first couple of wells, most cells do not have any puncta. In fact, the median value for well D02 is 0, while the mean is 1131. Not only are these very different numbers, the mean is actually larger than the 75th percentile—essentially most of the signal for this well is based on a minority of the cells. It is clear that the signal in this well is dominated by a few cells. Whether these are outliers, a subpopulation or artifactual signal from debris, cell aggregates or other confounding material is not clear at this time. Going back and looking at the images and image analysis algorithm would be important here, but frequently these situations are the result of heterogeneity in the cells [5] and cannot be eliminated through refinement of the image analysis algorithm. This assay would be one where you might want to look at the data a little closer. We will do this in Chapter 10.

9.3 PLATE AND CONTROL NORMALIZATIONS OF DATA

The first step in analyzing well-level data will be to normalize the data. **Normalization** of screening data attempts to remove systematic interplate variability, making features comparable across multiple plates (note: plate-to-plate normalization of data is different from "normally distributed data;" refer to Chapter 10 for more discussion). This is critical for experiments such as high throughput screens, which produce raw data from multiple plates. At the simplest level, this is usually implemented by scaling each data point to the negative control. The choice of normalizing to within-plate negative controls or all within-plate samples or wells largely depends on the experimental design. Normalization to within-plate samples is generally appropriate if an assay lacks well-defined negative controls and if the majority of the samples are expected to have a null biological effect. The use of all within-plate samples may be problematic for normalization techniques that use nonrobust descriptive statistics sensitive to outliers (e.g., mean, standard deviation; discussed further in the next chapter). It is generally preferred to normalize to within-plate negative controls if they exist, as it is reasonable to assume negative controls remain stable across different plates (unless there are other experimental factors contributing to heterogeneity). The calculation of assay statistics and hit analysis generally assume that the data are normally distributed (discussed in depth in Chapter 10), the normalization process should result in data that are normally (or nearly-normally) distributed. Recent reviews describe the derivation and use of these methods [2, 3], Bray and Carpenter discuss these and additional methods specifically within the context of HCS data [6]. These references provide in-depth treatment of the data analysis; we will mention the most common methods here.

9.3.1 Ratio or Percent of Control

The most common approach favored by biologists for its simple calculation and interpretability is to divide each observation by the mean (or for a robust alternative, the median) of the controls within each plate. This method however does not incorporate any information regarding the dispersion or variability of the control distribution.

9.3.2 *z*-Score or Robust *z*-score

A *z*-score is calculated as $z_i = (x_i - \bar{x}_{control})/s_{control}$ for each data point and is interpreted as the number of standard deviations from the mean of the control sample. When the control sample follows a normal distribution, the resulting *z*-score follows a standard normal distribution with mean $= 0$ and SD $= 1$. z-Scores have the advantage of incorporating information regarding the variability of the control distribution and are frequently used in high throughput screens to indicate the extremity of the observation in relation to the control distribution. An alternative to the z-score which is less sensitive to outliers is the robust z-score, which substitutes the control mean and standard deviation with the control median and median absolute deviation (MAD) [7].

9.3.3 *B*-score

The previous methods correct for systematic variations across multiple plates, however systematic variations may exist within a plate. The *B*-score adjusts for intraplate variability, effectively normalizing observations to all within-plate samples, and is appropriate for data that exhibit row and/or column effects. Its calculation is more complex as it uses an iterative "median polish" algorithm (a type of ANOVA—discussed in Appendix C based on medians rather than means), to adjust for row and column effects. For details surrounding the derivation and exact calculation of the *B*-score, see Reference 8.

9.3.4 Mixed Effects Models

There are times when these simple approaches to normalization are not appropriate, as the screen has significant variability along a particular axis—for example, day-to-day or plate-to-plate variability. In these settings, a Mixed Effects Model can be used for normalization. Very briefly, a mixed effects model is a statistical model (a close relative of an ANOVA), which accounts for both fixed effects (such as treatments) and random effects (such as a plate-to-plate and/or a day-to-day random effect). Usually the fixed effects are the "catch" that you are looking for, and the random effects are an undesired byproduct. Of course, the magnitude of the plate-to-plate and day-to-day random effects can be monitored as a quality control metric as they are direct readouts of the reproducibility of the assay (more about this in Chapter 11). In this case, a statistician will need to be consulted for the experimental design as the calculation of the different fixed and random effects requires specific experimental designs.

9.4 CALCULATION OF ASSAY STATISTICS

Metrics have been developed that assess the robustness of a screen. Of the available methods, the Z' statistic is far and away the most common calculation [9]. It measures the mean and variance of the positive and negative controls only, so it is reproducible and easy to understand. The V factor is also used by many investigators as it has the advantage of greater sensitivity to the data near the EC50 of a dose–response curve. We compare the methods in Figure 9.1. Performance of controls becomes both the foundation and the flaw in the calculation of assay statistics. Surrogate positive controls (e.g., toxins to substitute for general compounds or RNAi reagents) can create a false set of expectations. This is particularly relevant when morphological

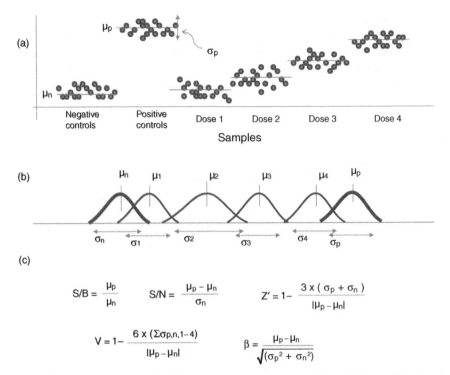

FIGURE 9.1 *Calculation of assay screening statistics.* Data collected during validation of an assay typically produce many replicates of test conditions that will be anticipated in the screen. (a) Plotting of control values. Positive and negative controls are usually essential, and are required for most assay validation and robustness statistics. Replicate values produce a mean value, designated as μ_n for negative controls and μ_p for positive controls. Mean values for the positive and negative controls and the standard deviation for the positive controls are highlighted. Other mean values for a dose response curve are highlighted by the horizontal bar for each dose. The variability measured through several statistics, usually the standard deviation, designated σ_p for the positive controls, in the figure. (b) Data are replotted as distributions with mean values along the horizontal axis, the variance is the spread in the individual values, as noted in the figure. (c) The equations used to determine assay robustness. S/B, S/N, Z' statistic, V factor and SSID (β) are shown.

changes are the endpoints. Occasionally, if a strong but biologically irrelevant positive control is used, the Z' statistic will be satisfying, but the screen itself will appear to produce no hits. Optimizing an assay for the "best-case" positive control may lead to "worst-case" performance in practice.

9.4.1 Signal to Background Ratio (S/B)

Signal-to-background ratio is the simplest method for calculating assay performance, taking the ratio of the mean value of the positive control responses to the negative control responses.

9.4.2   Signal to Noise Ratio (S/N)

Signal-to-noise ratio takes into account the variability of the negative controls. Here, the signal is not defined by the mean of the positive controls but by the difference between the positive and negative controls; this difference is divided by the variance of the negative controls.

9.4.3 Z' Statistic

The Z' statistic was the first well-recognized measure of assay performance [9], and it is still the most widely cited and accepted measure. It is calculated as the ratio of the difference between the means and the variances of the positive and negative controls. It is very stringent because it is intended to determine the likelihood that a single data point (compound or RNAi treatment) could be classified as a hit. Using a Z-score is appropriate when the screen will run as a single-point determination. In cases where a dose–response will be calculated, or where replicates will be run, there are more sensitive options which we discuss later. A perfect screen has no variance and a high signal window, so its Z' statistic reduces to zero. Good screens have values of 0.5 or greater and if the value falls to 0 or less, the assay is too variable to run as a screen (0.3 is a common cutoff for many screening facilities).

9.4.4   V Factor

The V factor complements the Z' statistic in that it integrates the mean and variance of the values along the dose–response curve into the measure of assay performance. Thus, it not only takes into account the performance of the maximal and minimal values (the positive and negative controls, respectively), but how the assay reports values as concentrations near the EC50 [10], a critically important variable for drug discovery. In these cases, the V factor can be preferred because the EC50 value is a critical parameter. The Z' is more common in HTS, where the ability of the screen to identify a hit is the paramount concern. The relationship between the V factor and the Z' statistic is clear from Figure 9.1c.

9.4.5   Strictly Standardized Mean Difference

The SSMD is used in some RNAi screens (which in turn are frequently run as imaging assays) [11]. There are several forms of the equation, based on how the screen is run, but it is essentially a modification of the S/N calculation, except that it anticipates unequal variance in either the positive or negative controls (this can be confirmed by comparing the formulas in Figure 9.1). This is a common occurrence in RNAi screens, as reagents that target the same gene will have variable results, and this can occur in the positive or negative controls. Weighting the variance of the positive and negative controls prevents an anomalous pattern in either one from dominating the calculation.

9.5 DATA ANALYSIS: HIT SELECTION

As noted, the process of data analysis for microtiter-based assays has been well discussed in the primary literature, as well as in several how-to guides and best-practices reviews [1–3, 10]. In point of fact, there are many methods for hit identification in a screen (Bhinder and Djaballah identified 33 published methods to identify hits from RNAi screens in addition to describing their own [12]). A few are widely used and several are modifications of commonly used methods. In essence, all methods define a distance from the mean or the median that takes into account the variance. Since we are discussing the analytical process of HCS data, we will stay with this concept, but as a practical matter, hit selection will frequently be coupled with orthogonal methods to classify the hits. In small molecule screens, the hits will be pooled by structural properties to identify common scaffolds. In RNAi screens, additional databases may be consulted, either protein–protein interaction or gene ontology databases, again with the idea that binning the hits into functional families can help prioritize the follow-up studies. In cases where such integrated analysis will be performed, the most robust method of hit selection available should be used.

9.5.1 Rank Order

Inelegant, yes, but effective in certain circumstances. In particular, in cases where a few hits are desired to help explore a novel phenotype. Testing the most effective perturbants in follow-up experiments can lead to novel connections and new research projects, but it places no assurance on how many connections may be valid in the entire data set (which is relevant for pathway analysis of the hits and other integrative approaches). It is common to calculate rank order data from the z score or B factor normalizations, which minimizes egregious artifacts in the data, as discussed in Sections 9.3.2 and 9.3.3. One common example of rank order is characterizing hits from a small molecule screen. Frequently, hits are selected on the basis of an activity threshold (e.g., "80% inhibition at 10 μM"). Such a threshold is set by practical considerations, such as the potency necessary for a tractable medicinal chemistry effort to begin. All compounds passing this threshold may advance to secondary assays.

9.5.2 Mean +/− k SD

The simplest method of selecting set of reasonably robust hits is to calculate the standard deviation (SD) of the dataset and select hits that are two or three SD above or below the mean. Depending on the normalization method and the performance of the controls, the mean may be calculated against the negative controls or the entire plate of samples (which is common for small molecule screens where the vast majority of compounds are negative).

9.5.3 Median +/− k MAD

Directly analogous to the method described above, except that the median of control or plate values and the median absolute deviation (MAD) is calculated instead of the SD.

This is used when some skewing of the data can be observed, particularly the presence of outliers, which in this case could result from difficulties in running the assay.

9.5.4   SSMD

In addition to its role in defining how robust a screen is (introduced in 9.4.5), SSMD values calculated on individual wells is a recent alternative to the z-score that allows for flexibility in how the controls can be used [13]. Its use in hit selection for RNAi screens derives from its accounting for the phenotypic variability that can occur for either the hits or controls in RNAi screens, which can confound other hit analysis methods.

9.5.5   *t*-Test

This is less common for screens, as it requires replicate values and most screening methods are based on a single data point for each sample, but it is used by some groups and greatly increases the confidence in individual reagents (compounds or RNAi) [14, 15]. The *t*-test is covered in greater detail in Chapter 10 and in Appendix C.

9.6   IC 50 DETERMINATIONS

9.6.1   Overview

The **dose–response relationship** describes the change in effect on whole cells caused by differing doses (concentrations) of compounds after a given exposure time. This may apply to individual cells or to cell populations. One assumption in this relationship is that there is almost always a dose below which no response (or limited response) occurs or can be measured. A second assumption is that once a maximum response is reached, any further increases in the dose will not result in any increased effect. As simple as these assumptions seem, there are numerous cases where they are not observed, and a back-up plan is needed when these assumptions fail.

Understanding the dose–response relationship is a necessary part of understanding the cause-and-effect relationship between compounds and its target cells. As Paracelsus once wrote, "The right dose differentiates a poison from a remedy." It should be noted that dose–response relationships will generally depend on the exposure time; quantifying the response after a different exposure time may lead to a different relationship and possibly different conclusions. Similar changes may be observed when changing the number of cells subjected to a given compound concentration. This limitation is caused by the descriptive nature of the approach and highlights the need for understanding the biological pathway and appropriate assay development.

The **half maximal inhibitory concentration (IC_{50})** is a measure of the effectiveness of a compound in inhibiting a biological function. Often, the compound in question is a drug candidate. This quantitative measure indicates how much of a particular compound is needed to inhibit a given biological process by half. In other words, it is the half maximal (50%) inhibitory concentration (IC) of a substance

(50% IC, or IC_{50}). It is commonly used as a measure of drug potency in pharmacological research. EC_{50} is more general, as it can be useful for activators in addition to inhibitors, and is more relevant for cell-based assays, where factors such as penetration into the cell can affect the measurement.

A standard dose response curve is defined by four parameters.

- The baseline response (Bottom)
- Maximum response (Top)
- Slope
- EC_{50}, which is the halfway point between bottom and top (not necessarily the 50% activity value. For example, if the bottom is at 20% and the top is 90%, then the EC_{50} is the concentration of drug that evokes a response of about 55%), see Section 9.6.2.3.

A dose–response curve should have enough data points to calculate all four parameters, in practice, this is frequently difficult. When problems occur, it typically affects the calculation of the slope, as this requires several data points in the middle of the response curve. Most drug discovery programs track the EC_{50}, so it is common to fit a curve to a simplified three-parameter formula, outlined in Figure 9.2. The graphical relationship is described in Figure 9.2a. The formula for this simplified case is shown in Figure 9.2b. This particular equation assumes a standard slope with the response going from 10% to 90% of maximal as x increases over about 2 log units but does not actually measure this change from the data.

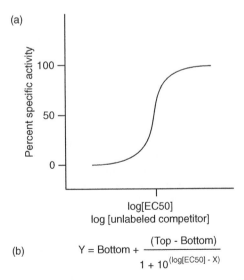

FIGURE 9.2   *Example of a dose response curve with a three-parameter fit.* (a) Dose–response cures typically have a sigmoidal shape, marked by clear plateaus at the high and low doses. In a three-parameter fit, the EC50 can be calculated, but the slope through the EC50 is not. (b) The formula used to generate the curve and to calculate the IC50.

9.6.2 Challenges of Fitting Data to a Standard Dose–Response Curve and Potential Biological Insights from Imaging Data

Fitting data to a standard dose–response curve can be very tricky in practice. Frequently, a suboptimal assay shows more variation in some runs and the data show significant scatter, or chemical properties of some compounds may contribute to the response. In other cases, there may be biologically relevant differences that cause these situations. Alternative algorithms (exploring different metrics) may help. It can also be helpful to understand how the response is occurring—at the 50% response level, are all the cells showing a 50% response, or are half the cells showing a 100% response? Having the image data available can help to explain unanticipated complexity.

9.6.2.1 Failure to Clearly Hit the Top or the Bottom of the Curve   Curve-fitting calculations depend on responses plateauing at the high and the low doses. For compounds of unanticipated potency, one or both ends of the dose range may still show dose-dependent effects for the range tested. In these cases, the algorithm may not be able to fit the curve. One solution is to "fix" the top or the bottom, to feed into the equation the maximum or minimum response values based on the available data (essentially, telling the algorithm that the first or last point is to be taken as the max or min value). This practice is common in screening and is somewhat risky because it, in effect, assigns the top or bottom of the dose range to an arbitrary concentration. Whenever possible, it is better to rerun the experiment. That said, it is common in drug discovery to run dozens of dose–response curves in a single run, and a few compounds will invariably show anomalous behaviors which can be provisionally evaluated and flagged, should they be selected for follow-up studies.

9.6.2.2 The Hill Slope   Although it is common to drop the calculation of the Hill Slope, it indicates important properties of a treatment when it can be calculated, so we return to this parameter. The complete, four-parameter equation is described in Figure 9.3, with example Hill Slopes shown in Figure 9.3a. This equation allows for a variable slope. The Hill slope (or Hill coefficient) is a dimensionless parameter that describes the steepness of the curve. If it is positive the curve increases as the concentration increases. If it is negative, the curve decreases as the concentration x increases. A standard sigmoid dose response has a Hill curve of |1.0|. The curve will be more shallow with Hill slopes closer to 0. With Hill slopes > |1.0|, the curve is steeper. Traditionally, a Hill slope much different than 1 is an indicator that the protein exists in a multimeric state, and that binding to one subunit affects binding by another subunit (either positive or negative cooperativity). In drug discovery, steep Hill slopes may indicate aberrant inhibitory effects, including compound aggregation and nonspecific binding, so a steep Hill slope is an early indication of nonphamacological properties for the inhibitor or the assay conditions [16]. For cell-based assays, there is also a potential that the nature of target can influence the Hill slope, such as targets that show high variability or feedback regulation of activity [17]. In these cases, the ability

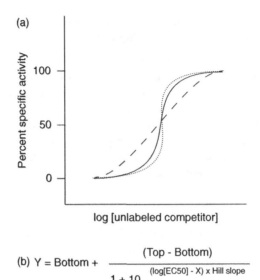

(b) $Y = \text{Bottom} + \dfrac{(\text{Top} - \text{Bottom})}{1 + 10^{(\log[\text{EC50}] - X) \times \text{Hill slope}}}$

FIGURE 9.3 *Effect of Hill slope on curve fits and assay values.* (a) Changes in the Hill slope do not impact the determination of the IC50, but it can affect how easily the min and max response levels can be determined. For shallow Hill slopes (< 1.0), demonstration that the full response range has been reached, making the EC50 calculation difficult, dashed line. (b) The dose–response curve calculation that includes the Hill slope.

to reanalyze the images at the cell-level can explain complex behavior in a standard dose–response curve.

9.6.2.3 Response Plateaus That Differ From Max and Min Values

It happens that after running many compounds in a validated assay, a compound will show a clear max or min value, through a stable and sustained response that is markedly different from the rest of the compounds and the controls. Again, this is common in drug discovery. In standard well-level assays, there are no options for evaluating why this happened. The alternatives are to report an **absolute IC50**, the 50% inhibition concentration based upon the max and min responses of the control compounds, or **relative IC50**, the 50% inhibition value that is calculated directly from the curve fit for that compound. However, in an imaging assay, there are several viable options. Recalling the discussion in the previous chapter on distinguishing what feature will be used as a metric, it may happen that the previous controls affected only a portion of the cells, whereas the new behavior results from the compound hitting all of the cells (or vice versa). Or perhaps the new compound causes cells to lift off the plate in addition to inducing apoptosis, limiting the maximum response. Looking at images from various compound treatments allows an imaging assay to diagnose these possibilities and can allow a more informed decision on how to score such situations.

9.7   CONCLUSIONS

Good data practice begins with reviewing the quality of the data, much like good image analysis begins with reviewing the images and segmentation. If the data do not show significant problems, it can be analyzed with standard well-level methods that are common for other assay formats. If not, alternatives for data analysis exist, and are covered in the following chapter.

Visualization of overall plate data to detect sporadic and systematic biases (e.g., edge effects, row/column effects) is important for normalization. Do the data indicate that there were problems in the experiment? A dramatic (visual) phenotypic event that produces a modest shift in the numerical response could indicate a problem in the algorithm (such as inaccuracies in segmenting cells or thresholding the image capture such that the intensity in one channel is underexposed or overexposed). This is not always easy to detect, as viewing images on a computer monitor often includes some level of brightness or contrast adjustments. Knowing how to limit these adjustments is important in viewing images. It could also happen that the feature being examined may not be reporting the event as observed. Recall that even for simple events, a handful of features are recorded and will not necessarily be equally sensitive to problems. Features such as average intensity, total intensity, and variation in intensity all report on changes in treatment groups, but will do so with different dynamic ranges.

KEY POINTS

1. Standard assay data analysis options are available for HCS data, but they are performed for a single cellular feature. Alternative features can be compared, but considering different features should be based on biological relevance and not primarily as a way to improve the assay robustness.

2. Summary statistics are widely used by statisticians because they provide a relatively concise view of the data. For the experimental biologist they provide information on assay or screen performance that is not available in other formats, and are an opportunity to gain more control and understanding of one's work.

3. The ability to review images and to reanalyze data is one of the most powerful aspects of HCS, including the power to examine and resolve complex behaviors in assay responses.

FURTHER READING

In addition to specific references within this chapter, a strong general appreciation of assay development and validation can be gained from the NIH/Eli Lilly Assay Guidance Manual (http://www.ncbi.nlm.nih.gov/books/NBK53196) which presents many examples of assay development, validation and data analysis for cellular and biochemical assays.

REFERENCES

1. Birmingham, A., et al. Statistical methods for analysis of high-throughput RNA interference screens. *Nature Methods*, 2009, **6**: 569–575.

2. Malo, N., et al. Statistical practice in high-throughput screening data analysis. *Nature Biotechnology*, 2006, **24**(2): 167–175.

3. Zhang, X.D. Illustration of SSMD, z score, SSMD*, z^* score, and t statistic for hit selection in RNAi high-throughput screens. *Journal of Biomolecular Screening*, 2011, **16**: 775–785.

4. Høyer-Hansen, M. and Jäättelä, M. Autophagy—an emerging target for cancer therapy. *Autophagy*, 2008, **4**(5): 574–580.

5. Haney, S.A. Rapid assessment and visualization of normality in high-content and other cell-level assays, and its impact on experimental results. *Journal of Biomolecular Screening*, 2014, **19**(5): 672–684.

6. Bray, M.A. and Carpenter, A.E. Advanced assay development guidelines for image-based high content screening and analysis, in *Assay Guidance Manual*, 2012, National Institutes of Health, Bethesda, MD.

7. Chung, N., et al. Median absolute deviation to improve hit selection for genome-scale RNAi screens. *Journal of Biomolecular Screening*, 2008, **13**: 149–158.

8. Brideau, C., et al. Improved statistical methods for hit selection in high-throughput screening. *Journal of Biomolecular Screening*, 2003, **8**: 634–647.

9. Zhang, J.-H., Chung, T.D.Y., and Oldenberg, K.R. A simple statistical parameter for use in evaluation and validation of high throughput screening assays. *Journal of Biomolecular Screening*, 1999, **4**: 67–73.

10. Ravkin, I. and Temov, V. Comparison of several classes of algorithms for cytoplasm to nucleus translocation. Tenth Annual Meeting of the Society for Biomolecular Screening, 2004.

11. Zhang, X.D. A pair on new statistical parameters for quality control in RNA interference high throughput screening assays. *Genomics*, 2007, **89**: 552–561.

12. Bhinder, B. and Djaballah, H. A simple method for analyzing actives in random RNAi screens: introducing the "H Score" for hit nomination & gene prioritization. *Combinatorial Chemistry and High Throughput Screening*, 2012, **15**: 686–704.

13. Zhang, X.D., et al. The use of strictly standardized mean difference for hit selection in primary RNA interference high throughput screening experiments. *Journal of Biomolecular Screening*, 2007, **12**: 497–509.

14. Pan, J., et al. A kinome-wide siRNA screen identifies multiple roles for protein kinases in hypoxic stress adaptation, including roles for IRAK4 and GAK in protection against apoptosis in VHL$^{-/-}$ renal carcinoma cells, despite activation of the NF-κB pathway. *Journal of Biomolecular Screening*, 2013, **18**(7): 782–796.

15. Whitehurst, A.W., et al. Synthetic lethal screen identification of chemosensitizer loci in cancer cells. *Nature*, 2007, **446**: 815–819.

16. Shoichet, B.K. Interpreting steep dose–response curves in early inhibitor discovery. *Journal of Medicinal Chemistry*, 2006, **49**: 7274–7277.

17. Fallahi-Sichani, M., et al. Metrics other than potency reveal systematic variation in responses to cancer drugs. *Nature Chemical Biology*, 2013, **9**: 708–714.

10

ANALYZING CELL-LEVEL DATA

Steven A. Haney, Lin Guey, and Arijit Chakravarty

10.1 INTRODUCTION

Up to this point, our discussions of data analysis have focused on selecting a strong cellular response and using it to evaluate treatment conditions on cells. This works well in many settings. However, sometimes more complex relationships between samples are important. In these cases, analysis of data at the cell level, including the distribution of individual cells within a well, need to be evaluated, rather than relying on the average values for the well. Population-level analyses require a deeper grounding in statistics than is necessary for standard cell-based screens. So bear with us as we wade through some of the statistics that you probably learned in college (quite?) a few years ago. In this chapter, we will present the concepts, and their application to high content data. Following the extraction of cellular features, as described in Chapters 4 and 8, the analysis of the summary statistics (described in Chapter 9) is an important step. This chapter presents many of the statistical foundations for assessing the skewness of the data as well as the paths for analyzing cell-level data that are normally and non-normally distributed. Although these steps are typically performed by a biostatistician or computational biologist, it is critical that biologists understand the rationale and basic concepts behind the statistics being utilized. Such an understanding can sharpen the formulation of hypotheses and improve communication between the experimenter and the analyst. Understanding the fundamentals of statistical hypothesis testing is key to being able to successfully interpret analytical results.

To go a step further, we will introduce very simple methods for applying these statistical tests to HCS data, through the use of the R statistical analysis package.

An Introduction to High Content Screening: Imaging Technology, Assay Development, and Data Analysis in Biology and Drug Discovery, First Edition. Edited by Steven A. Haney, Douglas Bowman, and Arijit Chakravarty.
© 2015 John Wiley & Sons, Inc. Published 2015 by John Wiley & Sons, Inc.

R has emerged as the language of choice for bioinformaticans for several reasons, the most relevant being that is easy to use for analyzing data and has a very active community of developers that continue to extend its capabilities. A bit of background on R, including how to download it and start using it is included in Appendix B (by the way, it is free). It should not be necessary to repeat these exercises to understand them, but they have been set up so that they can be repeated for those that want to actively explore the data. Discussion related to understanding HCS data and the options for analyzing such data is presented here, the process of extracting, plotting, and testing of the data using R is presented in the appendix.

So what is the difference between analyzing experiments at the well level versus at the cell level? Well-level data is one data point per well, replicate wells give you two, three or more data points. The challenge is to determine whether these few points are different from one or more data points for a different treatment condition. We reviewed these processes in the previous chapter. Cell-level analysis looks at hundreds of data points for each sample. The power here is that even if the mean values do not change very much and the samples would fail in basic measures of screening robustness (such as the Z' statistic), changes in distribution patterns can be rigorous and definitive demonstrations of an experimental effect.

10.2   UNDERSTANDING GENERAL STATISTICAL TERMS AND CONCEPTS

We will take a minute to clarify some of the important terms and concepts we will be using. Some of these concepts were introduced in Chapter 9. Many people have intuitive understandings of several elements of the following discussion, but they become critical to handling cell-level data, so we will take our time here.

10.2.1   Normal and Non-normal Distributions

A normal distribution is the classic bell-shaped curve, and results from data generated when a large number of independent and identically distributed random variables are measured. Rolling marbles through an array of evenly spaced pins will result in a normal distribution of marbles in the collection area, because each time the marble hits a pin, there is an equal chance of bouncing left or right. The mean and the median for a normal distribution are the same (recall that in Chapter 9, we noted that comparing the mean to the median is a quick way of checking whether a distribution is overweighted on one side, or skewed). A graphic comparison of a normal and non-normal distribution is made in Figure 10.1. The normal distribution is shown in the inset. This is compared to a skewed distribution, where the distributional shape has strong effects on the summary statistics, particularly the mean and median.

Non-normal distributions result from events that have an unequal chance of varying from the mean value on one side versus the other. A noncellular example of how a skewed distribution occurs and the challenge it presents to comparisons: throwing footballs from the end zone 20 yards will result in most of them lying near

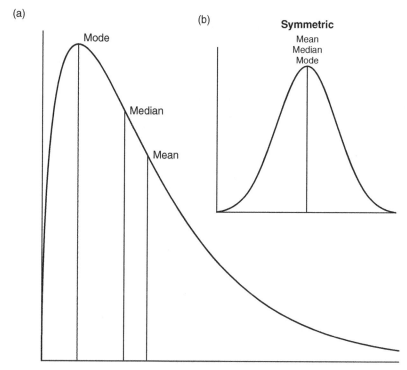

FIGURE 10.1 *Distribution shape and summary statistics.* This figure shows the relationship between the mean, median, and mode for a positively skewed distribution. A comparison with a symmetric (normal) distribution is shown in the inset.

the 20-yard line, but every so often, there will be one that hits with a strong forward roll and lands much further away. It will almost never happen that a football will land with a strong backward roll and land equally far short of the 20-yard line. Over a series of throws, this will generate a skewed distribution. Now imagine you had two people throwing footballs. The challenge then, is could you determine that another person was throwing to the 30-yard line, based upon two non-normal distributions? What about the 22-yard line?

Many biological processes will yield non-normal distributions. The most well-known is the distribution of DNA content per cell. Most cells have a DNA content of 2N, a complete set of paired chromosomes, because they are not dividing. Under normal circumstances, a smaller number of cells will have 4N the amount of DNA, because they are about to divide, and a few will be between these extremes (S phase), as they have begun the process of dividing. A histogram of DNA content per cell presents a profoundly non-normal distribution, with two pronounced humps for the 2N and 4N peaks. Historically, a common way to measure changes in these distributions is to fit normal distributions to the three segments (2N, S, and 4N) and show that one or more of them have changed. However, there are also statistical tests designed to work with non-normal distributions, and these provide a measure of reliability regarding whether a change in distribution for a test condition is indeed significant without parsing the data into subsets of normal distributions. These methods are essential for cases where the data are less well-characterized or familiar.

10.2.2   Normalizing or Transforming Data

There are two common uses of the phrase "normalizing data": (i) adjusting all of the primary data values to eliminate plate-to-plate variability and (ii) fitting data that do not fall into a normal distribution into one that does through a mathematical transformation. We discuss both meanings in this book, the former concept was important for analyzing screening data in Chapter 9, the latter is what we are discussing in this chapter. To avoid confusion, we will generally use the phrase "transforming data" when referring to the conversion of data to a normal distribution.

Transforming data is a common way to treat data, as it increases the options for applying a statistical test, including more powerful tests and tests that are more commonly used (and therefore better understood when reporting results). Raw data may be non-normal but it can often be transformed in a way that makes biological sense. An example here is measuring changes in the nuclear localization of a transcription factor. If it is equally distributed throughout the nucleus and cytoplasm, it has a ratio of 1. If 80% of it is localized to the nucleus, the ratio is 4, but if 80% of it is localized to the cytoplasm, the ratio is 0.25. Higher ratios yield numbers farther from 1 than lower ratios for the same degree of change. Transforming the data by taking the log of the numbers produces a new set of numbers where direction of change does not affect the degree of change. We discuss additional methods later in this chapter. An example of transforming data is presented in Section 10.3.2, on data visualization.

10.2.3   Robust Statistics

We had introduced the use of robust statistics briefly in Chapter 9. Standard statistical tests assume the signal results from events in a group of samples that are relatively equivalent, but it can also happen that such a measurement is influenced by some outliers. Robust statistics reduce the effects of such a set of bad actors in a group, for example, by using the median instead of the mean for a distribution, because the median is largely unaffected by a group of outliers, whereas the mean is. These approaches are common in HCS data because problems in sample measurements are common, such as clumped cells in a few wells, and in RNAi screening, where samples may show pronounced edge effects [1].

10.2.4   Parametric and Nonparametric Statistics

Parametric statistics refer to statistical tests for distributions that follow a recognized pattern or shape. The normal distribution is the most well-recognized, but the Poisson distribution is also a recognized parametric distribution. In these cases, we can discuss the data in terms of the parameters for that distribution, typically the mean and the standard deviation, but the median, standard error, and other measures can be used. Significance can be assessed when a change in these parameters exceed well-defined thresholds. Parametric tests are powerful, because they assume an underlying shape to the distribution of the data. Naturally, if this assumption is wrong, then the p-value provided by a parametric statistical test will be misleading. Nonparametric tests (surprisingly enough) are designed to test whether the change in a distribution is significant without presuming the distribution of the data fits a pattern or shape; significance is assessed without comparing two samples through parameters such as the mean or the median. On the one hand, if the assumption of normality is violated, parametric tests are likely to provide misleading answers. On the other hand, nonparametric tests are less powerful than parametric ones *if the data actually fit a parametric distribution*; they will underestimate the statistical significance of an effect if used with normally distributed data. It is therefore important to evaluate whether a distribution conforms to the distributional assumptions of a statistical test to decide whether a parametric or nonparametric method is most appropriate.

10.3 EXAMINING DATA

10.3.1 Descriptive Statistics

As discussed in Chapter 9, the first step in data analysis is to reduce a large number of data points into a number of concise, informative summary statistics. Table 10.1 presents some of the most descriptive statistics of location and dispersion (or spread) of the distribution. We present these with a bit more definition than we did previously.

The mean and standard deviation (SD) are the most popular summary statistics of location and dispersion, respectively, and are commonly reported as mean $\pm$ SD. The coefficient of variation (CV) is another measure of dispersion that is normalized to the sample mean. The CV is thus a unitless measure and is useful for comparing dispersions across different datasets with differing units. However, the CV is only defined for nonzero means and becomes highly sensitive to minute changes in the mean as the mean approaches zero. Although the mean, standard deviation, and CV are among the most commonly reported descriptive statistics, they are not robust as they are highly sensitive to outliers and, consequently, are most appropriate for symmetrical distributions.

In contrast, the median, interquartile range (IQR), and median absolute deviation (MAD) are statistics derived from rank order statistics, in which observations are considered in terms of their ranking within a distribution. These order statistics are robust to outliers and thus appropriate for skewed data. A robust alternative of reporting the mean $\pm$ SD to summarize data is to report the median $\pm$ IQR or median $\pm$ MAD. The mode is another useful measure in assessing whether a distribution is unimodal (having one local maximum) or multimodal (having more than one local maximum). It is also the most suitable measure of central tendency for nominal data, defined as data consisting solely of categories (e.g., "heads" and "tails"). These were

TABLE 10.1   Common Summary Measures of Location and Dispersion

Descriptive statistic	Formula	General comments
Measures of location		
Mean	$\bar{x} = \left(\sum_{i=1}^{n} x_i \right) \Big/ n$	The arithmetic mean is the most commonly reported summary statistic to describe the central tendency of a distribution; however, it is not robust as it is sensitive to outliers and is not appropriate for highly skewed distributions.
Median	$\tilde{x} = x_{(n+1)/2}$ if n is odd $\tilde{x} = \left(x_{n/2} + x_{(n/2)+1} \right)$ if n is even	The median is a more robust measure of location, particularly for highly skewed distributions.
Mode	The most frequent value in the distribution	Distributions with two local maximums (or peaks) are called bimodal and those with more than two local maximums are multimodal. This measure is most appropriate for nominal or categorical data.
Measures of dispersion		
Variance	$s^2 = \dfrac{\sum \left(x_i - \bar{x} \right)^2}{n-1}$	The variance is defined as the square of the standard deviation, the more commonly reported summary measure of dispersion, and is highly sensitive to outliers.
Standard deviation (SD)	$s = \sqrt{\dfrac{\sum \left(x_i - \bar{x} \right)^2}{n-1}}$	The SD is the most common reported summary statistic to describe the dispersion of a distribution; however, similar to the mean and variance, it is not a robust statistic.
Coefficient of variation (CV)	$s/\bar{x}$	The CV is a measure of dispersion normalized to the mean and is usually reported as a percentage. The CV is only defined for nonzero means and becomes highly sensitive to minute changes in the mean as the mean approaches zero.
Interquartile range (IQR)	Q3–Q1, where Q3 is the 75th percentile and Q1 is the 25th percentile respectively	These quartiles are calculated in a similar manner to the median. The IQR is a robust measure of dispersion and more appropriate for highly skewed distributions.

TABLE 10.1   (*Continued*)

Descriptive statistic	Formula	General comments
	Measures of location	
Median absolute deviation (MAD)	$\text{median}_i(\lvert x_i - \text{median}_j(x_j) \rvert)$	The MAD is the median of the absolute value of the deviations from the sample median. Similar to the IQR, the MAD is another robust measure of dispersion but is more commonly used in high content screening applications than the IQR.
	Measures of shape	
Skewness	$\sum_{i=1}^{n} (x_i - \bar{x})^3 \Big/ (n-1) s^3$	Skewness measures the amount of asymmetry in a distribution with positive values indicating that the data are right-skewed (having a long right tail) and negative values indicating that the data are left-skewed (having a long left tail). Symmetrical distributions have a skewness of 0.
Kurtosis	$\sum_{i=1}^{n} (x_i - \bar{x})^4 \Big/ (n-1) s^4$	Kurtosis is a measure of the center of a distribution relative to the shoulders and tails of a distribution. High kurtosis values indicate heavy tails and a narrow peak; low kurtosis values indicate thin tails and a flat peak. Normal or Gaussian distributions have a kurtosis of 3.

introduced in Figure 10.1. For perfectly symmetrical distributions, the mean, median, and mode are exactly the same.

Other useful summary statistics that describe the overall shape of a distribution are skewness and kurtosis. Skewness measures the amount of asymmetry in a distribution and its sample estimation is given in Table 10.1. Distributions with positive skewness values have long right tails and are referred to as positively-skewed or right-skewed (Figure 10.1); conversely, distributions with negative skewness values have long left tails and are referred to negatively-skewed or left-skewed (the opposite shape to Figure 10.1). For comparison, the normal distribution and the relationship between the summary statistics, is shown in an inset of Figure 10.1. Perfectly symmetrical distributions have a skewness of 0. Kurtosis, on the other hand, is a measure of how heavy the tails of a distribution are. Distributions with high kurtosis have a narrow peak and heavy tails while distributions with low kurtosis have a flat top at the mean and thin tails. The sample kurtosis estimate is often centered at 3 as it is known that normal (Gaussian) distributions have a kurtosis of 3. Skewness and kurtosis can be used together to quantitatively infer characteristics regarding the shape of a distribution. These measures, along with all descriptive statistics, should always be complemented with a plot such as a histogram, which delineates the shape of a distribution.

10.3.2 Data Visualization

Descriptive statistics are useful for summarizing characteristics of a data set but should never replace graphical visualization of the data. A histogram can be plotted as a first pass in assessing normality as it outlines the general shape of a distribution, indicating whether the distribution resembles the familiar bell-shaped normal curve or whether the distribution is highly skewed or has heavy tails (Figure 10.2a, c). A

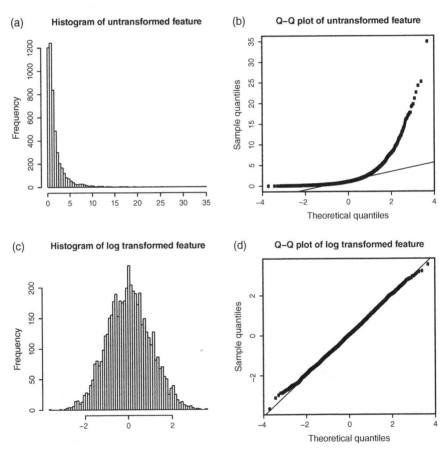

FIGURE 10.2 *Histograms and Q–Q plots of transformed data.* (a) A histogram of raw data reveals the distribution to be highly right-skewed. (b) A Q–Q plot of raw data shows clear departures from normality. (c) The log-transform of the feature distribution shows that the distribution is more symmetrical as shown by the histogram. (d) The log-transformed data also closely resembles a normal distribution, as shown by the linearity in the Q–Q plot.

histogram is often used as an estimate of the **probability density function**, defined as the theoretical probability for each possible data point. A more powerful visual technique to assess normality is the quantile–quantile (Q–Q) plot (Figure 10.2b, d), which plots the quantiles of the theoretical distribution by the quantiles of the empirical or observed distribution. Quantiles are various points taken throughout the cumulative distribution function where the cumulative distribution function is the cumulative sum of the probabilities for each data point. Thus, if quantiles of an empirical distribution perfectly matched the quantiles of a theoretical normal distribution, all data points would fall onto a straight 45° line ($y = x$). Substantial deviations from the 45° line, particularly at the center of the line, indicate non-normality. One advantage of the Q–Q plot is that it is much easier to judge whether a plot approximates a straight line than to judge whether the shape of a histogram approximates a normal density curve. In the example shown in Figure 10.2, the original data are clearly not normally distributed, as shown in Figure 10.2b, but is nearly so when transformed, as shown in Figure 10.2d. Also, although Q–Q plots are usually used to compare a test distribution to a normal distribution, they can be derived for other distribution patterns to assess whether an empirical distribution fits any theoretical distribution. For example, plotting the data in Figure 10.2a against a theoretical Poisson distribution would yield a straight line in a Q–Q plot if it was in fact a Poisson distribution.

Data visualization and summary statistics are very effective for uncovering abnormalities and outliers that may suggest experimental artifacts such as edge effects or auto-focusing errors. Cytological features believed to be independent of the experimental design (i.e., randomly distributed across treatments or wells) can be used for quality control monitoring. Other graphical tools include barplots, scatterplots, and boxplots, to name a few. Scatterplots are especially useful for visualizing correlations between replicates to assess reproducibility. Boxplots display multiple descriptive statistics such as the median, IQR, and skew simultaneously and can help identify outliers within a distribution (Figure 10.3). Boxplots are favored by statisticians because they report on the feature values per well (much as a plate heat map does), but also provide data on the **dispersion**.

Outliers can be the most informative observations, as they can be indicative of an experimental or image processing error, or indicative of a truly extreme data point due to a particular treatment. Outliers identified due to experimental artifacts should be excluded from all subsequent data manipulations and analyses as they can bias results. Note that this is different from assuming that all outliers are "bad data" and disregarding them—this is common practice, and the other name for it is cherry-picking.

10.3.3   Transformation of Data

Many cytological features do not follow a normal distribution and may need to be transformed to better approximate one. A data transformation performs a mathematical operation on each data point. If the transformed data conform to a normal distribution, the transformed data can be used in subsequent parametric data analyses

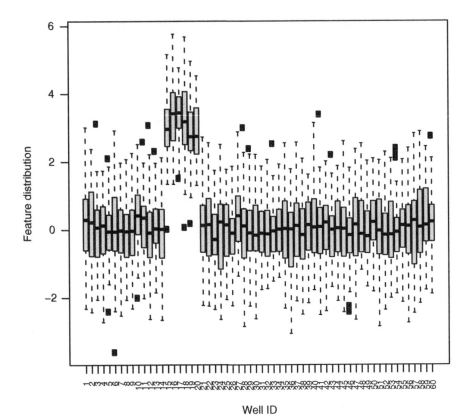

FIGURE 10.3 *Data content in a boxplot.* Boxplots of data from each individual well from a hypothetical plate. Boxplots plot the median and interquartile range, along with whiskers that extend from the 1st and 3rd quartiles. Data points that fall outside of the whiskers can be considered outliers. Wells 15–20 show clear departures from other wells.

rather than analyzing the original data with a nonparametric test. Some common methods are described in the following sections.

10.3.3.1 Log Transformation   One of the most common transformations used in statistics is the log transformation ($y = \log(x)$), which compresses larger values of a distribution closer together and spreads smaller values farther apart (as demonstrated in Figure 10.1). It is therefore useful in making positively skewed distributions or distributions with large outliers more symmetrical. However, it is not suitable for negatively skewed distributions and can exacerbate the skewness of these distributions. The log transformation is also useful in converting a feature from a multiplicative scale, such as a ratio of nuclear intensity to cytoplasmic intensity levels, to an additive scale (since $\log(a/b) = \log(a) - \log(b)$). Another utility of the log transformation is to stabilize variation between comparison groups as statistics such as the *t*-test and analysis of variance (ANOVA) assume equality of variances between groups. Finally, note that the log transformation cannot be applied to values of 0. If the data contain zeros (such as no speckles per cell), then the entire dataset needs be increased so that the lowest value is not 0, typically set to 1.

10.3.3.2 *Square Root Transformation*   The square root transformation ($y = \sqrt{x}$) is used primarily to transform count data (such as nuclear or speckle count), which tend to follow a Poisson distribution. For positively skewed distributions, the square root transformation will give a more normal distribution to the data, but do so less dramatically than the log transformation.

10.3.3.3 *Power Transformations*   Power transformations, typically square or cube transformations, are the opposite of the square root transformation and can be appropriate for the opposite data issue—significant left or negative skew.

10.3.3.4 *Reciprocal Transformation*   This transformation ($y = 1/x$), is useful for extremely right-skewed distributions. It also requires that data do not contain 0 values. The biological interpretation of the reciprocal data is an important factor in deciding whether it is a relevant transformation. Variability of staining could be converted to a measure of uniformity in staining; if the inverse measure is acceptable for the purpose of comparing two conditions, then the effect of the transformation on the data could be considered.

It is worthwhile to explore a variety of different transformations before deciding which transformation yields the best result. These are just examples; they can be applied to any feature and can be checked with a Q–Q plot for their effects. However, when considering a new method for transforming data, it is important to discuss the problem with a *bona fide* statistician. This is particularly true when data variance changes in the dataset. Most statistical tests require that the variance remain constant or proportional throughout the test conditions. It has been shown that variance can change through the test conditions [2], and can even progress from rightward skew, through equal variance to negative skew as compound treatment increases through a dose–response curve. This change in variance for samples within an experiment (**heteroskedasticity**) can eliminate some statistical tests from consideration, particularly the *t*-test and ANOVA. The effect of data transformation needs to be evaluated across the entire dataset.

10.4 DEVELOPING A DATA ANALYSIS PLAN

Now that we have presented some background on distribution patterns and how they affect analysis options, we turn to what to do when you have a high content dataset that requires a statistical analysis, at the single cell level breaking the process down into a set of (mostly) sequential steps. Note that a discussion of the data analysis plan should occur between the biologists and statisticians before the studies begin—scope of the intended analysis has a strong effect on experimental design, which cannot be added on *post hoc*.

10.4.1 Review the Summary Statistics

As noted previously, a method for generating a complete set of summary statistics for a single-plate assay using R is presented in Appendix B. Since we will be introducing methods for analyzing cell-level data in R, familiarizing yourself with R could enable you to recreate some of these examples and apply them to your own datasets. Most commercial data analysis packages do not include tools for analyzing cell-level data (other than in "canned" cell cycle modules), so we turn to R to help us view and analyze our data at the cell level. For a dataset that appears to be highly skewed when looking at the summary statistics, it becomes necessary to further interrogate the data using additional tools, mostly graphical but can also include some statistical tests. We discuss these options next.

10.4.2 Determine the Distribution of the Data

The next step in analyzing the data is to look at the distributions of the data in the samples. The shape of the data distribution determines which statistical tests are appropriate (or rather, which ones are inappropriate), and offers clues to how well the experiment went. For example, when past datasets have been normally distributed but the experiment you are analyzing currently suddenly is not, this is an immediate red flag that there is a problem with the assay, either on the experimental (wet lab) side or the image capture and quantitation side. A bias in the collection of the data has occurred, such as a compression of data on one side of the mean.

What are your options if you are unsure from a summary statistics table whether the data is excessively skewed? As we introduced earlier, an increasingly popular way to look at the distribution in data is by a Q–Q plot, which plots each point in a sample by its quantile (e.g., percentile) value, if it followed a parametric distribution (usually the normal distribution). Going back to the example of the FYVE data we discussed in Chapter 9 and presented in Table 9.1, we have plotted data from some of the wells in density distribution plots (essentially, smoothed histograms that are based on the probability density function mentioned earlier) and in Q–Q plots to show how they compare to each other. Examples of Q–Q plots are shown in Figure 10.4. Two wells are compared, one for a well with a mean value that is low (D02) and one where the mean value for the well is higher (D07). In both cases, Figures 10.4a and 10.4c, the data show rightward or positive skewing. The extent to which each distribution follows a normal distribution is shown in remaining panels in the figure. As can be seen, HCS data are frequently non-normally distributed, and can be highly skewed. Values at the higher end of the distribution are higher than what would be expected for a normal distribution, as seen by the rise above the Q–Q line. The distributions also highlight why formal statistical tests and visualizations are important. Depending on the experiment, switching from a parametric test to a nonparametric test or transforming the data to a normal distribution may be required.

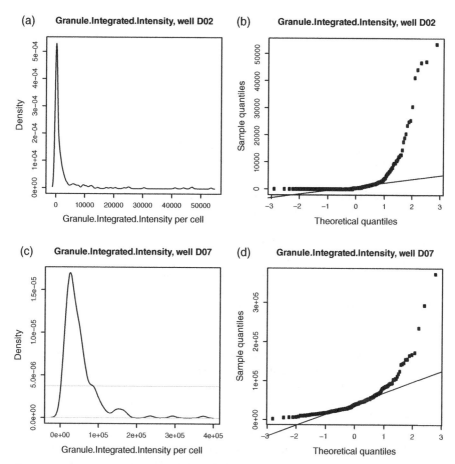

FIGURE 10.4 *Density and Q–Q plots for FYVE granule formation.* Probability density plots (a,c) and Q–Q plots (b,d) for two wells reporting integrated intensity per cell. Wells with lower (D02) and higher (D07) signal at the well level are compared. Note that not only does well D02 show lower average signal, but the majority of signal at the well level is the result of a very strong right skew, which may be considered as outlier cells, depending on the study and the analytical process.

In this case, you will need to consider the differences in distribution shape between two samples that will be compared.

10.4.3   Consider Transforming Non-normal Data

Transforming the data enables parametric statistical tests to be used, which are more sensitive and can be easier to interpret (since the general scientific audience is more familiar with statistical tests based on normal distributions). In the ratio example above, strong increases in nuclear localization result in high numerical changes, but strong decreases result in modest numerical changes. A log transformation helps restore some biological relevance to the data, as it can be just as important if the exclusion from the nucleus increases by 3 fold. Data transformations should be explored whenever possible, but they should be interpreted cautiously, and many statisticians will look for results from transformed data to be replicated, or at least trending towards the same conclusion, in the raw data.

10.5 CELL-LEVEL DATA ANALYSIS: COMPARING DISTRIBUTIONS THROUGH INFERENTIAL STATISTICS

Inferential statistics contrasts with descriptive statistics. The latter looks at a dataset to assess what properties it possesses (is it normally distributed?), but makes no conclusions about any of the samples. That is the provenance of inferential statistics. We have already discussed inferential statistics in Chapter 9, during the sections on hit selection and dose–response curves. Indeed, these are procedures we use frequently. However, as the earlier discussions of normality and parametric statistical tests have indicated, the process is more nuanced when discussing cell-level HCS data.

We revisit the process here because when looking at the cell-level data, there are more options for how to summarize the data within each well and how to make these comparisons. Additionally, HCS data are different than the data that many inferential statistical processes were developed for, in particular, the number of observations can be 10–1000 times larger, making significance measures much greater (i.e., very small p-values), but the difference between them in these cases can be small (a modest **effect size**). Citing a low p-value as the sole measure of the robustness of an experiment can be misleading. This is where discussions between a biologist and a statistician become critical. Therefore, we return to the basics of inferential statistics. The mean and median of a sample are **estimators** of that sample, attempts to identify a "typical" value for the well that can be used to compare wells. The appropriateness of using these as estimators has direct bearing on getting a result from a statistical test that reflects the actual biology or pharmacology. Other measures may need to be used to accurately describe the sample, or it may be that using estimators is impractical (such as in cases where samples show heteroskedasticity), and a nonparametric test is more appropriate.

Although rarely invoked directly in cell-based assays, the process of determining whether a treatment has affected cells is embedded in the concept of the **null hypothesis**. A statistical test formally evaluates whether the difference in two samples is based on variation inherent to the data collection process ("accepting the null hypothesis") or that the difference does in fact result from the treatment and that the two samples are reporting authentically different values ("rejecting the null hypothesis"). A complete discussion of this is included in Appendix C, but for the current discussion, we will leave it as the significance of a test is the chance that the null hypothesis was accepted or rejected incorrectly, and that for most studies, a 5% chance that this may have occurred is considered acceptable. The challenges in HCS are to understand the difference between a statistically stable response (significance) and a biologically relevant response (importance), and to relate these standards to data that can be non-normally distributed, something many experimental biologists may not have considered previously.

10.6   ANALYZING NORMAL (OR TRANSFORMED) DATA

If you have been paying attention so far, you know where we are heading next: if the data are normal, we can use standard parametric methods to analyze them. These include t-tests for a comparison between two groups, or a one-way ANOVA for comparisons between three or more groups with a single treatment. These tests were introduced in Chapter 9, what follows here is a short description of the approach and when it is most appropriate for your high content datasets. In this section, we introduce the most common parametric inferential statistical tests, and some of the major assumptions made when using them. We describe them only here, they can be implemented easily in most statistical analysis and spreadsheet programs (including Excel^{TM} and R), and present additional theoretical background in Appendix B.

10.6.1   The t-Test

There are several choices to make when you are running a t-test in practice on your data. For starters: if you have lots of outliers, the t-test is not the right test for you. In HCS, you will typically have a large sample size and it may be close to a normal distribution, but you may have lots of outliers. Every experiment is different, so you need to look. Should your t-test be one-tailed or two-tailed? Two-tailed, almost always. Unless you are absolutely certain that the variable being examined is only going to move in one direction. Choosing a one-tailed test without a good rationale has the effect of making your p-value look better (twice as good), and but may be a violation of good practices (more than twice as bad). What about the question of equal or unequal variances? Depends on the data. The best way is to examine the histograms of the data, and do a Levene's test or an F-test to test whether the two populations have an equal or an unequal variance, and then choose appropriately. Do not choose unequal variance blindly, as it costs you degrees of freedom (equivalent to losing statistical power). Welch's t-test does not assume equal variances and would be used if unequal variances have been determined. Refer to the discussion in Appendix C on this topic.

10.6.2   ANOVA Tests

Again, everyone has heard of ANOVA. Whenever you have a three-(or more) way comparison with a single treatment, the correct test to use most of the time is a one-way ANOVA with a *post-hoc* test (a follow-up statistical test between two groups that is similar to a t-test). When is an ANOVA a bad idea? When the data are not normally distributed (in which case a Kruskal–Wallis test is more appropriate, discussed in the next section) or if the variances between groups are not equal, in which case a Welch's ANOVA is more appropriate. Welch's ANOVA is a direct extension of

Welch's unequal variances *t*-test, which explicitly allows for the treatment groups to have different variances. Again, the steps for a three-way comparison are very similar to those for a two-way comparison. Examine your data visually, check for normality with a histogram and a Q–Q plot, and then test variances between groups.

An ANOVA simply compares the variability between the different groups (between-groups error) to the noise in the experiment (within-groups error) and spits out whether your treatment effect was statistically significant. It does not tell you which groups are different from each other. Usually, an ANOVA is followed up with a *post-hoc* test (such as a Tukey's or Dunnett's tests) to compare specific groups against each other. These tests provide you with a *p*-value that is protected against the problem of multiple corrections (more about this later).

10.7   ANALYZING NON-NORMAL DATA

What happens when you decide that the dataset can not be transformed to a normal distribution and must be analyzed with a non-parametric statistical test? Often, one simply uses the Wilcoxon rank sum (the nonparametric equivalent of a *t*-test) or Kruskal–Wallis test (the nonparametric ANOVA). Nonparametric tests are also known as distribution-free tests because they make no assumptions about the underlying distribution from which the data are drawn. This makes these tests more robust and sometimes simpler to use (also, harder to misuse). There are fewer choices to make with a nonparametric test, and they are often one-step procedures. This increased simplicity and robustness comes at a cost, however: in cases where a parametric test would be appropriate, nonparametric tests have less power. In practical terms, the *p*-values are typically larger for a nonparametric test on a given dataset than for the parametric equivalent.

From an implementation standpoint, nonparametric tests are a bit of a headache for a beginner as you will not find them in Microsoft Excel (unlike the *t*-test and the one-way ANOVA). These tests are available as part of simple statistical packages like PRISM and JMP. Also, there are a number of online resources that provide the tests as web-based forms. As it turns out, they are also available in R. We discuss the tests here, and follow up with additional background material in Appendix C.

10.7.1   Wilcoxon's Rank Sum

Also known as a Mann–Whitney U test, this is the simplest nonparametric hypothesis test, which compares whether the ranked sums across multiple groups differ significantly. Use it when comparing two groups that are not normally distributed. The only choice to make is one-tailed or two-tailed test.

10.7.2   Kruskal–Wallis ANOVA

This is a nonparametric alternative of a one-way ANOVA and is a direct extension of the Wilcoxon ranked sum test to three or more groups. It is identical to a one-way analysis of variance with the data replaced by their ranks. The null hypothesis for this test is that the populations from which the samples originate have the same median. Since it is a nonparametric method, the Kruskal–Wallis test does not assume a normal population, unlike the analogous one-way analysis of variance. However, the test does assume an identically shaped and scaled distribution for each group, except for any difference in medians. The Kruskal–Wallis test does *not* assume that the data are normally distributed; that is its big advantage. It does, however, assume that the observations in each group come from populations that show **homoscedasticity**; if not, the Kruskal–Wallis test may give inaccurate results [3]. These changes in distribution do occur in high content experiments [2], so it is necessary to check them before beginning this test.

10.7.3   The Kolmogrov–Smirnoff (KS) Statistic

The KS statistic has been used in several studies to compare nonparametric high content feature data [4, 5]. Its popularity derives from the fact that it is very easy to perform, and is very sensitive to changes in the distribution of two samples. The test compares two populations where the values have been rank-ordered and scaled from 0 to 1 (each value is measured for its fractional contribution to the total value of the samples). A change in distribution will cause the curve to change, and the gap between these curves at the widest point is taken as the KS statistic. The KS test is easily applied in R.

10.7.4   Bootstrapping

Bootstrapping is the practice of estimating properties of an estimator (such as its variance) by measuring those properties when sampling from an approximating distribution. In practice, this is usually implemented by repeatedly resampling the observed dataset with replacement, to generate "new" datasets of the same size. These datasets can then be used to compute, for example, the variance of the mean for an unusually shaped distribution.

At first blush, sampling with replacement does not seem intuitive at all. The logic behind this is that the sample we have collected is often the best guess we have as to the shape of the population from which the sample was taken. For instance, a sample of observations with two peaks in its histogram would not be well approximated by a Gaussian or normal bell curve, which has only one peak. Therefore, instead of assuming a mathematical shape (like the normal curve or some other) for the population, we instead use the shape of the sample. The key principle of the bootstrap is to provide a way to simulate repeated observations from an unknown population using the obtained sample as a basis.

While the implementation of bootstrapping can be cumbersome (and may require that you write your own code to do it!), it is a powerful technique that is worth mentioning here. Basically, pretty much any statistical test can be applied using bootstrapping. In situations where no standard statistical test exists for the problem at hand, bootstrapping will often provide a reliable and statistically sound way to answer the question at hand. When in doubt, bootstrap!

10.8   WHEN TO CALL FOR HELP

A biologist should always work in consultation with a trained statistician. Statistical tests are easy to use and the results sound convincing. This can lead to a false sense of security that data are being analyzed appropriately—it can be hard to tell when one has strayed from accepted practices. In this chapter, we have highlighted several specific cases when a biologist should consult a statistician for assistance. There are other situations. One situation in particular is the case of multiple comparisons. In statistics, the multiple comparisons or multiple testing problem occurs when one considers a set of statistical inferences simultaneously [1]. For example, if one relies on a significance level of 0.05 for a single test, but then compares 20 different tests (independent features or endpoints) in a single experiment, then one should expect one test to be scored as significant every time. Several statistical techniques have been developed to prevent this from happening, allowing significance levels for single and multiple comparisons to be directly compared. These techniques generally require a stronger level of evidence to be observed in order for an individual comparison to be deemed "significant", so as to compensate for the number of inferences being made.

10.9   CONCLUSIONS

We have introduced the options available when considering how to analyze data. The task becomes one of reviewing the data and determining the appropriate tests to use. We touched on this situation when we introduced the use of summary statistics to evaluate the fit of the data to a classical, normally distributed distribution. In cases where the data are a fit, or when the data are affected by the presence of a small set of outliers, the methods discussed in Chapter 9 are fully adequate. In cases where the fit is less clear, methods for more critical appraisals of the data were introduced in this chapter, and we will demonstrate their use in assessing high content data shortly.

KEY POINTS

- Most HCS data are non-normally distributed. In many cases, these departures from normality can be managed through robust statistical tests or data transformations.
- Non-normal distributions, particularly highly skewed distributions, occur naturally in HCS data and in biology in general. Ignoring these patterns (such as through routine elimination of skewed tails) may not be appropriate for the biology behind the assay. Nonparametric methods of analyzing data are available (as are parametric methods that are based on distributions other than the normal distribution), and should be used in these cases.

- Analyzing data at the cell level is certainly more time-consuming than simply taking the well means or medians as assay metrics, but doing so enables a broader range of scientific questions to be addressed.
- A few methods for analyzing cell-level data are both simple and powerful, and could be considered as standard data analysis tools, even by scientists who expect to analyze most of their data at the well level. Detailed summary statistics, the Q–Q plot for individual wells and the Boxplot for a plate are good examples.

FURTHER READING

There is much additional information in Appendix C, including more detailed background and examples on how to apply the tests described in this chapter. Our intent here is to introduce these concepts and sensitize an experimenter to situations where they are relevant. When the need arises, the operational details can be found in the appendix. Additional sources for background and examples of statistical tests in biological research include Statistical Methods in Research by Domenico Spina, Chapter 26 in volume 746 of *Methods in Molecular Biology* (2011).

Many online resources are also available, and a good starting place is http://statpages.org/.

REFERENCES

1. Birmingham, A. et al. Statistical methods for analysis of high-throughput RNA interference screens. *Nature Methods*, 2009, **6**: 569–575.
2. Haney, S.A. Rapid assessment and visualization of normality in high-content and other cell-level assays, and its impact on experimental results. *Journal of Biomolecular Screening*, 2014, **19**(5): 672–684.
3. Fagerland, M.W. and Sandvik, L. Performance of five two-sample location tests for skewed distributions with unequal variances. *Contemporary Clinical Trials*, 2009, **30**: 490–496.
4. Giuliano, K.A., Chen, Y.T., and Taylor, D.L. High-content screening with siRNA optimizes a cell biological approach to drug discovery: defining the role of P53 activation in the cellular response to anticancer drugs. *Journal of Biomolecular Screening*, 2004, **9**: 557–568.
5. Perlman, Z.E. et al. Multidimensional drug profiling by automated microscopy. *Science*, 2004, **306**: 1194–1198.

SECTION IV

ADVANCED WORK

We now begin to enter the iterative phase of HCS. Having made initial progress on mastering the instrumentation, assay design, image acquisition, and data analysis, we begin the process of improving, refining, and extending this work. We begin with Chapter 11, taking a critical appraisal of assay development, both the wet-bench phase that includes cell culture and staining, and the computational phase. Once an assay starts to show some promise, it will frequently move to a screen or routine assay. In Chapter 12, we discuss the process of how this transition opens a host of complications that need additional attention. Finally, in Chapter 13, we take a look at *in vivo* studies. HCS assays performed in cell-based assays offer insights into signaling mechanisms, but such studies performed *in vivo* provide important preclinical context to the manipulations. In all of these cases, initial progress in developing a quantitative cellular imaging assay may still need significant development to achieve the level of robustness necessary for their intended uses.

An Introduction to High Content Screening: Imaging Technology, Assay Development, and Data Analysis in Biology and Drug Discovery, First Edition. Edited by Steven A. Haney, Douglas Bowman, and Arijit Chakravarty.
© 2015 John Wiley & Sons, Inc. Published 2015 by John Wiley & Sons, Inc.

11

DESIGNING ROBUST ASSAYS

Arijit Chakravarty, Douglas Bowman, Anthony Davies, Steven A. Haney, and Caroline Shamu

11.1 INTRODUCTION

Eventually, every high content assay reaches this point: you identify the biology you are interested in, select a metric that should work, design an assay around this metric, run pilot experiments and everything looks fine. Then you scale up, and the first time you run the assay, the images and/or numbers do not look right. Is it the biology? Is it the bench work? Or is it the analysis? Welcome to troubleshooting mode! We hope your stay here is short.

In this chapter, we will talk a little about how trouble usually manifests in the high content setting. Then we will take a look at how to design an assay to minimize the chances of trouble occurring in the first place, and to maximize your chances of spotting trouble when it does occur. Finally, we will provide a troubleshooting guide, with a special emphasis on image processing.

11.2 COMMON TECHNICAL ISSUES IN HIGH CONTENT ASSAYS

In theory, every stage in the operation of a high content assay is a potential source of technical failures. For example, cell culture can lead to variations in cell density that in turn lead to sampling errors or differences in biology. Immunofluorescence (IF) staining protocols can cause technical issues when antibodies are not sufficiently optimized. Issues surrounding image acquisition can also lead to poor data quality when wells are out of focus, or when insufficient exposure times lead to a poor

An Introduction to High Content Screening: Imaging Technology, Assay Development, and Data Analysis in Biology and Drug Discovery, First Edition. Edited by Steven A. Haney, Douglas Bowman, and Arijit Chakravarty.
© 2015 John Wiley & Sons, Inc. Published 2015 by John Wiley & Sons, Inc.

signal-to-noise ratio. Image analysis protocols can further contribute to the variability of an assay, typically due to thresholding and segmentation errors.

In this section, we will look at the common sources of failure in high content assays, and discuss some of the patterns associated with these sources of failure. Often, there may be more than one underlying cause, and more than one potential solution. For example, when working with a very weak antibody, refinements in staining, image acquisition and processing can all be employed to improve or resolve experimental issues.

11.2.1 At the Bench

11.2.1.1 Biology is Not Robust   In some cases, the underlying biology is at fault. This happens when a biologist walks in to a high content assay development setting looking to develop an assay based on an effect observed in a few images or a small test sample, and this effect is not representative of the totality of the data. This happens more often than you would think in high content assays, perhaps because of the tendency of the human eye to gravitate towards a pattern in a dataset or a few interesting exceptions at the microscope. From the standpoint of a high content person, who is not necessarily an expert on biology, it is typical to take the problem statement at face value, only to find many weeks later that the biological effect was much weaker than first expected. As you can imagine, when defining a high content assay in the earliest stages, it is really important to differentiate between images and quantitative data, between opinions and facts. The best way to do this right at the beginning of the exercise is to ensure that a randomly selected set of images is scored in a blinded manner for the phenotype/effect of interest. This topic will be covered in more depth later in this chapter.

11.2.1.2 Poorly Reproducible Staining   The signal-to-noise ratio of assay samples will fluctuate with day-to-day, or sample-to-sample, variations in the quality of staining. This may not always affect the performance of the assay. For example, an assay that evaluates a specific substructure for morphological changes (for example, spindle bipolarity or neurite outgrowth) may be very resilient to changes in signal intensity. On the other hand, an assay that uses a global threshold to identify objects of interest is likely to be highly vulnerable to fluctuations in staining intensity. It pays to understand the impact of variability in staining intensity on the final output of your assay. If the final output is sensitive to these changes, and if your signal-to-noise ratio fluctuates from run to run, consider redesigning your assay to minimize the effect of these variations on the final output or include a control for normalization.

11.2.1.3 Autofluorescence
Causes of autofluorescence. The dictionary definition of autofluorescence is "a self induced fluorescence". A more comprehensive definition would be a tendency of a sample to exhibit an unwanted or unexpected fluorescence which is either the result of naturally occurring ligands or as an artefact resulting from sample preparation. In essence, this autofluorescence is the cellular background that must be factored out of an image that is labelled with a fluorescent reagent. In some cases, it

does not interfere with the experiment, typically when the target to be labelled in the cell is expressed well, such as overexpression of a green fluorescent fusion protein or labeling an abundant protein such as tubulin. In other cases, autofluorescence can be an issue, for example, labeling an endogenously expressed protein that is not highly abundant, certain cell types (e.g., primary hepatocytes) can accumulate high levels of autofluorescent compounds, and when the labelling reagent is weakly effective, such as some antibodies. In these cases, it may take time to reduce the background fluorescence of the cells. The factors that contribute to autofluorescence include:

> *Natural fluorescence.* Natural fluorescence in cells and tissues is due in a large part to molecules like flavins, porphyrins, and lipofuscins. These molecules absorb and reemit light in the same way as the fluorophores described above and in Chapter 2.
>
> Scientists who work with adult primary cells, especially cardiomyocytes, hepatocytes or neurons, will be familiar with lipofuscin fluorescence. Lipofuscin fluorescence often manifests itself as punctate intracellular spots, and can be a persistent problem even in samples that have been extensively processed with fixatives and organic solvents. The spectral characteristics of Lipofuscin are such that it will fluoresce when exposed to light wavelengths 360–650 nm [1].
>
> *Fixative-induced fluorescence.* As discussed in Section 11.2.1, fixatives fall into two categories, (i) organic solvents such as alcohols and (ii) cross-linking agents such as aldehydes. Aldehyde fixatives include formaldehyde, paraformadehyde and glutaraldehyde, and these fixative agents can react with amino acids and proteins resulting in fluorescence. Formaldehyde and paraformaldehyde are commonly used as fixatives in high content IF microscopy and often do not cause an unacceptable level of autofluorescence. The use of glutaraldehyde however commonly results in a high level of autofluorescence and consequently should be avoided where possible [1].

Addressing autofluorescence. There are several approaches that can be taken to reduce the problem of autofluorescence: (i) avoid, (ii) quench, and (iii) exclude or subtract during image analysis.

> *Avoiding the problem.* The first and simplest means dealing with autofluorescence is to avoid it. This may be achieved by using narrow band filter sets and reducing intensity of light source. If you are able to determine the spectral characteristics of the autofluorescence, it may be possible to find appropriate excitation or emission wave lengths where autofluorescence is eliminated or reduced. You may also be able to avoid the problem by altering the wavelength of the fluorophore. Autofluorescence is typically widely fluorescent, such as a range of 500–600 nm, but this means that labeling cells outside of this range is not a problem. In this case, the 488/GFP channel on most imagers is still available.
>
> *Quenching and bleaching.* There are a number of well-described quenching and bleaching techniques which can be employed to reduce or eliminate autofluorescence emanating from different sources (see Table 11.1).

TABLE 11.1   Treatment Strategies to Reduce Autofluorescence

Source of immunofluorescence	Methods of quenching[a]
Lipfuscins	CuSO₄ or Sudan Black
Collagens and elastins	Pontamine Sky Blue
Unbound or excess fluorescent conjugates and general background	Evan's Blue and Trypan Blue
Aldehyde fluorescence	Sodium Borohydride
	Pre-bleach treatment long term (12–48 h) exposure of the sample to wavelengths of light UV to 633 nm)

[a]Selected references:

Schnell et al. *J Histochem Cytochem,* 1999, **47**: 719.

Werkmeister et al. *Clin. Materials,* 1990, **6**: 13.

Merino, J.M. et al. *Exp Eye Res,* 2011, **93**: 956.

Excluding or subtracting autofluorescence during image analysis. Depending on the nature of the autofluorescence, it may be possible to exclude or subtract the autofluorescent signal using the image analysis or data analysis tools available with many of the high content analysis platforms. For example, you may find that the localization or pattern of autoflorescence may be such that you can exclude it during image analysis (i.e., if the target of interest is located in nucleus and autofluorescence is mainly cytoplasm). Additionally, if the intensity of the autofluorescent staining is sufficiently low (i.e., does not saturate the camera) it may be possible to simply subtract the autofluorescent intensity background from unlabelled controls.

11.2.1.4 Scale-up Issues   Another common source of technical issues in a high content assay is the transition from a pilot assay to a full-scale assay. Scale-up issues arise when the conditions tested in the pilot are different from the ones used to run the assay. Although this can sound like an obvious "gotcha," it can still trip up perfectly reasonable people! Sometimes, the discrepancy between pilot and production-scale assay is obvious, such as when a pilot was developed using one batch of reagents and the production-scale assay uses a different batch. Other times, the differences may be far more subtle, such as when the assay plates sit at room temperature for much longer with the production-scale assay than with the pilot, leading to differences in signal-to-noise ratio between the two versions of the assay.

11.2.2 During Image Analysis

11.2.2.1 Acquisition   Many problems in high content assays start here—to trot out a cliché, Garbage In, Garbage Out. Poorly executed image acquisition can make images unusable for further processing. An example of this is when the images are overexposed because of some change in preparation or instrument settings. Over-exposure leads to a loss of information in the fine structure of images, and can obscure differences across a population (especially for intensity-based segmentation or intensity features), in effect leading to a reduction in the dynamic range of the assay. Another example is when images are poorly focused, which often may lead to degradation in performance of segmentation algorithms in image processing. Out-of-focus objects can also look larger, making cellular area metrics inaccurate. Other more subtle problems are related to insufficient resolution: either due to incorrect objective (e.g., magnification, correction collar setting not correct for specific plate type), camera binning set too high, or low exposure times.

In the end, image quality drives the quality of the numbers you see, and there are many situations where the numbers may "look okay" on a given day, but the fundamental data, the images, do not. In these cases, a high content assay can behave a little like a ticking time bomb. Eventually, a poorly functioning assay *will* provide badly behaved numbers, and the problem then becomes much harder to fix. Therefore (not to belabor the point), a well-designed assay should always have a quality control (QC) step built into the workflow where someone looks at the images themselves. Acquisition systems where this is not easy to do are prone to work poorly in the long run.

11.2.2.2   *Image Processing*   When developing an image analysis routine, it is worth checking the quality of your image processing and segmentation, as the final numbers will only be as good as the underlying image processing. Over or underseg-mentation can lead to reduced accuracy of the quantitative metric. In extreme cases, incorrect segmentation or masking can lead to artifactual results. For example, if segmentation errors cause the wrong population of cells to be selected or the data are being normalized by cell count and the cells are mis-segmented. The inappropriate use of a global threshold will also lead to unpredictable results if only a subpopulation of the desired cells is identified correctly. On most high content software platforms, image processing routines for segmentation will typically provide a visual readout of the quality of segmentation, for example, outlines for each identified object. One way to approach the image processing for a high content assay is to think of it as cooking rather than baking. In contrast to baking, where a recipe followed exactly leads to a predictable result, cooking requires occasional monitoring (tasting) and slight tweaks to keep things on track. Every once in a while, be sure to check in on your images, and on the steps involved in extracting metrics from raw images. If you do not, you may be unpleasantly surprised when the final data are served up!

11.2.2.3   *Poor Assay Design or Wrong Metric*   A closely related problem occurs when the assay performs poorly because the selected metric is not statistically robust, or does not appropriately capture the biology. As an example of the former scenario, consider what happens when you have a metric that requires normalization of the intensity of marker X with marker Y. If the intensity of marker Y is highly variable, such normalization may actually add noise rather than remove it. As an example of the latter scenario, where the selected metric does not accurately capture the biology of interest, consider what would happen if an assay designed to identify compounds causing mitotic accumulation used 4N nuclear DNA content as its

readout. Such an assay would correctly identify compounds causing mitotic accumulation, but would also be likely to inappropriately identify compounds causing nonmitotic phenotypes—such as delays in G2 or the formation of tetraploid cells. The "false positives" of the assay would not manifest during the operation of the assay, but would lead to a reduced overall rate of success of the assay. Clearly, as we have discussed previously in Chapter 8, the metric being measured defines the question being asked.

11.2.2.4 Lack of Adequate Statistical Power   In some settings, the inherent biological variability of the assay system requires a higher replicate number than the one that was selected, or the replicates are deployed in such a manner (adjacent wells on the same plate, for example) that falsely understates the true variability of the assay. In such a case, assay results will appear fairly robust at first blush, but will be (mysteriously) variable from run to run of the assay. To guard against this problem, it is usually advisable to spread replicates across the experimental axis that has the most variability. Although the results may appear noisier, the answers that the assay provides will be more reliable in the long run.

11.3   DESIGNING ASSAYS TO MINIMIZE TROUBLE

Sometimes the best way to avoid issues with the performance of an assay is to design for robustness up front. In this section, we will discuss ways in which to design high content assays to minimize trouble and maximize chances of success.

11.3.1   Choosing the Right Antibodies

Primary antibodies are critical to the success of any experiment, so their characterization is essential. In particular, it is dangerous to assume that the antibody correctly recognizes the antigen as stated by the manufacturer, particularly for less widely used antibodies. It is critical to fully characterize the primary antibodies as they are a key aspect to any experiment. When using an antibody that someone else has used, be sure to determine what work was done to validate it, and make sure that you have the right lot (batch) of the antibody. Lot-to-lot variability can be a significant factor in the success of a screen. When working with a new antibody, consider running a positive control targeting the protein with siRNAs and showing a concomitant loss of antibody staining. If the antibody is specific for a modified form of the protein (phosphorylation, acetylation, etc.) and the extent of modification can be manipulated through the activation of a signaling pathway or culture condition, these should be tested as part of the primary antibody characterization.

When selecting an antibody, it is essential that the immuno-reagent is thoroughly checked. Here are some questions worth asking when you are evaluating the quality of your antibody:

- Is the binding affinity of the antibody to its target sufficient for the needs of the assay?

- How specific is the antibody? Is the epitope specific to the target molecule that you wish to study?
- How pure is your antibody sample? Antibodies are typically affinity-purified before use. Antibodies that are not purified may potentially have more lot–lot variability.
- Is the antibody suitable for the experiments? The chemical and structural characteristics of proteins can change depending on fixation methods. This is not a concern for polyclonal antibodies, but assays using monoclonal antibodies will occasionally run into performance problems for this reason.
- Was the antibody tested under the right conditions? (e.g., SDS PAGE western blotting, conducted under denaturing conditions, tells you nothing about the way an antibody will perform in immunofluorescence, which is nondenaturing)

11.3.2 Optimizing Your Antibodies

Trying to optimize an antibody for use in a cell-based assay may at times resemble a chicken-and-egg scenario. How do you know you have found the optimum concentration of antibody, when you have not optimized blocking, permeabilization and fixation steps? Indeed, most commercially available antibodies will be supplied with detailed standard operating procedures (SOPs) covering the steps required for preparation and use of the antibody. It is typical to use this SOP as a starting point for further exploration of the space of all possible combinations of experimental conditions. This SOP should also be included in the assay SOP (See Table 11.1).

A critical and common tweak is to optimize the concentration of the primary and (if used) secondary antibody. For this, the most useful scheme is to optimize both simultaneously as a matrix of conditions, in a cross-titration assay. For example, this may involve setting up a 96- or 384-well micro plate with cells treated with a positive control (i.e., a treatment that will elicit a maximal cellular response), a negative control (i.e., a treatment that will elicit a minimal cellular response) and a resting control (un-treated cell). The second step will be to then perform serial dilutions of the primary and secondary antibody as shown in Figure 11.1. In this case, technical triplicates allow for better quantification of the magnitude and specificity of the staining. A no-primary antibody control is essential for all optimization experiments, as a low level of secondary antibody staining or autofluorescence can frequently be detected in the absence of primary antibody. Although specific staining with most primary antibodies gives a substantially higher signal, in cases where the primary antibody does not recognize the epitope, the residual staining/autofluorescence can be confused for a low level of specific staining with the primary antibody if a no-primary antibody control is not included. The information gained from this experiment will allow for the correct antibody concentration to be ascertained. Performing this experiment will also give a strong indication of the dynamic range of the assay and potentially save money in reagent costs if a lower concentration results in equal assay performance.

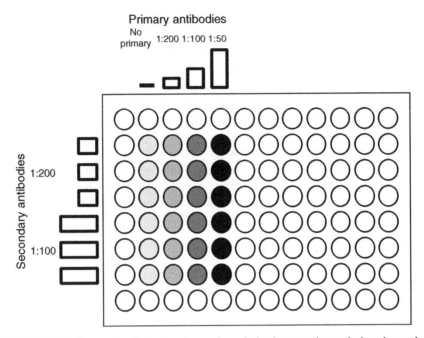

FIGURE 11.1   *Protocol optimization.* A sample optimization experiment designed to understand the differences in primary and secondary concentrations. Once images are acquired and analyzed, the signal-to-noise and cellular phenotypes can be analyzed to determine the best protocol.

11.3.3   Preparation of Samples and Effects on Fluorescence

Somewhat obviously, the different steps in an IF protocol have the potential to affect the quality of the final signal. What follows is a short discussion of some key points to consider in this respect.

11.3.3.1   Fixation and Permeabilization   Ideally a fixative should immobilize cellular antigens but at the same time preserve fine cellular structure. The fixative should also permit unrestricted access to all cellular and subcellular compartments, therefore allowing antibodies to reach their target antigens. There is a wide range of fixatives available for use with cells and tissues, but ultimately the choice will depend on the chemical nature of the antigen under examination and the antibody used. The most commonly used cell and tissue fixatives generally fall into two categories: (i) organic solvents such as alcohols and (ii) cross-linking agents such as formaldehyde.

Organic solvents (methanol or ethanol) dissolve lipids and dehydrate the biological sample resulting in precipitation of the cellular proteins. This method is especially useful for studying the cytoskeleton. Cross-linking reagents are generally considered to be preferable if the general cellular structure needs to be preserved. However, cross-linking fixatives may damage or chemically alter the antigen to the extent that the antibody will no longer recognize the epitope and hence, fail to bind its target. It should also be noted that it is often necessary to permeabilize cells when using a cross linking fixative, this additional step ensures that the internal cellular spaces are fully accessible to antibody reagents.

11.3.3.2 Blocking   This step is required to prevent nonspecific binding of the primary and/or the secondary antibody. A blocking step is almost always incorporated into immuno assays. As the name suggests, a blocking agent will prevent the interaction between the diagnostic antibody and any "sticky molecules", by binding preferentially to these sites and hence preventing the subsequent binding of the primary and secondary antibodies. Blocking is important procedure, as without this step, the assay readout may suffer for a low signal to noise due to high background. It is for this reason that blocking should always occur before and during the antibody incubation step. The most popular and widely used blocking agent is serum albumin, which can either be used in a purified form, or a part of serum, for example, fetal calf serum or goat serum. In addition to the serum albumin, many assay scientists may add a mild detergent. An ideal blocking buffer will prevent nonspecific binding of antibodies to the growth substrate (i.e., bottom of microscope slide or micro plate), as well as the cells or tissues under examination.

11.3.3.3 Washing   Multiple washing steps are essential for the removal of excess diagnostic reagents and the reduction of background. In general, a wash buffer will consist of physiological saline solution such as Phosphate buffered saline (PBS) and a mild detergent such as 0.1%–0.3% Triton X100, which aid the removal unbound reagents. To prevent the loss of the block during washing, it is advisable to add blocking substance to the wash solution. Washing may also result in the loss of cells, so care should be taken during this process.

11.3.4 Planning Ahead with Image Analysis

Many of the problems with assay quality can be pinned on poor assay design, and design problems that surface during the running of an assay are difficult at best to fix at that stage, as they make comparisons between assay runs impossible. Planning ahead for a robust assay is easier than "tightening up" an assay once it is running!

11.3.4.1 Pick the Best Descriptor of the Biological Effect to Measure   Selection of an appropriate metric is a good place to start with designing a robust assay. Careful consideration should be given to the noise and statistical properties of assay metrics. This topic has already been covered in some detail in Chapter 8.

11.3.4.2 Determine Application/Algorithm or Write a New One   The phenotype can be reduced to one of a few types, and this will lead to the use of an existing algorithm or the development of a custom algorithm. At this point, the analytical options unique to each imaging platform drive the options available. For example, most vendors provide highly developed ("canned") analytical routines which are designed for specific assays: count cells, analyze cytoplasmic-to-nuclear translocations, or to count and measure spots/puncta. Using such pre-existing algorithms can certainly be a fast start to image analysis, but they often depend on following a specific protocol during the assay set up, such as using specified cytological markers as counterstains, and can be limited in the ways they can be modified to adapt to different cell types or different stains. For example, some preset image analysis algorithms may have difficulties with multinucleated cells or cells that form higher order structures, such as which are observed in skeletal muscle myotubes. Others are highly flexible systems for establishing custom image analysis solutions, but lack such canned guides. They may require more time to generate solutions for a scientist just beginning their work in HCS, but learning these systems can be achieved with one or two studies. Regardless of what system is being used, an analysis solution needs to be developed that measures and reports the event(s) under study. This stage of the experiment will be absolutely dependent on the platform and analysis solutions being used.

11.3.4.3 Optimize Segmentation Parameters While Viewing Masking and Overlays
All of image analysis is about the conversion of visual data (an image) into numerical data, specifically, a table of numbers. To do this, the cytological structures within the cells, as well as overall numbers of cells, need to be calculated. This is almost always an iterative process where thresholds for objects are set (refer to Chapters 4 and 8, for both background and practical examples of setting object criteria), the resulting algorithm tested. The first thing to check is an assessment of the masking and segmentation. Setting thresholds too low will enlarge the area around objects, in essence, creating larger objects and cause nearby objects to be merged into single objects. Setting the thresholds too high will have the opposite effect, and can cause some objects to be split in two. Although thresholding affects segmentation, most image analysis systems can utilize the shape of the object or the absolute size of the object(s) to determine if it represents one or two objects. Most image analysis software have the ability to turn this segmentation overlay on and off, which allows for easy assessment of the algorithm's accuracy.

11.3.4.4 Checking Positive/Negative Controls for Dynamic Range
As a practical test of the algorithm, the performance of positive and negative controls should be compared. A low signal-to-noise ratio is often the first sign of trouble, and while it is tempting to try and push the envelope with tweaks to image processing and analysis routines, at some point it is actually easier to take a step back and fix the underlying problem that gives rise to a low signal-to-noise ratio. More about this later.

11.3.4.5 Understand/Determine Metrics to Help Define Cellular Phenotype
The benefits and challenges of working with image analysis are that a large number of features are determined for each cell, and each feature may be algorithmically combined with other features. Furthermore, statistical measures of each feature (standard deviation, etc.) are also available for each feature. As such, it takes some effort to determine which of the features best represents the biological events. Starting with a simple example of the nuclear accumulation of a transcription factor, measures of the amount of the transcription factor in the cytoplasm and nucleus will be reported, as can the ratio and the sum of the two pools within the cell. Subtracting the cytoplasmic from the nuclear will give a negative value for cells where more than half of

the target is in the cytoplasm, using a ratio will give a number less than 0.5 for the same cells. However, it may be that the biological events triggered by the translocation are driven solely by the amount of the transcription factor in the nucleus, and therefore the absolute amount of the transcription factor in the nucleus may be the most biologically relevant feature; the amount or proportion of the transcription factor in the cytoplasm could be irrelevant. Determining which features to use require of how the raw data will be processed and analyzed, including which statistical tests are important, and the underlying biology. To stay focused on the biology, having a second event measured in the same assay, such as colocalization of a cofactor or an event closely related to the key biological question, such as phosphorylation of a target that is regulated by the same receptor or expression of a gene controlled by the transcription factor can be crucial in knowing which features are most relevant, even if the multiplexed assay will not be used later (such as in a screen). The depth of biological understanding is paramount here, as it may be important to correlate two biological events that may be difficult to set up in a single assay. For example, signal transduction through a pathway and the resulting changes in gene expression can be temporally distinct events.

11.4   LOOKING FOR TROUBLE: BUILDING IN QUALITY CONTROL

HCS assays, like everything else, change as they age; light bulbs fade, cell lines grow a little differently, new problems crop up. Some of these changes may have little effect on the assay, while others may significantly impact the assay. QC is the best way to preserve data quality as your assay matures. The two aspects of assay performance that QC focuses on are bias and variability. QC is one of those things that everybody agrees should be done, but no two people do in the same way. People who run assays often have their own distinct ways of performing quality control. With that said, there are some common general themes that we will touch upon in the following section.

11.4.1   Using Controls for QC

Building in the right controls is the first step in QC. Although it is sometimes hard to do (especially with the limited real estate on a 96-well plate), it is worth having a positive and a negative control wherever possible for a high content assay. Running a reference compound with each batch of plates is a common QC measure. A drastically different IC_{50} for a reference compound may signal a bias in the assay.

11.4.2   Uniformity Plates

A second approach to detecting biases is to run uniformity plates on an occasional basis. A uniformity plate is one in which the entire plate consists of repeats of a single sample. This QC technique is surprisingly effective at detecting a wide range of spatial effects in assay conditions or image acquisition such as assay automation issues (e.g., leftover compound in robotic pipetting), microscopy problems (e.g., illumination, focusing or stage issues), or even incubator problems (plates drying out in the incubator, leading to pH changes in the outer wells). Uniformity plates can also be used to examine variability, more in a relative than an absolute sense. A statistical technique known as variance components analysis (a close relative of the ANOVA test) can be used to dissect the degree of variability in your platform (for instance, comparing the variability from well-to-well effects against row-to-row, plate-to-plate, day-to-day or operator-to-operator variability).

11.4.3   Monitoring Assay Statistics

Standard deviation-derived measures (such as Z'): The Z' statistic, a commonly used screening statistic, is useful in getting a sense of the signal-to-noise ratio in an assay. The Z' is the difference in IC_{50}s between positive and negative control divided by the standard deviation of the assay. A reduction in the Z' ratio is indicative of problems with the variability of an assay.

11.4.4   Monitoring Meta-data

Some aspects of metadata (data about data) provide useful clues to assay performance. These include stage depth (Z-axis depth, especially useful if you are having trouble with your autofocus) or stage temperature (very valuable information to log for live-cell assays!). While this information is not always useful in determining whether a problem has occurred, having it on hand (being logged continuously in a fixed location, for instance) can help the troubleshooting process.

11.4.5   Visually Inspect Images via Montages or Random Sampling

We have said this before (and we will say it again)—*look at your images* from time to time. The space of what can go wrong in a high content assay is infinite, and the best time to find and fix a problem can often be before it has resulted in corrupt data. Your assay workflow and platform should make it easy to perform random sampling and visual review of samples on an ongoing basis. Visual inspection serves two purposes, early detection of trouble, and as a diagnostic tool for when you do have a problem. In both cases, you will need to ensure that your assay workflow and platform can let you connect your images with the metric of interest.

As we mentioned earlier in this section, QC approaches vary from person to person and from situation to situation. Hopefully the ideas in this section will help you design a robust portfolio of QC approaches that work for you, keeping in mind that too much QC is better in the long run than too little!

11.4.6   Lock down Standard Operating Procedures (SOPs)

Once your assay has been developed and is working well, the next step is to standardize its performance by putting together an SOP, a user's manual for the assay. For assays that are run over the long term, the SOP goes several steps beyond the experimental

TABLE 11.2   Typical Information Detailed in a SOP (Standard Operating Procedure) to Enable Reproducibility for an Assay That is Performed by Multiple Users in Different Settings or at Different Times

Standard Operating Procedure	
Sections	Description
Objective	Summarize biology and assay
Scope and responsibility	Define who will be using the SOP and who is responsible for oversight of any procedures, including calibrations
Definitions and references	Define any terms and detail any external references required for SOP
Reagents	Define all reagents including product numbers and sources
Tools and equipment	List all tools, instruments, and other equipment needed to perform assay
Procedure	Detailed procedure describing all aspects of assay
Appendix	Include any additional information, flowcharts, etc.

protocol in ensuring consistency and reproducibility, especially if there are different users running the assay.

An SOP typically includes a background to the assay, the protocol for the assay and for cell culture, as well as a list of reagents and plastic-ware (especially manufacturer and batch number). A well-written SOP can also be useful in troubleshooting an assay or minimizing the chances of trouble cropping up later. For instance, if there are steps in your protocol that give rise to specific problems (e.g., a secondary antibody that is sensitive to degradation, or a critical wash step that, when omitted, causes background problems), it is worth making a mention of it in the SOP. An SOP should also specify key positive and negative controls, and QC measures (as described in the previous section). See Table 11.2 for a template for a typical SOP, to provide you with a place to start if you are looking to write your own SOPs.

11.5   CONCLUSIONS

This chapter strives to bring some real-world perspective to the ambiguous problem of improving assay performance. This can be a difficult phase. Expectations can be high when reporting early successes during the assay development process. Data may be less impressive once the assay is in production mode. Making things worse, it may not be clear where the problem is coming from. This chapter is intended as a survey of potential sources of problems, as well as strategies to mitigate them or avoid them in the first place. However, unique situations arise all the time, so this is best handled through a healthy contribution from others in the lab through brain-storming.

KEY POINTS

1. Understand the biological response and choose the right metric for the biological phenotype.
2. Define controls and plate layout early in the process to standardize analytical tools.
3. Understand the sources of noise and variables associated with each step of the workflow. Iteratively optimize individual steps: staining protocol, then instrument, then analysis algorithm.
4. Develop QC procedure to monitor assay performance.

FURTHER READING

Many basic principles of assay design in a wide variety of formats are discussed in the NIH Assay Guidance Manual [3]. Some methods are discussed here, such as the use of uniformity plates, but the most of the material is geared towards general assay development and scale-up issues. Much of the material in this chapter was developed through HCS-specific validation exercises that were reviewed in Chakravarty et al. [4]. Some additional discussion on automation-specific assay development and troubleshooting can be found there.

REFERENCES

1. Baschong, W., Suetterlin, R., and Laeng, R.H. Control of autofluorescence of archival formaldehyde-fixed, paraffin-embedded tissue in confocal laser scanning microscopy (CLSM). *Journal of Histochemistry & Cytochemistry*, 2001, **49**(12): 1565.
2. Kelloff, G.J. and Sigman, C.C. New science-based endpoints to accelerate oncology drug development. *European Journal of Cancer*, 2005, **41**: 491–501.
3. Sittampalam, G.S. et al. (eds). *Assay Guidance Manual*. Eli Lilly and the National Center for Advancing Translational Sciences, Bethesda, MD, 2012.
4. Haney, S.A. (ed.). Chakravarty, A. et al., *Developing robust high content assays, High Content Screening: Science, Technology and Applications*, John Wiley & Sons, Hoboken, NJ, 2008, pp. 85–110.

12

AUTOMATION AND SCREENING

John Ringeling, John Donovan, Arijit Chakravarty,
Anthony Davies, Steven A. Haney, Douglas Bowman,
and Ben Knight

12.1 INTRODUCTION

Automation and screening are often thought of as redundant descriptions of the process of scaling up an assay to greater throughput. Indeed, automation is essential to screening, but there is more to it. Screening invokes concepts of standardization, and automation enables it, but enabling standardization can be very relevant to general laboratory assays regardless of scale. This chapter will explore these concepts.

12.2 SOME PRELIMINARY CONSIDERATIONS

12.2.1 Assay or Screen?

The difference between an assay setting and a screening one is commonly distinguished by scale or throughput. Assays are artisanal, screens are mass-produced. Assays are designed to test a small set of reagents for their effects on cells, and all of the compounds, and therefore all of the results, may be used in subsequent studies. A classic example is the structure–activity relationship (SAR) for a series of compounds, where systematic changes within a series of compounds are assessed to determine which regions can be changed without interfering with the interaction with the target protein; even compounds showing no discernible effects provide important information. Assays can afford to be more complex, since the lower throughput generally

An Introduction to High Content Screening: Imaging Technology, Assay Development, and Data Analysis in Biology and Drug Discovery, First Edition. Edited by Steven A. Haney, Douglas Bowman, and Arijit Chakravarty.
© 2015 John Wiley & Sons, Inc. Published 2015 by John Wiley & Sons, Inc.

means that metrics can be more complex (including profiling experiments that use many features) and samples can be run in replicates, to provide a statistical test for each compound and concentration.

A typical screen, especially in a drug discovery setting, will process samples at a throughput that is one or more orders of magnitude higher than the corresponding assay. Screens are designed to rapidly assess a large panel of reagents and eliminate most of them (indeed, almost all of them) from further consideration. As such, screens tend to rely on simpler, but very robust, measurements. Data analyses from both assays and screens rely on common elements of statistics, but typically, the methods used in screening are different from those used in HCS assays, particularly those that use cell-based data and morphological features. These topics are presented in the data analysis chapters, in an earlier section of the book. This chapter will address the logistical challenges that occur when an assay is used to develop a screen, including the impact of automation on the assay, and the choice of metric that may be necessary when scaling up for higher throughput. In keeping with this shift in throughput, the screening mindset is very much a manufacturing one. Standardization becomes key, and the level of attention paid to questions of reproducibility and quality control can often make or break a screen. High content techniques are often mentioned in the context of screening. Up to this point in the book, we have discussed the fundamentals of these techniques, and their smaller-scale application in an assay context. For many situations, high content assays might end up being sufficient for your needs. In some situations, however, you may decide to go the screening route. This chapter discusses the transition from assay to screen, and the unique challenges that you will face in this transition. It is worth pointing out that many of the challenges in this context are ones of scale, reproducibility, and quality control.

12.2.2 To Automate or Not?

Automation is essential for screening, but it also has uses in general assays as well. Cases where automation for assays is helpful include:

- The use of liquid handlers, robotic addition, and removal of buffers and media (during the experimental phase) or the addition of fixatives and labeled antibodies (in the plate-processing phase) can be helpful when cell lines adhere poorly to the plate matrix. Automated liquid handlers can work at slow and consistent speeds, and can even be set to work at the edge of the well rather than the middle.
- Barcodes can be used in multiuser facilities, where individual users can drop off their barcoded plates and have them processed in an overnight queue.
- Automated plate handling can manage the plates during a nightly run. The data can be automatically pushed to the individuals' accounts for analysis. Automated liquid handlers can also be used to standardize higher density plate formats. The use of higher density plates (384-well plates versus 96-well plates) has advantages in increasing the number of wells available for replicate conditions

(four or more replicate wells for each dose of a dose–response assay), increasing the robustness of any assay. There is typically no trade-off in terms of cells per well, as 384-well plates are usually seeded at 1000 cells or greater, and this is enough for almost any HCS assay, even phenotypic profiling.

Other examples can be listed, but the general point is that automation can be very valuable to a research lab, and ought to be considered at the time the HCS instrument is purchased. Often, automation is purchased through the HCS vendor (particularly plate-handling robots); and somewhat unintuitive, certain instrumentation options are mutually exclusive (such as brightfield illumination and automated compound addition). These situations were discussed in Chapter 5. In the following sections we will discuss some of the key choices when deciding whether or not (and how) to automate your high content platform.

12.3 LABORATORY OPTIONS

12.3.1 Workstation versus Fully Automated Systems

There are two approaches to automating multistep assays: workstation and fully automated. With a workstation approach, steps including reagent addition, removal, plate washing, and incubation are performed on individual automated equipment and the assay plates transferred between them by the person performing the assay. There are advantages to this approach which need to be carefully weighed with the disadvantages. Often, a workstation approach can be less expensive because it does not require a robotic arm, an integrated platform and specialized scheduling software with associated drivers. A workstation approach can also offer greater flexibility where equipment may be free for multiple simultaneous users and multiple assay types. This equipment may also be more easily repurposed and moved for future needs. The disadvantage to this approach is that it still requires significant user intervention and attention. The most important consideration is that timing for individual steps will be user-dependent. In this scenario, there is a risk of plate-to-plate, user-to-user, and day-to-day differences in data.

The other approach to automating high content assays is to use a fully automated system. These systems use scheduling software and a robotic arm (or arms) to move plates between different stations. It is possible to fully automate high content assays on such systems from cell plating to imaging. There are multiple commercial vendors that build such system for the biopharma industry but given appropriate in-house engineering resources, such systems can also be self-assembled. The advantages of fully automated systems mirror the disadvantages of the workstation approach. If properly programmed, maintained, and set up, a fully automated system will require minimal user intervention and should provide consistent assay performance. The disadvantages can be the cost and some loss in flexibility in the use of the individual components. It is also true that if a system is highly customized for a particular assay format, repurposing it can be challenging.

12.3.2   Liquid Handler/Reagent Dispenser/Plate Washer Requirements

When reduced to their essence, most biological assays (either cell-based or biochemical) are exercises in moving liquids—adding, removing or moving reagents to, from, and between microplates. High content assays are no exception and in fact require a great deal of liquid handling since they typically have a large number of steps and different reagents and buffers, as described in Chapter 7. It is imperative therefore to select the appropriate liquid handling equipment when considering automating the workflow. An important consideration is the reagent volume requirements for the assay. Commercially available liquid handlers and reagent dispensers can transfer or dispense volumes from the picoliter to milliliter range and greater (although none presently can cover this entire range). Assay volume requirements often are dependent on the microplate format (96, 384, or 1536). It is wise to invest in liquid handling devices that are capable of working with multiple formats so you will have the option to move to higher density plate formats. Precision and accuracy of the liquid handling equipment is also important, for example, the addition of the reagent (most often the experimental compounds) that produces the signal of interest. For other reagents, such as wash buffers, this is less important. Dead volume is also important when considering liquid handling devices. Reagent dispensers with tubing can require milliliters of extra reagent. While this is not necessarily an issue for inexpensive reagents, it is very important for expensive antibodies. There is a wide array of commercially available solutions for assay liquid handling (with a wide price range as well). Vendor sales and application teams are good sources of information, but it is important to identify and talk to actual customers who use these instruments. Having a vendor bring in an instrument for a demo prior to purchase is also an option. System integrators or vendors who integrate fully automated multivendor systems are always good resources as they have likely worked with many of the instruments on the market, and therefore they may know some of the strengths and weaknesses of the instruments.

Plate washing devices are important instruments for high content assay automation, as it is typically necessary to completely remove and wash away a reagent before adding the next reagent. Because cell types vary greatly in their ability to adhere to microplate surfaces, it is important to have control over aspiration and dispense speeds, and the locations within the well where these events occur. Some commercially available washers also have associated reagent dispense capabilities. For lower throughput needs, these instruments can serve as a nearly fully automated assay platform.

12.3.3   Barcode Reading Requirements

The use of barcodes on high content assay plates is an important consideration, especially for higher throughput needs. For both workstation and fully automated approaches, barcodes can ensure that your imaging data are traceable back through the assay to the cell and experimental compounds. When purchasing an automated system or equipment, it is important to discuss your barcode requirements with the vendor to ensure that their product is capable of accommodating your data-tracking needs. This is also a good time internally to consider the requirements and perhaps advantages that can be gained by linking automation to your assay request system, image acquisition, data analysis, and finally to data storage and management systems. For a high value and expensive assay format like high content screening, this will enable rapid troubleshooting should there be data quality issues.

12.3.4   Vendor Selection Issues

Deploying an automated HCS system will be a costly and time-consuming process. It is therefore important to select not only the equipment and systems that can perform your assay but the correct vendor. You will do best with a vendor that is established, reliable, and has a reputation for delivering what they promise and supporting their products. Identify other customers of the vendor's systems and equipment and spend time discussing their experience. These connections can be made through the vendor's sales force but independently identifying them through other means can be even more valuable. As with any scientific instrument, it is important to get a good sense of the vendor's service organization. Even the most robust equipment benefits from periodic preventative maintenance. After the initial purchase and installation of the system, it will be valuable to build a relationship with members of the service team. After the standard warranty expires, a service contract is desirable to keep the system operating continuously. Finally, vendor pricing on systems and equipment can often vary widely. It may be important to stay within an established budget, but buying equipment or systems from unreliable vendors can be a mistake. The ability of the vendor to deliver a product that is usable and performs your assay reliably should be the primary deciding factor for your purchase.

Another important aspect of choosing a vendor (for fully automated systems in particular) is the software that the operator will use to schedule experiment runs. This software will also be interfacing with the many different instruments that make up the workflow. Although there are often similarities in how vendors approach the problem of communicating with devices, there are remarkably different approaches vendors take in creating an interface for a user to create and run automated methods. Some software is very complex and can require an expert user while others are designed to be used by nearly everybody. There can be advantages to both types of software interfaces. Complex software often gives the expert user a high degree of flexibility and control while simpler software can sometimes feel limiting. Simpler software interfaces are a better choice for an environment that lacks automation expertise. Have the vendor provide a detailed software demonstration and have them show you how a method to perform your assay would be created and run. At this point, you can have the vendor demonstrate that the system is capable of meeting your assay and throughput needs via simulation. It is also helpful to get a good sense of how a user will resolve problems that inevitably occur during an automated run. A built-in reliable error notification and recovery mechanism is desirable for high content screening assays which can stretch over days and involve expensive reagents. Automation software has improved greatly over the last 10 years and most vendors include error recovery and notification mechanisms. One can also use webcams at various points in the workflow to allow an operator to remotely and visually monitor the instruments and plate handling throughout the process.

12.3.5   Highly Customized versus More General Systems and Software

High content assays are complex, and have special demands for automation which often require customized solutions. But, there are potential problems with overly customizing automated systems. As mentioned earlier, it can be difficult or even impossible to repurpose a highly customized system. This might also be true of repurposing it for a slightly different high content assay format. Another potential problem is relying on a specific individual at the company who created the solution, as these individuals may leave the company, and the ability of the company to fully support the system may be compromised.

12.3.6   Managing Expectations About Automation

Finally, it is important that all involved in automating a high content screening assays clearly understand the actual value of automation. Automated systems and equipment should produce more consistent data quality as compared to manual workflows performed by multiple individuals over time. Automated systems, combined with the use of barcodes, will also provide a clear traceable path between the assay plates and the data. The value of freeing highly skilled individuals from the tedious tasks involved in high content assays should not be minimized. A senior scientist who spends most of their time moving liquid around is not being properly utilized. Automation can also obviously enable greater throughput for high content assays. At some point, automation is essential, as it would be impossible for an individual to manually perform assays that require very short incubation times for fixation and permeabilization. A mistake that people often make when thinking about automation though is that it will speed up an assay. This is seldom the case, though automation does allow for overnight, unattended runs which increase overall throughput.

12.4   THE AUTOMATED HCS LABORATORY

12.4.1   Setting Up the Automated Laboratory

Automation applied to HCS is presented diagrammatically in Figure 12.1, in which two situations are overlaid. In the first, core stations are highlighted. Setting up a laboratory as a set of independent core stations is less expensive and more flexible, but potentially lowers throughput. Flexibility is increased because scientists bring their plates to individual work stations to perform each task. This also includes revisiting a station more than once during a single experiment. Using a liquid handling station to manage both the experimental treatments as well as the post-fixation plate processing can be done with a single liquid handling workstation, but each situation

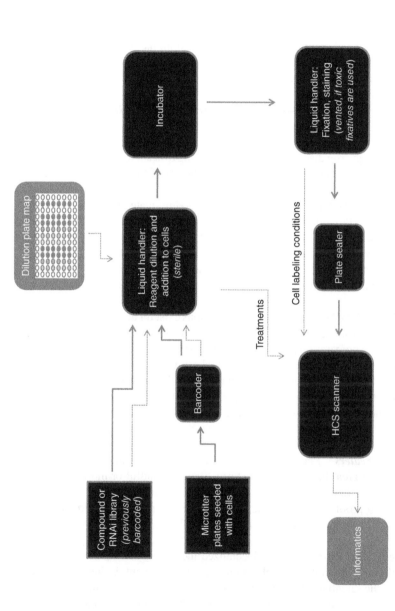

FIGURE 12.1 *Layout of an automated screening laboratory.* Two paths are described. The path of the plates themselves are shown in solid lines, the path of the information regarding the experiment is shown in the stippled lines. Workstations are depicted as rounded rectangles, whether these are physically linked through plate transfer robotics, or serve as individual stations does not affect their logistical relationship in a screening laboratory. In some cases, data may be explicitly fed into individual workstations, but it is also possible that the specification of plate treatments may be coded for each instrument individually and then recoded into the informatics and analysis package.

requires specific considerations. On the other hand, an integrated automation solution is essential for truly high throughput screening. Extending throughput comes at the expense of flexibility. The transition from assay to screen usually involves a change in the equipment used. This change may include exploring alternatives with the automated microscopy setup depending on throughput, as well as weighing one's options with specialized equipment for assay automation, including liquid handlers and barcoders. Many of these items were introduced earlier, this section represents something of an annotated checklist.

12.4.1.1 Selecting Equipment—Liquid Handlers, Barcoder, Microscopes   As a general rule, the equipment used to automate an HCS assay can be expensive and specialized. As such, it is often confined to industrial labs (pharmaceutical and biotech companies) although drug discovery research is moving to academic labs as well. This discussion will be brief, but will introduce the key pieces of equipment to consider.

12.4.1.2 Barcode Readers   As noted above, barcoding plates increases not only throughput, but also flexibility. A barcode reader is relatively cheap and easy to use. It is required in screening to link the cell data with the compound (or RNAi) library plates. In an assay lab, barcoding can be used to load imaging protocols (developed by investigators previously) automatically for individual experiments. Barcoding enables the instrument to read plates and push the data back to the investigators.

12.4.1.3 Liquid Handling   A wide range of options exist for automated liquid handling. Small plate washing workstations can help an investigator through the tedious but important plate preparation process. Fully automated liquid handling stations can manage dilutions and reformatting of reagent stocks, timed compound addition steps, and complicated immunofluorescence staining protocols. Such instruments can be maintained in sterile environments for experimental treatments or housed in controlled airflow hoods to minimize exposure to fixatives during the post-experimental processing.

12.4.1.4 Plate Sealers   Plate sealers are another relatively simple piece of equipment that can be a great help to a scientist processing a handful or more plates at a time. A plate sealer fixes a piece of cellophane onto the top of a plate to ensure a tight, evaporation-proof seal, increasing the shelf life of a plate. This eliminates the need for a plate lid, which can simplify robotic workflows.

12.4.1.5 Cell Culture Incubators   Specialized incubators can be used in a fully automated laboratory to culture the plates during the experiment. Using a fully integrated system, the scientist can deliver cell culture plates to the screening laboratory, and they do not have to handle the plates again until the screen is completed. This is useful for both fixed end-point and time-lapse experiments.

12.4.1.6 Software   Computer software can be a hidden expense for an automated laboratory. All instruments that track individual plates (bar coders, liquid handling stations, HCS instruments, etc.) have software that can accept data on the plate and track the plate and well information. Software to run reagent dilution and reformatting steps are essential, other steps can be handled offline in a modular lab format (such as plate sealing and plate washing). Individual software packages for each instrument almost always have interfaces that enable them to be integrated into packages that track the automated screening process and can report experimental results (Pipeline Pilot™ is a well-known commercial product, but open-source alternatives exist). The challenge is in maintaining full integration and upkeep of the software as the screening laboratory matures.

12.4.1.7 Robotic Plate Handling   Robotic arms move plates from stacking stations to an instrument, and then offload the plate to a separate stack once the plate has been processed or read on the instrument. Most HCS platforms have an option for purchasing a robotic arm and a plate stacking array to load and unload plates from the scanner. In most cases, the specific choice should not limit other automation options, as industrial standards regarding equipment integration have been specified. Robotic plate handling is the key to connecting individual workstations and creating an integrated screening laboratory.

12.4.2 Connecting the Components

A diagrammatic version of a screening laboratory is shown in Figure 12.1. Perturbants (compounds or RNAi libraries) are aliquoted and formatted before entering the screening lab, while cells may be dispensed in the screening laboratory, using basic liquid handling equipment (including the use of multichannel hand pipettors for small libraries, say 5000 wells or less). Direct addition of screening plates to cell culture plates can be handled with modest liquid handling devices. Automated dilution series, or reformatting 96-well reagent plates into 384-well screening plates requires more sophisticated automation. In all of the plate handling steps, options range from individual workstations and the transfer of the plates by laboratory personnel, to fully automated systems where the transfer between each station is handled by conveyors or robotic arms. In the workstation mode, most stations will still require plate stacking and automated plate handling, to allow batch processing of plates. In a workstation setting, multiple liquid handling steps can be handled by a single station.

The informatics pathway feeds into the plate processing pathway at distinct steps. A platemap needs to be designed, so that compounds can be selected and diluted appropriately. Each compound plate has been previously barcoded to identify each well in each plate. The platemap will specify what needs to be done to each compound plate: which compounds will be chosen and how they will be rearrayed. The conversion from 96- to 384-well plates means that each compound may be added to four wells in the experimental plates. Barcoding the cell plates is needed to associate each experimental plate with the compounds used from each of the library plates. Most HCS instruments are capable of tracking plate and well information, and they can easily be fit to read plate barcodes, so the package of information (feature data and metadata) can be passed to the informatics server and database.

12.4.3 Reagent Considerations

Changing batches of reagents midway through a screen can be bad news. As with any other screening situation, it is useful to think through your reagent needs for a screen beforehand and stockpile appropriately. This step—the last step in scale-up from assay to screen—is tedious and often quite time-consuming, so it is tempting to skip it altogether. Do not. The failure to stockpile reagents appropriately is a common source of screen failure, and is very hard, if not impossible, to rectify if it hits you along the way. The time spent in this step will be more than paid off when the screen runs smoothly.

The biggest challenge in this space is knowing what reagents to control for. The answer is everything—cells, cell culture media, fixation reagents, PBS, antibodies, plates, troughs, pipette tips (for every single pipetting step). It is common to plan ahead and bank cell lines being used so as to control the passage number quite tightly (usually two or three passages at most).

12.4.4 Planning Ahead with Informatics

As distinct from the automation software, informatics is generally defined as the point where sample tracking ends and data analysis of the treatment conditions begins. In general, the process groups information by treatment (compound dose, gene targeted by one or more siRNAs) and evaluates the effect on the metrics used as assay or screen endpoints. We discussed the approaches that can be used in data analysis in Chapter 9, and the topic of informatics has been covered already in Chapter 6, but it is worth highlighting a couple of key points here in the context of the transition from assay to screen.

A critical, often overlooked step in assay scale-up is thinking through the archiving and data access needs of the screen. In the design stages of a screen, it is important to have a sense of what data will need to be archived, how long it will need to be archived for, and how accessible each individual piece of data needs to be. For example, for a high throughput screen that generates an IC_{50} number, a minimal set of data would obviously have to include the individual data points (the metrics derived from each well) as well as the IC_{50}s for each compound run. Less obvious is the choice of what to do with the images. These are typically not saved in a screening setting, because the sheer volume of the data makes it prohibitively expensive. Although data storage costs are dropping by the day, accessing the data can still be fairly challenging, and the number of situations where the raw images could be useful are limited, especially for a mature screen. It then becomes a cost-benefit calculation weighing the actual effort spent archiving images from screening against the potential benefit of being able to access the images if the need arises. If you do decide to archive the images, there are potential benefits to the choice (such as the ability to troubleshoot), but only if the link between the numerical data and raw images is robust (persistent over time) and easily navigated.

A second dimension to consider for the informatics associated with screens is the duration of time for which the data need to be retained. Table 12.1 provides a quick

TABLE 12.1   A Checklist for an Automation-Friendly Dose Response Assay

☑ For *each* reagent being pipetted, are accuracy and precision known, and satisfactory?

☑ Is the assay robust to the timing constraints of incubations? (i.e., is a 30-min drug incubation crucial to the end result or is there a "grace period" which will allow relaxation of automation conditions.)

☑ Have the dead volume requirements for your "precious" reagents (e.g., primary antibodies) been minimized?

☑ Is the throughput capacity of your automation appropriate to your needs?

☑ Has the automation routine been defined with appropriate liquid handling control? (e.g., pipet slower or at edge of well as to not disturb cells)

☑ Has the assay been engineered and *tested* for robustness? (need to be able to rely on instrumentation to process plates day in day out with minimal equipment downtime)

☑ Have user interaction points been designed to maximize "walk-away" portions of the assay?

☑ Have you input appropriate Quality Controls? If the assay "breaks," will it fail catastrophically (a Good Thing), or will bad data become mixed in silently with good (not a Good Thing)?

☑ Do you have a Standard Operating Procedure (or would you prefer to wait until you *wish* you had one?)

☑ Have you tested for sterility? Do you have ongoing quality control measures to evaluate sterility on an ongoing basis? (e.g., for cell plating, long-term drug addition)

☑ Has your scheduling software been tested, especially for larger runs?

---

checklist of questions worth asking when planning your informatics needs for a high throughput screen.

12.4.5   Designing an Automation—Friendly Dose–Response Assay

12.4.5.1   *Replicates*
When designing dose response assays (or any other imaging assays for that matter), you will need to strike a balance between robust solid assay data and a point of diminishing returns. In other words, establish at which point it does not make sense anymore to add more replicates and or take more images/well, since this will directly affect how long it will take to run such an assay, cost, as well as throughput. Typically, what we do is run a new assay with up to 10 replicates and up to 36 images per well. (More replicates than we would ever consider running in a real assay!). Once the data are acquired, we look at the way in which the variability of the assay is dependent on the number of replicates.

12.4.5.2   *Positive/Negative Controls*
Plate-to-plate variations are often the bane of high content assays, as they can lead to considerable variability in the final readouts. To account for plate-to-plate variations, consider running a maximum inhibition control (100% or positive control) on each plate. It is often a good idea to also include DMSO or other relevant negative controls on each plate to normalize the data or do a background subtraction. Before settling on a positive control, test a few different

compounds for their usability. Sometimes, if your positive controls are too potent, you could end up with an undesirable situation where none of the test compounds will yield complete inhibition, making a good curve fit difficult. If you encounter this problem, you can (sometimes) fix it by titrating the positive control down, but thinking ahead with the positive control may reduce the pain you experience. Other considerations for good positive controls are cost, availability, and toxicity.

12.4.5.3 Number of Dose Points and Range   When fitting a dose–response curve, as most of us know, there are three key pieces of information to "nail down," the top, the bottom, and the slope of the assay. If your experimentally determined dose response curve misses the point at which the underlying relationship "picks up" or "levels off," the fit will be poor, and the estimate of IC_{50} will be inaccurate. Garbage in, garbage out. Even worse, your experimentally determined points may transition from "no response" to "full response" within one dose group, or miss the top or the bottom of the curve altogether. When faced with these outcomes, your best option is to rerun the entire curve, and hope that the new dose ranges that you have selected do a better job of identifying the top and bottom of the curve. Refer to Section 9.6 for an extended discussion.

However, at the outset, you would like to plan your experiment to maximize your chances of success, so having a rough idea of potency of the compounds to be tested will provide you with a better shot at choosing the right dose range. Even with a 15-point curve, the assay may not hit the low and high plateau if the top dose is off by several orders of magnitude. (Here our experiences as a rough guideline for you: for secondary assays to assist with SAR, we typically start at 25 µM using a three-fold dilution series. We typically use a nine-point dose response, which allows us to have the lowest concentration at 3.8 nM. Once a project reaches Hit-to-Lead, we typically lower the top dose to 10 µM or even 1 µM, which gives us 1.5 nM or 0.15 nM respectively. We have found that based on the available "real estate" in 96- or 384-well plates, in order to utilize well space most efficiently, even employing a 10-point assay will allow for proper controls.)

12.5   CONCLUSIONS

Automation and screening can be unexpectedly challenging for those who have viewed it from afar—in fact, it is one of the most specialized areas of drug discovery. Automation is something that can be used by any HCS laboratory, regardless of scale, to minimize technical complications that can be more acute in imaging assays, due to the number of steps involved.

KEY POINTS

1. Understand all parts of the process (instruments, software, metadata tracking) before automating parts of the workflow or the entire workflow

2. Think through your options carefully, considering the tradeoffs before deciding how tightly integrated you want your workflow to be

3. Establish robust process control over experiments and lock down SOPs once assays have been developed and validated.

FURTHER READING

Society for Laboratory Automation and Screening (www.slas.org) is a multidisciplinary community focusing on advanced laboratory technologies.

Laboratory Robotics Interest Group (lab-robotics.org) is a special interest group focused on laboratory automation.

13

HIGH CONTENT ANALYSIS FOR TISSUE SAMPLES

KRISTINE BURKE, VAISHALI SHINDE, ALICE MCDONALD, DOUGLAS BOWMAN, AND ARIJIT CHAKRAVARTY

13.1 INTRODUCTION

Conceptually, it is pretty easy to see that high content techniques are just as applicable in a tissue-based content as they are in a cellular context. Tissue-based high content analysis is a rapidly growing field, and the extension of basic biology derived from an *in vitro* setting to the *in vivo* setting can often be of critical importance, for mechanistic work as well as for drug discovery and development. Current work on tissue HCS is most commonly applied to pharmacodynamics (the study of drug action *in vivo*), but can also be applied to areas of general immunohistochemistry, including the analysis of genetically modified animals, other model systems (such as zebrafish), and human pathology (digital pathology). In a drug discovery setting, the extension of high content techniques to tissue samples provides a powerful means of assessing the pharmacodynamic effects of a given drug, especially when a drug targets a specific subpopulation of cells. There are a number of challenges inherent to tissue-based HCS, for instance, the three-dimensional (3D) architecture, irregular outlines and histological heterogeneity of tissue sections [1].

The staining technology commonly used for tissue-based HCS is immunohistochemistry (IHC), where an antibody is combined with chemical staining procedures to characterize the presence or change in a protein. These procedures are used to define a cell type by the protein expression (e.g., CD31 expression to indicate endothelial cells), the pathology of a protein that accumulates or localizes abnormally, or a

An Introduction to High Content Screening: Imaging Technology, Assay Development, and Data Analysis in Biology and Drug Discovery, First Edition. Edited by Steven A. Haney, Douglas Bowman, and Arijit Chakravarty.
© 2015 John Wiley & Sons, Inc. Published 2015 by John Wiley & Sons, Inc.

biomarker showing a response to a therapeutic treatment. Immunofluorescence (IF) is becoming more common for tissue-based HCS where there are limited samples available (multiple markers on single slide) or where colocalization of multiple markers is important.

In this chapter, we focus on the challenges commonly encountered in developing high content assays for an *in vivo* setting, as well as potential solutions for these challenges. We also discuss general principles of high content assay design for tissues. We will go over some of the technical aspects of IHC as well as IF for tissues, and discuss the pros and cons of both technologies. Finally, we review some of the technical components of a tissue imaging system including slide scanning hardware and image analysis software.

13.2   DESIGN CHOICES IN SETTING UP A HIGH CONTENT ASSAY IN TISSUE

Developing an IHC or IF assay is a complicated process with many technical choices and "gotchas" that are not obvious to biologists used to running HCS assays on plastic. We will attempt to cover some of these gotchas in this section, and for a more detailed coverage of the same topic, please refer to the DAKO (Dako, an Agilent Technologies Company) handbook [2].

13.2.1   IF or IHC? When and Why

IHC and IF are two different staining modalities which are used for the localization and quantitation of proteins in a tissue setting. Both methods use antibodies as the primary tool and experimental design can be quite similar if not identical. The fundamental difference between IHC and IF is antibody detection.

In IHC staining, visualization is performed by means of an enzyme (typically horseradish peroxidase or alkaline phosphatase) conjugated to the primary antibody that can catalyze a color-producing reaction. IHC is colorimetric and can be easily reviewed under a light microscope. Although IHC is a relatively mature technology, dating back to the early 1940s, protocols are constantly improving, and better techniques with higher sensitivity have developed since then, including the advent of autostainers (instruments capable of automated wet lab staining protocol). There are multiple chromogen substrates to choose from and a variety of systems to enhance detection. Signal enhancement, however, can also lead to higher background. Depending on the chromogen of choice and the experimental tissue, there is also the possibility of endogenous substrate that increases background and decreases assay sensitivity. Special treatments are often required to deal with this issue when it arises.

While multiple IHC stains can be done simultaneously, co-staining with IHC is challenging for the beginner. The difficulty lies in distinguishing between chromogens that overlap in tissue, especially in cases where the protein of interest is in the same cellular compartment. While a regular brightfield color camera cannot separate out multiple chromogens well, specialized techniques utilizing hardware (multispectral imaging [3]) or software (e.g., a color deconvolution algorithm post acquisition) can spectrally separate multiple immunostains.

IF requires the use of a fluorescence microscope and has a couple of major advantages over IHC. It has a higher dynamic range, provides better subcellular localization, and is relatively easier to devise image processing routines for (due to the monochromatic nature of each stain). IF is the preferred method for colocalization studies due to the variety of available fluorochomes and the simplicity of the experimental design. Cellular structures and antibody colocalization are much clearer and permit the exploration of 3D architecture *in vivo* (using either a confocal microscope or a Z-stack of consecutive images through a series of focal planes, followed by deconvolution).

With all of those advantages, why is IF not more commonly used in tissue-based HCS approaches? There are some major drawbacks to IF, both scientific and cultural. From a technical standpoint, IF typically stains specific cellular compartments. This means that tissue morphology is difficult to assess, and co-stains may be necessary to identify which cellular structures are positive. Another technical drawback of IF is tissue autofluorescence. As with IHC, special treatments, protocol changes, or image processing steps are often required to deal with this issue. Arguably, the major reason why IF is not more commonly used in tissue-based setting is cultural. Traditionally trained pathologists are substantially more comfortable with IHC, as it allows them to leverage their training in histology and assess tissue morphologies which are preserved by the use of a counterstain.

So to sum up—if you are looking to identify a specific cell type or work with tissue histology, go with IHC, especially if a pathological evaluation of tissue is required. In situations where IHC provides sufficient resolution and dynamic range to provide the answer you are looking for (tissue-level questions), it is better to use it, as the methodology provides additional information that is accessible to a trained pathologist. On the other hand, if the question being asked requires precise colocalization, or higher resolution (cell or organelle-level questions), IF is the better technology. When you are trying to quantitate subcellular morphology or levels of a specific protein in tissues, it is probably IF you want.

13.2.2   Frozen Sections or Paraffin?

The choice of tissue collection and fixation method is the next important step. The two common tissue collection methods are frozen and formalin fixed paraffin embedded (FFPE).

FFPE tissue is a widely used method for sample preservation and archiving. By one estimate, over a billion clinical tissue samples, most of them FFPE, are being stored in various medical laboratories across the globe. FFPE sample preparation is tedious, to put it mildly—samples are collected commonly in formalin, and then processed by dehydrating through a series of soaks in various concentrations of ethanol, and xylene (some protocols require up to 11 changes of soaking liquid, each an hour in length!). Tissue samples are then oriented by the histotechnician and manually embedded in paraffin. Sectioning requires the use of a slicing device known as a microtome, which takes skill and experience to use properly. The cut paraffin sections are floated on a warm water bath to remove the wrinkles and then picked up on a glass slide. Slides are then usually baked to further adhere the tissue to the slide.

For some antigens, an additional series of steps (epitope unmasking) is required for optimal processing. Antigen retrieval is a simple method of boiling paraffin-embedded tissue sections in an aqueous retrieval solution to enhance the signal. Antigen retrieval reagents break cross-links between proteins which are formed during tissue fixation, thus significantly improving tissue immunoreactivity. For FFPE slides, antigen retrieval is done after deparaffinization and rehydration. In formalin, or other aldehyde fixed paraffin embedded tissue samples, the protein cross-links can be broken using multiple antigen retrieval methods. Each differs in the type of reagent used and temperature.

Although the process and embed protocol is painful, FFPE has some substantial advantages. FFPE blocks are indefinitely stable—they can be stored at room temperature for years, and tissue morphology is very well preserved with this method. One limitation of FFPE-processed samples is that DNA and RNA contained within them is not as readily accessible for molecular analysis, as nucleic acids in this method are heavily modified and trapped by extensive protein–nucleic acid and protein–protein cross linking. Although a number of commercial kits are currently available for nucleic acid extraction from FFPE samples, this is another (tedious) step to look forward to.

Frozen samples lie at the other extreme of the effort–reward spectrum. Samples are snap frozen using dry ice or liquid nitrogen. Samples are then embedded in OCT (Optimal Cutting Temperature) embedding medium prior to sectioning on a cryostat. Frozen tissues can be either pre- or post-fixed (rapid fixation is essential to keep certain antigens intact). The method is simple and requires little extra work, although it does require a cryostat, for storing the samples at a stable low temperature. Frozen tissue samples can also be easily used for other assays (e.g., microarray profiling or RNA characterization.) The downside of this method of tissue collection is that the morphology is not as well preserved, freezing artifacts are fairly common and frozen tissues may also have a higher tendency for background staining. Length of antigen retrieval is crucial in frozen sections; longer incubation in acetone can result in loss of tissue morphology due to tissue shrinkage which can interfere in the imaging and analysis of the tissue section.

13.2.3   Primary and Secondary Antibody Choices

Staining protocol design choices are remarkably similar to those of cell-based assays. A choice may need to be made between a monoclonal and polyclonal primary antibody. Once again, the same rules that were discussed in Chapter 11 apply to tissue assays. Polyclonal antibodies react to multiple epitopes of a protein, and can be more straightforward to make work (few epitope-masking issues). Conversely, monoclonal antibodies react to a specific protein epitope, thus possessing higher specificity, with fewer chances of cross reactivity between proteins with high levels of homology, although monoclonal antibodies theoretically have a higher risk of epitope masking, and may require more tweaking to optimize. From the standpoint of scaling up, monoclonal antibodies have a potentially limitless supply as they are derived from a hybridoma cell line (as opposed to polyclonal antibodies, which can suffer from batch-to-batch variations).

Again, analogous to the cell-based setting, primary antibodies used for slide-based assays can be used directly or indirectly. The direct method refers to antibodies that are directly conjugated to either a substrate that can interact with a chromogen or a fluorochrome which is directly detected. Indirect staining methods amplify the signal and require the use of a primary antibody targeting the antigen of choice followed by a secondary antibody that is directly conjugated to a substrate and targets the primary antibody. The direct method is a cleaner staining method but depending on the antibody, may not be easily detected. The indirect method allows for signal amplification but can also amplify background staining. As indicated above, if a specific tissue is being assessed, nonspecific background from the detection needs to be considered.

The actual staining process can be achieved either by an autostainer or by manual staining. The advantages of using an autostainer include good reproducibility, less manual interruption, shorter run time, bulk staining capability, and ease of optimization. On the other hand, the advantage of using manual staining is flexibility, allowing the user to run multiple staining protocols simultaneously. Generally speaking, the tendency is to automate whenever possible, and use manual staining only when the situation specifically calls for it.

As with cell-based HCS assays, once the assay has been chosen, antibody conditions need to be optimized. The antibody should be titrated to determine the optimum concentration of the antibody where there is the highest level of specificity and the lowest level of background. Close attention should be paid to the diluent used as this one factor can result in drastically varying outcomes. For further information on antibody titrations and incubation temperatures, refer to the DAKO handbook. As discussed in the earlier chapters, protocol optimization is an iterative process that involves staining, scanning, and image analysis to develop a robust assay to measure both cellular/tissue morphology and protein expression.

13.3   SYSTEM CONFIGURATION: ASPECTS UNIQUE
TO TISSUE-BASED HCS

Although conceptually very similar to standard HCS, tissue HCS has developed from the pathology and histology fields rather than the cell-based assay development world. The nature of the sample (a tissue section) and data analysis (with a strong role for the context of the cell in addition to its actual state) introduce many differences to the two approaches and to the development of the relevant image collection systems. This section will review the system configuration necessary for high content tissue acquisition. An overview of the hardware, software, and data management system will help guide the setup of an automated microscope system. Also a review of whole slide image systems (WSI) will give the reader an understanding of new technology that might be a good alternative to stand-alone microscope setups. The following information is geared toward researchers setting up a tissue-based image analysis system and pathologists who might be looking to migrate over to digital slides for pathology reviews.

There are many commercially available slide-scanning systems that are relatively hands-off and capable of scanning a large number of slides in an automated manner. While these "black box" optical platforms exist, it is necessary to understand the fundamental components of the imaging system. Understanding the central components of an imaging system can be useful if you are building/purchasing a microscope-based system or trying to compare different instruments.

13.3.1   Optical Configuration

As with HCS systems, resolution is determined by the optical configuration of the microscope. It is important to understand which objective magnification, numerical aperture, and type of digital camera is necessary to image a specimen at the optimal resolution for a specific assay. The proper configuration of the microscope is also essential. For brightfield microscopy, Köhler illumination is a technique for correctly aligning the microscope for optimal contrast and resolution [4]. Prior to image analysis, Köhler illumination should always be checked. For fluorescence microscopy, a high numerical aperture is desired. Because of the large areas of the slide that are scanned, immersion objectives are typically not used.

13.3.2   Digital Camera

Two types of camera are typically used in tissue scanning systems: line-scan and area-scan. Line-scan cameras scan the sample in strips to create a whole slide image, while area-scan cameras scan the sample in tiles. These tiles are stitched together to create a whole slide image. Color (RGB) CCD cameras are typically used for IHC assays, while a monochrome CCD camera (the standard for cellular HCS systems) should be used for IF assays because of the increased sensitivity.

13.3.3   Stage Accessories

Due to the variability (thickness, morphology) of the sample across the slide, the software needs to focus on each area to make sure the entire sample is in sharp focus. On a microscope-based system, a motorized XYZ stage would allow the user to mark multiple field of views (FOV) which the software can store and then acquire the images in a hand-off mode. For example, if imaging mitotic cells throughout a large tissue section, you can mark the XYZ coordinates of multiple cells and then have the software automatically scan all of the points. The addition of a slide loader to the microscope enables the microscope for automatic slide handling with potential capacity of upward of 200 slides (the corollary to the microtiter plate holder). The loading and unloading of slides is controlled with software. For example, the user can configure the software to load slides and then select an acquisition script which would image specific areas of the slide while autofocusing on each area to maximize focus.

13.3.4 Software

Software is a critical component on both the commercial slide-scanning systems and a custom-built microscope-based system. Two key factors for creating an efficient workflow from digital imaging to analysis are automation and integration. Obviously, the software must control all hardware components in an efficient manner, but the acquisition workflow and flexibility is also important. The system must define the area of the slide to be scanned, either by manual intervention or some type of automated "sample detection" routine. The system should have an automated naming scheme, either by scanning a barcode on the slide or by using a number system based on day of scan. Depending on the thickness of the sample or 3D aspect of the sample, it may be useful to collect a 3D stack (Z-stack) in order to collect multiple planes of focus. The resulting stack contains 3D information about the specimen and allows the pathologist to emulate the manual focusing normally performed when viewing the glass slide on a microscope.

As with plate-based HCS, many of the instrument vendors supply image analysis algorithms with the imaging instrument. There are also a number of third party software platforms which are compatible with the many different instruments and file formats used by WSI systems. These parallel the cell-based assay platforms, ranging from those with canned modules for specific applications to the development environment, enabling the user to develop custom solutions.

13.3.5 Whole Slide Imaging System

Whole slide imaging systems, much like the available HCS systems, are fully integrated systems designed for high throughput slide-based scanning. WSI scanners differ from automated microscopes in that they are a closed system that can use various scanning methods for image stitching or tiling to create a seamless digital slide. These are highly automated "black-boxes" that enable the user to load many slides into some type of cassette and then set up an automated acquisition of all the slides in a walk-away mode. The required scanning time of a WSI scanner to scan a whole slide is minutes, whereas it would take automated microscope much longer to complete. The WSI system also includes a data/image management system, visualization software, and image analysis software capable of handling large image files. The data/image management system is discussed in more detail in the next section. Visualization software acts like a virtual microscope, allowing the user to adjust magnification, pan, and zoom, and compare different stains. There are a number of advantages of this digital "virtual microscope" including the ability to view multiple samples simultaneously and the ability to annotate areas of interest. Instead of selecting individual FOVs, the user can now choose to image the whole tissue of interest or select regions (Figure 13.1). WSI enables a much better understanding of the tissue morphology and heterogeneity by capturing the entire sample. In additional

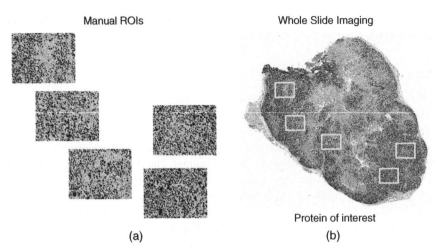

FIGURE 13.1   *Field of View versus Whole Slide Imaging.* (a) A typical acquisition protocol on a standard microscope-based system involves the selection of five manually selected individual fields of view captured by the imaging system. (b) Whole slide imaging creates an image of the entire sample. In addition to the analysis of many cellular measurements, additional features (percent viable tumor, percent necrosis) can be measured to understand overall tissue morphology and heterogeneity.

to the individual cellular measurements, tumor morphology can be characterized. For example, with a WSI of a xenograft, you can analyze the percent viable tumor versus necrotic area.

13.3.6   Data Management

Similar to plate-based assays, WSI also generates large volumes of data. As discussed in Chapter 6, creating a digital storage system is important for integrating metadata to the digital images and easy retrieval of data for future data analysis. This data management system can be based on a relational database or a simple file/folder-based structure. If using a simple folder-based structure, a file naming convention for images and a folder structure for easy retrieval of images should be established. Adding metadata such as assay name, dose, tissue type, and time point to the file name will enable easy retrieval of the images.

The best solution for acquiring and digitally archiving tissue-based images is to use an image management system. Automatic integration of metadata from a LIMS (laboratory information management system) system to the image management system is very important and can be done in several ways. Barcodes can be utilized to uniquely identify each slide. This barcode would then link the slide data with the LIMS sample data. Many of the WSI platforms include their own proprietary databases to store both the image and sample data. There are a number of alternatives for the microscope-based systems. One example, as discussed in Chapter 6, is OME [5], an open-source initiative to create a platform for storing both image and sample data which support both commercial and home-grown systems.

13.4   DATA ANALYSIS

Although it has taken us a while to get here, this last section is the most critical part of the whole exercise! While the execution of high content assays in tissue is daunting enough, the analysis of images derived from tissues poses a set of unique challenges. One major image analysis challenge is the heterogeneity of the tissue samples. In contrast to cell culture samples, tissue samples possess histological variations in morphology due to the differences in cell type, and pathological responses of tumor tissue (such as necrosis). Tissue heterogeneity poses significant challenges to the image analysis algorithms used to quantitate protein expression levels because of the wide range of signal levels and morphological characteristics. Heterogeneity may also introduce bias if specific regions are incorrectly excluded (or included) in the analysis. For example, xenograft tumors include areas of necrotic tissue, which arise due to the poor vascularization of the tumor mass, as well as due to the immune response to tumor inoculation. It is therefore important that any quantitative effect is measured only in specific regions of the sample. Clinical samples likewise may contain areas of tumor and nontumor, or an assay may be relevant to only the epidermal layer of a skin biopsy, that is, the PD effect is region-specific. Clinical samples are faced with a second level of heterogeneity as their tumor tissues are often from different tumor types and frequently differ in histological grade, leading to variability in morphology. The development of a robust image analysis method *for in vivo* samples is typically more challenging and time-consuming than the *in vitro* analysis described in earlier chapters. The image analysis algorithm must first identify the region of interest to analyze, then perform the specific cellular analysis within this region.

There are a number of strategies one can employ to address the issues of heterogeneity. A direct method is to stain the tissue sample with an additional biomarker that enables the selection of the regions of interest in an automated manner. For example, tumor tissue may be differentiated from normal tissue by the use of a tumor-specific biomarker (Figure 13.2). Later-stage clinical trials, which are typically focused on a single indication, may be more amenable to this approach than early clinical trials which place few restrictions on the type of tumor beyond broad characterization (e.g., solid tumor or hematological malignancy). Staining with a tissue-specific biomarker has the advantage of providing a precise definition of region boundaries, as the link between the biomarker and the tissue of interest is tight. The image analysis algorithm can then use this tissue-specific biomarker to gate the segmentation and analysis. Of course, a molecular biomarker capable of differentiating the region of interest from other regions of the tissue may not exist. Even if a biomarker does exist, it may not be feasible or cost-effective to optimize it for use in an assay setting. The use of an additional biomarker may also bring with it the need to reoptimize staining protocols or redesign the assay.

A second method to address tissue heterogeneity without using a molecular biomarker for the tissue of interest is to manually draw regions of interest on the image, thereby limiting the acquisition and/or analysis to user-selected regions. Hand-drawn regions have the advantage of transparency, where every field of view being analyzed is visually inspected before the analysis algorithm is run. However, this approach is subjective, and may introduce greater variability as well as systematic

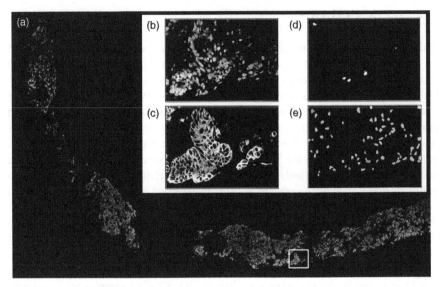

FIGURE 13.2 *Utilizing immunofluorescence to define tumor region.* (a) Needle biopsy containing both tumor and host tissue. Enlarged view shown in upper panel. (b) Nuclei (c) PanKeratin (d) phospho-histone H3. (e) Image analysis segmentation results, measuring nuclei only (bright grey), tumor marker positive nuclei (dark grey), and mitotic cells within tumor region (medium grey).

biases into the final analysis. Therefore, user-specified regions on images from histological stains (such as hematoxylin–eosin) should be defined by a trained pathologist. In addition, user-specified regions work best when region boundaries are either fairly smooth or tolerant of error. Image processing routines can be designed to work on a preprocessed version of the original image to facilitate the reproducible marking of user-defined regions. If an entire tissue sample is being imaged, it is often more efficient to first acquire a low magnification image of the entire tissue. The user may then mark the areas of interest, and have the acquisition system collect images of marked regions at higher magnification to minimize acquisition time. Alternatively, if the acquisition system in use permits walk-away, it may be desirable to acquire the entire tissue at high magnification, and once again provide a view of the entire tissue for the user to mark the regions of interest for the analysis.

Finally, new machine learning algorithms are being developed to automatically detect the regions of interest with the entire images. The user first trains the software to identify regions of interest, for example, viable versus necrotic, by manually drawing representative areas within a set of images. Once trained, the algorithm is applied to the entire sample and only the detected regions of interest are further analyzed with the cellular analysis algorithm.

In practice, staining with a tissue-specific biomarker and drawing regions by hand are not mutually exclusive methods, and may in fact have to be combined. We used such an approach to identify tumor tissue within liver punch biopsies. Since the majority of the samples that we were dealing with consisted of liver metastases from carcinomas, we elected to use an epithelial biomarker, cytokeratin, to distinguish tumor sample from liver tissue (Figure 13.2). The image processing algorithm can identify this marker and use it as a mask to gate the measurement. This biomarker was able to correctly identify tumor regions without the need for further manual intervention. However, when the approach was applied to other tumor types, the cytokeratin biomarker in use was not sufficient to identify the tumor in every case. In addition, in some cases, the cytokeratin biomarker stained adjacent tissues (poorly differentiated hepatocytes), albeit with a distinctly different subcellular localization. This led us to develop other biomarkers that were used on tumor sections immediately adjacent to the assayed samples. Hematoxylin–eosin and Ki-67 stains were eventually used to identify tumor regions, which were then drawn by hand. The region of interest was further filtered using the cytokeratin-positive region. This combined approach enabled us to implement a system that was sufficiently flexible to adapt to the fluid sample characteristics of a Phase I clinical trial.

Many of the image processing algorithms developed for HCS were originally developed for cell-based assays. Adapting these analysis algorithms to tissue samples is particularly challenging, since the properties of individual cells in tissue differ significantly from the properties of the same cells *in vitro*. While cell density can be easily controlled in plate-based assays, cell density is dependent on tumor and tissue type and can cause numerous problems for image analysis. The cells also reside in a 3D architecture as opposed to lying flat in a multiwell plate. This altered context leads to difficulty in identifying cellular and nuclear boundaries, and often leads to the misidentification of individual cells due to over- or under-segmentation. For example, the physical sectioning of the sample leads to varying sizes of nuclei in a single field of view depending on the relative location of the cell in the 3D architecture. Many of the "canned" segmentation algorithms include size parameters to help identify individual cells. One common problem is to accurately segment the small fragments of nuclei, the algorithm will commonly over-segment individual larger nuclei. In addition, "canned" segmentation algorithms optimized for tissue culture cells, with their uniform shape, often fail with cells in tissue, which are far more variable in their shape due to both the orientation of cells in tissue and their tight packing. An alternate, more simple approach is to use an area measurement instead of nuclei count. The algorithm simply identifies pixels that are greater than a specified intensity and scores these as positive pixels. As discussed in Chapter 4 it is important to understand the differences between a pixel-based approach versus an object-based approach (Figure 13.3).

There are a number of other potential sources of variability in a tissue-based setting that further impede the establishment of accurate segmentation routines. For example, signal-to-noise and background variability within a specific field of view may make it difficult to set a single threshold that is capable of accurately segmenting an image. Adaptive thresholding algorithms often account for this variability, but it may be necessary to adjust thresholding parameters across different samples. Similar variability may exist across samples, making it necessary to adjust segmentation parameters within a single experiment. Additionally, for IF assays, autofluorescence and

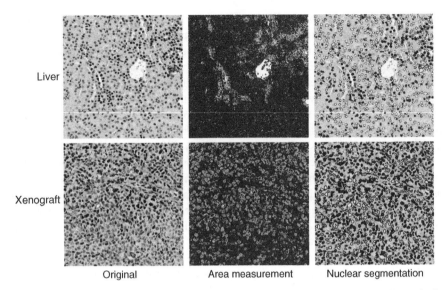

FIGURE 13.3   *Pixel analysis versus cell analysis.* Pixel analysis identifies positive pixels above a user-defined threshold and classifies (color coding) these pixels based on staining intensity to give a basic area measurement. Cell analysis identifies nuclei based on hematoxylin-stained nuclei, and classifies each individual nucleus based on intensity of DAB stain. These are normally marked in a color scheme (red-orange-yellow-blue for strong to weak) but are shown as 3 shades of grey in figure.

nonspecific fluorescence background may further reduce the overall signal-to-noise ratio, necessitating additional correction images to be acquired or additional preprocessing before quantitation. There may be steps to reduce background fluorescence in the staining protocol, and there are technologies such as spectral unmixing [3] to remove background fluorescence.

Assay designers commonly deal with issues related to segmentation by modifying image processing parameters, on a trial-and-error basis, to optimize the performance of their algorithms. While this is certainly a reasonable starting point, the intrinsic variability of tissue samples requires the identification of image-processing parameter settings that will work across a range of different signal-to-noise ratios. Even when such parameter settings exist, a gradual drift in the staining procedures over time (for example due to antibody degradation) may in turn result in the degradation in performance over time of image processing algorithms optimized for a particular set of conditions. For example, in an assay where a threshold is needed to separate specific populations from others, a decrease in the signal-to-noise will increase the likelihood that the segmentation algorithm will fail. For this reason, among others, it is important to develop the image analysis algorithm using a number of samples representing high and low responders as well as slides from different days. This includes both the ROI detection algorithm as well as the cellular segmentation algorithm. In an assay being run in production mode, a failure of the algorithm is typically silent, as it yields no obvious clues or error messages. However, such a failure can have catastrophic consequences for the assay, as it invalidates any numerical metrics that are derived from detected ROIs and specific cell segmentation. Thus, we recommend monitoring image segmentation performance on an ongoing basis through the visualization of segmentation overlays.

13.5 CONCLUSIONS

There are many similarities between cell-based (multiwell plates) and tissue-based (microscope slides) assays, including wet lab protocol, image acquisition, and image analysis. Both involve many different platform technologies which are important to understand in order to develop an efficient, fully integrated, workflow. And, both can benefit from the same quality control and robust assay development strategies highlighted throughout this book.

KEY POINTS

1. As with plate-based assays, it is important to understand the key parameters to optimize the assay during the development process. These include tissue processing, IHC and IF wet-lab protocol, image acquisition, and image analysis.
2. A good understanding of the imaging system components will enable you to select the appropriate slide-scanning system and optimize the setup to acquire sufficient resolution images for image analysis.

FURTHER READING

DAKO Immunohistochemical Staining Methods Education Guide. An invaluable online resource that reviews all aspects of immunohistochemistry, from antibodies to image acquisition and analysis.

There are many online microscopy websites which include diagrams and animated tutorials to educate the reader on all aspects of optics and microscopy, from objective specifications (magnification, numerical aperture) to advanced microscopy techniques. Olympus (http://www.olympusmicro.com/index.html) and Nikon MicroscopyU (http://www.microscopyu.com) are two examples.

IHC World (http://www.ihcworld.com/) is an online information resource for all things IHC.

The Human Protein Atlas Project (http://www.proteinatlas.org) is an online resource for information on proteins, searchable by gene or protein, that includes protein expression profiles and specific cellular localization.

Buchwalow, I.B. and Bèocker, W., *Immunohistochemistry: Basics and Methods*. Springer, 2010.

REFERENCES

1. Chakravarty, A. and Bowman, D. et al. Using High Content Analysis for Pharmacodynamic Assays in Tissue. *High Content Screening*, John Wiley & Sons, Inc., 2007, pp. 269–291.

2. DAKO Immunohistochemical Staining Methods Education Guide. http://www.dako. com/us/index/knowledgecenter/kc_publications/kc_publications_edu/immunohistochemic al_staining_methods.htm, accessed August 13, 2014.

3. Levenson, R.M. and Mansfield, J.R. Multispectral imaging in biology and medicine: slices of life. *Cytometry part A*, 2006, **69**(8): 748–758.

4. Nikon MicroscopyU: Kohler Illumination. http://www.microscopyu.com/tutorials/java/koh ler/index.html, accessed August 13, 2014.

5. Open Microscopy Environment. http://www.openmicroscopy.org, accessed August 13, 2014.

SECTION V

HIGH CONTENT ANALYTICS

Analytics, the mining of large datasets to uncover patterns or make predictions, is growing in many fields, including social media and business development. Biological data analytics has been a natural outgrowth of early omics technologies such as transcriptional profiling and proteomics. This trend is extending into HCS, as the amount of data generated in imaging experiments is seemingly comprehensive and readily available for mining. The chapters in this section introduce the two main approaches to data mining, exploratory clustering and factoring and supervised learning, which generates rules based on data trends that are applied to new samples.

There are more caveats than best practices in these approaches, so the next two chapters introduce the respective subjects, but largely within the view of introducing each in the context of prominent methods for each approach and considerations on experimental design and data processing than can improve the relevance and extensibility of a study.

No question about it, these methods have the capacity to alert researchers to unanticipated connections between perturbations and cellular effects, and to develop the next generation of biomarkers. However, this promise, combined with the "early days" nature of applying these methods to high content data, set up a challenging situation for researchers hoping to discover the barely discoverable. This is where the contributions of a team of bench biologists and data analysts really need to come together.

An Introduction to High Content Screening: Imaging Technology, Assay Development, and Data Analysis in Biology and Drug Discovery, First Edition. Edited by Steven A. Haney, Douglas Bowman, and Arijit Chakravarty.
© 2015 John Wiley & Sons, Inc. Published 2015 by John Wiley & Sons, Inc.

14

FACTORING AND CLUSTERING HIGH CONTENT DATA

Steven A. Haney

14.1 INTRODUCTION

It is becoming common practice to integrate the vast numbers of measurements made per cell in an HCS assay into a plot or dendrogram that shows the relationships between cell subtypes, target classes, or phenotypic responses. However, the results are frequently less than robust, or at least of questionable value. There are several reasons for this. First, the output is always dramatic, but quality measures for the classification are not obvious, and for a few methods, do not exist. Second, the methods apply significant computational pressure to a dataset, which frequently reduces a matrix of many treatments and features to a few properties or groups. This is often performed with a "black box" approach to the methodology, the average biologist has not studied the different approaches to explain trends in the data in detail. It is thus imperative to validate results both at a computational and a biological level. Third, there is no defined rulebook for these studies. Each experiment introduces its own strengths and limitations, and each study needs an analytical design that is "fit for purpose." Results are usually evaluated through more than one method or set of algorithmic adjustments. Finally, and crucially, large high content datasets are victims of the "curse of dimensionality." When the number of parameters being examined increases, the data points characterizing a biological response become more and more sparsely distributed. The amount of data points required to support a statistically reliable result can grow, in the worst case, grow exponentially. As a

An Introduction to High Content Screening: Imaging Technology, Assay Development, and Data Analysis in Biology and Drug Discovery, First Edition. Edited by Steven A. Haney, Douglas Bowman, and Arijit Chakravarty.
© 2015 John Wiley & Sons, Inc. Published 2015 by John Wiley & Sons, Inc.

result, the patterns observed in high dimensional data, while visually strong, can be completely spurious, a result of overfitting.

As such, it is most common to refer to these methods as "exploratory data analysis," and the findings are better thought of as hypotheses for follow-up experiments rather than definitive conclusions. Such exploratory analysis may yield an indication no stronger than "one of these compounds is acting funny." Follow-up experiments could take the shape of specific cellular assays for proliferation, stress, or candidate signaling pathways, or through another multidimensional assay, but one with defined quality measures. These would fall under the heading of Machine Learning (covered in the following chapter). A classic experiment in HCS unsupervised learning is to treat cells with many compounds and determine how many separate classes of compounds exist, based on phenotypic cellular responses. Such classes could then be analyzed to see if a toxic response is present in one of the groups, and used in subsequent rounds of compound testing to see if newer derivatives retain the characteristics of earlier ones. As such, if the caveats of unsupervised learning (UL) methods are understood, then they can be very helpful in identifying hidden relationships in data and can start the process toward defining unique properties of potential therapeutics, or of the common regulation of cellular processes thought to be independent.

There are numerous methods that are covered under the umbrella of "unsupervised learning." The methods themselves have been developed over decades in other sciences, including Psychology and Ecology. These methods gained popularity in the biomedical sciences and drug discovery with the advent of large datasets, particularly the "omics" technologies, such as transcriptional profiling and metabolomics. Their use in high content studies is an area of active interest, but is still very much in the development stage. In this chapter, we review several unsupervised learning techniques that have a direct application in HCS assays.

14.2   COMMON UNSUPERVISED LEARNING METHODS

The most common methods for unsupervised learning are introduced below. These methods treat data very differently, and return different information [1,2]. A graphical comparison of several of these approaches is shown for a 2D dataset in Figure 14.1. The methods are applied to high dimensional datasets (more than a hundred features, or dimensions, can be integrated) along the same guidelines that are depicted in the figure. The major methods are the following.

14.2.1   Principal Components Analysis

Principal components analysis (PCA) answers the question "what are the major trends in the feature responses and how would the treatment conditions be grouped by these trends?" PCA starts by defining an equation that runs through the widest spread in the data; this is the first principal component. There is a significant conceptual issue here, and understanding it is key to understanding PCA. The most commonly used phrase is that the first axis is that which "explains the most variance in the data."

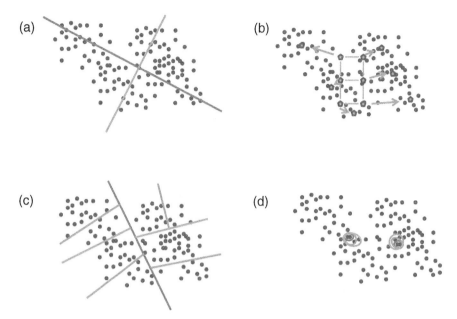

FIGURE 14.1 *Analyzing data using unsupervised learning methods.* A 2D dataset is analyzed by four approaches. Samples measured by two features are plotted by each in a traditional XY coordinate system (axes not drawn). (a) In PCA, a line is drawn through the greatest dispersion in the data (longest line). Reducing the dimensionality from two to one dimension would be achieved if data points are now compared by how they relate to this line instead of the original XY coordinates. Adding a second line, one that traverses the maximum dispersion remaining in the data (shorter line) allow a more accurate description of the sample relationships, but does so by replacing the original coordinate system with a different 2D system. (b) SOMs classify samples by seeding the dataspace with an array of points that will serve as represent groups or clusters. Placement is improved for each point by iterative cycles of distance measurements, reducing these for each group to a minimum. k-Means clustering works along a similar principle, but the number of points is not seeded in an organized pattern, and therefore no relationship between the cluster centers is presumed. (c) Divisive hierarchical clustering separates data into groups, first along the widest separation in the dataspace (longest line). Each successive cycle separates each cluster into two clusters (successive lines segmenting the samples), until all samples have been separated. (d) Agglomerative hierarchical clustering finds the closest pair of samples and groups them as a cluster (initial clusters shown in small circles for two sets of closely related points). Successive clustering is achieved by including samples one at a time until all samples or clusters are grouped (shown increasingly larger circles).

The general interpretation of this is the axis (or ordinate, or most simply, a line) that is plotted in the orientation necessary to run through the data at its widest spread. It answers the question "if you had to pick one coordinate that best represented the differences between the samples across all of the features, what would that line be?" In Figure 14.1a, the first principal component would be the line that runs along the widest spread in the data. All samples are then evaluated by where they fall along that line. To add a second principal component, a second line is drawn through the data that runs through the widest spread remaining in the data in a direction that is orthogonal (perpendicular) to the first. The process is repeated until all trends in the data have been evaluated (for a two-feature dataset, we are done after the second component is determined). The result is that a set of test conditions can be defined using these algebraic equations instead of the original feature data. Ultimately, this process has the potential to replace the original dimensionality of features with one of the same number of equations/dimensions/lines, but as it happens, most highly dimensional datasets can be approximated well in only a few equations/components. This is because of two reasons. First, recall that the first principal component is the one that best describes the differences between samples in a dataset. Each new component is the best measure of the remaining sample differences, and some features do not show much variance, and therefore do not differentiate samples well. Thus the principal components are rank ordered for their contributions to characterizing the dataset. Second, many HCS features are highly correlated. Adding a feature that is highly correlated to one that is already in the dataset yields little new information. Therefore, the effective amount of information that can be used to classify or differentiate samples in a dataset is far less than the number of features it contains. At some point, the gain in definition by including each successive component becomes vanishingly small, so a decision is made to stop adding components and just work with the most relevant ones. A complex dataset can be described well by a few principal components, frequently less than five. If the number of components can be reduced to two or three, samples can be evaluated in standard 2D or 3D graphs, plotting them along the axes using the component scores. A general method for determining the appropriate number of components, usually done graphically through a Scree plot, is discussed later.

14.2.2 Factor Analysis

Factor Analysis is an approach that is used widely in the social sciences, where there is a general problem trying to understand deep personal characteristics through observing surface behaviors. Thus, we can ask questions about likes and dislikes regarding daily life and try to determine the extent to which factors such as religion, political party affiliation, or gender play in the decision making process. This approach groups feature responses into a few factors, and then the samples can be evaluated on the basis of the factors (rather than on the individual responses or features). This is a very attractive method for HCS scientists, since most biological studies focus on altering the activity of signaling pathways and growth processes. These are conceptually similar to factors, and general responses to these perturbations can be measured by changes in the "feature space" for an imaging assay. Features that correlate could be characterized as being responsive to a common factor, samples can then be evaluated on the basis of a few factors instead of the complete panel of features. The challenge is ascribing algorithmic factors identified through feature change patterns to specific cellular processes such as a single signaling event. Novel associations between groups of features would certainly require explicit validation in follow-up studies. Although conceptually based in the idea that the factors defined are underlying drivers of the phenotypic changes, the process is similar to PCA, so much so that many consider it a form of PCA. This is not accurate, but stems from the reduction of the dataset to a few factors/components. The distinction is that PCA is strictly an algebraic reduction in complexity that rank orders the components, whereas FA is a grouping of features that are responding to an underlying driver (the factor).

14.2.3   k-Means Clustering

This method places an arbitrary number (k) of cluster centroids into a dataset, initially at random. Each data point is assigned to the closest centroid. The mean location of each cluster is then recomputed by averaging the positions of the data points in each cluster, and the centroids are relocated to these new cluster centers. This process is repeated until the centroids stop moving (within some tolerance). The number of centroids (k) that provide the best differentiation of the data is frequently determined empirically by trying several seedings, say sequentially trying two to six centroids, and comparing the results. The optimal number of centroids is that which accurately reports on the properties of each group. Too many centroids means that bona-fide biological clusters become separated, too few means that bona-fide clusters may be merged, and their properties become averaged over the samples of what should be considered as distinct groups. One difficulty in assessing results is that all samples will be clustered, regardless of how well the data supports such a linkage. There is no mechanism for excluding true outliers or even of specifying an "outlier cluster"— they will be merged with other samples into clusters, although the impact of outliers can be minimized through taking the median of the distances to define the cluster location, which is defined as k-medoid clustering, which has been integrated into the modified clustering algorithm partitioning around medoids, or PAM. Outliers can be separated from clusters in density-based clustering, introduced briefly below. Similar to the Scree plot for PCS, quantitative measures can be used to compare how well different numbers of clusters describe the data (discussed later).

14.2.4   Self-Organizing Maps

Self-organizing maps (SOMs) have been used in many analytical fields to reduce the dimensionality of a dataset into an easily visualized projection. Specifically, data with many parameters and samples can be reduced to a few major trends or events, and the map is a visual representation of these trends and the fraction of each trend in the set [3]. The number of maps generated is chosen to start the analysis, similar to k-means clustering, but the initial seeding of cluster centers (nodes) are based on an array of nodes. Each sample is measured against all of the nodes in the array, rather than the closest centroid (as in k-means clustering), and each node becomes associated with a sample because it is closest to one node among all the others. Originally, the output was a map of colored regions where each region was a cluster or trend in the dataset. The largest regions represented greatest contribution to the variability of the feature responses. This is another technique that has been used extensively in transcriptional profiling, particularly when the maps are presented as an array of line plots that allowed a quick understanding of the data trends, such as gene expression changes over time [4]. SOMs are similar to k-means clusters, in that they define subgroups as prototypical samples of common properties. The relevant difference is that the final clusters retain the array-based relationship, clusters nearer to each other in the array are more similar to each other, whereas in k-means clustering, no relationship between clusters is inferred.

14.2.5   Hierarchical Clustering

Another method popularized by its wide use in transcriptional profiling studies [5], hierarchical clustering assumes that all samples are related to each other, some are closely related and are therefore members of a cluster, similar clusters are themselves clustered, and all clusters are ultimately part of a single cluster of all samples [6]. There are two approaches to hierarchical clustering, a "bottom up" approach that finds samples, features, or perturbations which are most similar to each other, and then successively groups additional examples with others until the entire set is connected (agglomerative clustering), and a "top down" approach that divides the entire sample set into two clusters on the basis of maximal dissimilarity between the groups, and then repeats this process for each group until all of the samples have been separated (divisive clustering). The point at which these changes occur is represented as branch points in a dendrogram. Hierarchical clustering is one method that is prone to being misapplied. Two cases where this can occur is (1) when it is applied to samples that are not related to each other, and (2) when the features used to cluster samples are not capable of differentiating the samples. Examples of treatment conditions that may not be hierarchically related are presented in Figure 14.2. If the samples are indeed related in hierarchical terms, then there is still the issue of whether the features are appropriate to distinguish these relationships (discussed below and also highlighted in Figure 14.2). It can be difficult to tell how strong (or genuine) the clusters are, but (as with other methods described here) there are methods for checking how robust the classification is. For high content data, the grouping of samples can be hyperstratified, showing apparent separation into small clusters when the data actually does not support such fine distinctions. It is possible to limit the extent of clustering (referred to as pruning, following the characterization of a dendrogram as a tree) when the extent of separation in the dendrogram is not supported by the quality metrics of the analysis.

14.2.6   Emerging Clustering Methods: Density and Grid-Based Clustering

HCS data can make use of the classification methods discussed above, but it can be very challenging. The spectrum of feature relationships, from orthogonal to partial correlations to identical, is one reason, the inability to eliminate outliers is another. For samples that can be divided into higher density regions versus sparsely distributed samples farther away, density-based clustering methods such as density-based spatial clustering of applications with noise (DBSCAN), are alternative methods that

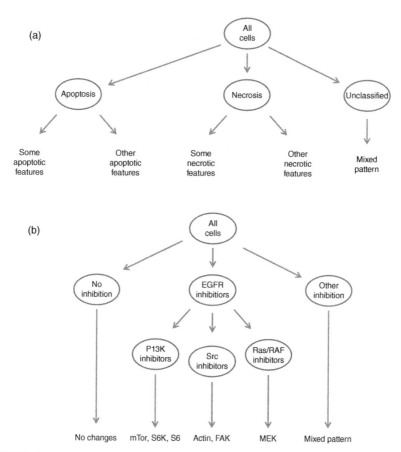

FIGURE 14.2 *Potential hierarchical relationships.* Two experimental situations that may be hierarchical in nature, but may be difficult to show by unsupervised learning. In (a), cell-death pathways, particularly apoptosis and necrosis, can be observed when cells are treated with cytotoxic reagents. The ability to detect a particular cell-death pathway is dependent on the features recorded, but furthermore, the ability to construct a hierarchical relationship is dependent on the features being linked by the cell-death pathway. If the ability to see minor and major cluster relationships cannot be provided by the features (such as few treatments that produce a complete apoptotic or necrotic phenotype), then the subgroups could separate, and it would appear that the apoptotic and necrotic subclusters could be mixed. This would really be a case of forcing a relationship between clusters that is weakly defined by the feature data. In (b), a similar situation could occur. EGFR is known to regulate several pathways in the cell, and therefore it could be presumed that inhibitors of EGFR and the pathways it regulate would also form a hierarchical relationship. However, observing this separation may be difficult as the pathways are complicated by feedback regulation and cross-talk. Even if the relationship between EGFR and its proximal targets is relatively clear, the phenotypic responses may be complex and lead to many 'mixed pattern' samples.

eliminates the confounding effects of the outliers [7]. Density clustering is based on a defined distance between nearby samples that will therefore be defined as nearby, a minimum number of cells linked by this distance may be required for the samples to be classified as a cluster. All samples that are linked by these criteria will be considered as a common cluster. All samples that do not conform to these criteria are not part of the cluster, and may not be classified into any cluster. Additional rules may be introduced in other density-based methods. Grid-based methods will calculate a multidimensional grid that is based on the dataset and fit samples into cells defined by the grid. Further definition to the dataset, such as linking nearby cells can be calculated on the cells rather than the samples, reducing computational time. In addition to providing flexibility with clustering decisions, these can also greatly reduce computational time needed to classify the samples.

14.3   PREPARING FOR AN UNSUPERVISED LEARNING STUDY

The output from unsupervised learning studies is abstract, so it is frequently impossible to catch problems *post hoc*. All unsupervised learning experiments need a clear design that is capable of supporting the analysis phase of the study. This can be somewhat similar to a clinical trial, one needs to determine the analysis plan, the anticipated results and then determine if the study is large enough to meet such thresholds (generally termed "powering a study"), and whether the metrics tracked in the study will allow for meaningful parsing of the data. Ironically, even studies described as "hypothesis free" need to go through this process, perhaps even more so. Hypothesis-free experiments, for example, trading the concrete hypothesis of "I believe this compound affects these cells in this way" with the fuzzier "I wonder if there are compounds out there that affect these cells in some interesting way," are an important type of study in the biological sciences. However, the trade-off of such a shift is that the possibilities are literally endless, so the parameters and cut-offs need to be considered at the start. What follows are some general guidance on the process, but each study needs to reevaluate each step as it relates to that study. This will not be an exhaustive review of steps in running such a study, but is intended to alert the reader to the process and particularly acute challenges. In this section, we introduce some of the considerations for developing a study. The principles are common to all UL studies, but we will focus on matters specific to HCS data.

14.3.1   Develop a Provisional Hypothesis

What is the purpose of the experiment? To cluster a panel of related compounds and determine which show off-target effects? To identify cellular events that correlate with the inhibition of various signaling pathways? Defining the goal is essential to determining the number of reagents required and what features need to be considered. It also helps define the analytical process. As discussed above, the methods for analyzing multiparametric data produce results that may look similar, they actually analyze the data very differently, and some methods are inappropriate for certain experimental designs. Therefore, after determining the goal of the experiment, the next step is finding the correct method for analyzing the data. More than likely, a pilot study is very valuable. A selection of cytotoxic reagents or well-characterized tool compounds, or siRNAs against several related signaling pathways, can be an important reality check for an anticipated study. The best place to start is showing that you can group treatments phenotypically, into perhaps two or three classes of reagents.

14.3.2   Select Your Cell Labels Carefully

What to measure? Selection of labels is critical to the study design. A study to contrast apoptosis and autophagy needs to have clear markers for both, and each needs to be well validated in a multiplexed system at the outset. A general phenotypic profiling study would include as many independent cytological markers as possible, including markers for the cytoskeleton and other cellular landmarks, such as the plasma membrane and organelles [8]. At present, HCS investigators typically have four channels to use when building a multiplexed assay. One will almost always be nuclei, both because they are relevant to so many types of studies (proliferation, toxicology, development, and others) and because many image analysis methods start with the identification of the nucleus. Beyond this, markers used in the study need to be firmly linked to question being asked. Do the structures being quantified respond to treatment conditions in relevant ways? Consider an example based on using the actin cytoskeleton to measure some important kinase signaling pathways. Src, Fak, and Rock kinases all directly affect actin networks, and there are many additional kinases that affect these proximal regulators of the actin cytoskeleton. In the case of cellular organelles, the shape and localization of the ER and mitochondria are affected by many signaling and metabolic pathways as well. A study to characterize treatment conditions on the basis of their effects on proliferation typically includes direct measures of energy balance, cell cycle, events that trigger mitosis, and DNA synthesis.

14.3.3   Establish Treatment Conditions

This can be a surprisingly complex decision. Do you treat all reagents at one dose or time point? Do you treat each at the IC50 or IC90 (for compounds) or at the time of maximal knockdown for each individual target (for RNAi). A complex but thorough approach applied by Perlman et al. [9] was to run each compound in a dose-response assay, and then align the phenotypic effects through their IC50s. Furthermore, they recognized that the distributions of the data were nonparametric, and so they used a nonparametric test to compare the samples. Nobody said these studies were easy. It does not get easier with RNAi reagents. In addition to a wide range of potencies and the extent off-target effects [10], mRNA decay rates vary widely, with half-lives ranging from seconds to hours, so the caveats to compound dosing pertain to RNAi screens as well, and more. Selecting treatment conditions is one of the best reasons to consider one or more small pilot studies before planning for the full study.

14.3.4   Collect and Analyze Images

Phenotypic studies are subtle. Relevant subpopulations can be a small fraction of the total number of cells analyzed, so generally speaking imaging of 1000 or more cells may be needed to capture such rare events in a statistically valid dataset. In this regard, phenotypic profiling can be similar to multiparametric flow cytometry [11].

14.3.5   Prepare the Data: Feature Selection

Which measurements to include? Many features reported through image analysis are redundant. In particular, there are many cases where two features measure the same thing, or as inverses, or are directly related through a third feature. An example of the latter is the relationship between total intensity, average intensity and area. Depending on the learning method used, these redundancies may overweight the analysis. PCA is a method that reduces dimensionality (the number of features), and is resistant to redundancy; highly redundant features would be compressed into a single component. On the other hand, hierarchical clustering usually assumes that all features are independent, and therefore equally relevant. A perturbation that changes several redundant features will have a greater influence on the classification than one that changes a single, unique, feature. Despite the fact that some methods are resistant to this overrepresentation, it is better to eliminate redundant features during this stage of data review. Frequently, several learning methods are compared, therefore cleaning the feature data at the outset is a more stable way to explore the data.

How then do you control for this overrepresentation? After the data has been collected, the file of the cell-level data can be opened and inspected (refer to Chapter 8 for a discussion on the feature data generated in an assay). Look at the columns; any features that do not represent relevant biological measurements can be eliminated immediately. Examples include columns for cell number (the ordinal identification of cells within a field), or, in many cases, the position of the center of the nucleus in (x, y) space within a field (which is not generally relevant to most studies (but may be, [12]). Following this, groups of columns can be compared in scatterplot arrays, as shown in Figure 14.3. Here, columns of feature data are plotted against each other pairwise in an array of scatterplots. This analysis is very simple to perform, but very important, this review allows for the quick visual identification of related features that may not be detected by a cursory review of the feature columns. The process used for this figure is discussed in Appendix B. Algebraically identical features can be identified as perfectly correlated, as shown in the figure for the three features shown in the lower right section, and can be deleted outright. Other features will be correlated, but less than absolutely, as highlighted in the middle row and column of the figure. Oftentimes such highly related features can be eliminated without material consequence to the study, but it may happen that there is a biological reason to retain them. In such cases, correlation should be noted, and used during review of the output to see how much of a role these feature groups play in forming any conclusions.

The plan outlined here looks at correlations in the data, but is a biologically based filtering method (essentially, hand selecting features for study). There are good

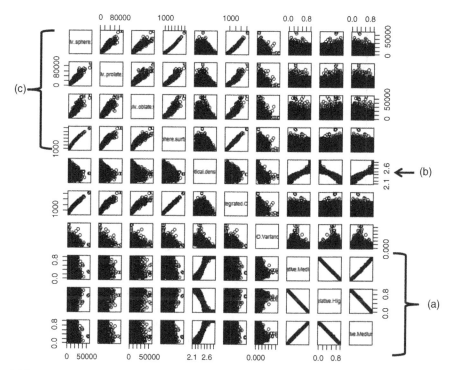

FIGURE 14.3 *Pairwise scatterplots of HCS cell-level data.* Cell-level data for the FYVE puncta assay (described in text) are plotted in feature × feature arrays. 10 of the 70 features are plotted here. While the data and the feature names are compressed in this figure, key relationships between the features that are important to feature selection can be observed. In the lower right quadrant of the figure (a) several feature pairs that are absolutely (completely) concordant can be detected by the strong diagonal lines formed by the data points. These patterns indicate that they are algebraically identical, and that two of the three can be eliminated without consequence to a study (and doing so will in fact improve most learning approaches). Another pattern is the highly correlated feature pairs indicated by (b). In this case, a clear relationship exists between the features (further accentuated in this figure because the relationship is repeated across the identical features described in a). Other features showing strong correlations are highlighted by (c). The features are measuring similar properties of the cells, but with different scaling. Eliminating the redundant features in b or c outright might help some methods, it may also happen that comparing different feature reduction strategies may be called for. Details on these features are available in the text.

reasons for doing this, including the importance of understanding the biological basis for the study. Understanding the features to be compared is part of staying linked to the biological basis for making comparisons. However, this is also an explicitly biased process. In some cases, an investigator may feel that an algorithmic alternative could be used (particularly for partially correlated features such as was highlighted in Figure 14.3b). One method is to use PCA to reduce features that are highly correlated to the same component, so that comparisons based on these features are not overemphasized through redundancy, or through generating a correlation matrix (discussed below) and eliminating features that are highly correlated (say, with coefficients greater than |0.7|).

14.3.6   Scale the Data

Raw data is not used for multivariate analyses, it must be processed. UL methods characterize the similarity or differences between samples or features, and the first step is to develop a matrix of the relationships between the features. This is challenging for HCS, as the feature comparisons are trickier for the kinds of studies that UL methods were developed for. In these cases (such as in ecology or psychology), data tends to be collected on a common scale across all the measurements (such as questions that ask a respondent to rate items from 1 to 10), and therefore do not need to be rescaled. In HCS, features of integrated area can range from 0 to greater than 10^5, whereas shape features may be from -1 to 1. Therefore, HCS data must always be rescaled before any multivariate learning algorithm can be applied to the data. The simplest method to scale is to convert all feature columns to a single range, such as from 0 to 1. This can be accomplished with the formula in Figure 14.4a, which is essentially the same as scaling an assay metric to the percent of control. There are other methods for scaling data, and one of these may be better for some data types or study designs. It is also possible to combine scaling with the next step in the process, as discussed below.

14.3.7   Generate the Similarity or Difference Matrices

In addition to rescaling, the relationship between the features needs to quantified. There are three options, the appropriate one to use depends on the UL method. Most methods will group samples through the similarity of the features, so either a **correlation matrix** or a **covariance matrix** needs to be calculated from the feature

(a)

$$x' = \frac{x - x_{min}}{x_{max} - x_{min}}$$

(b)

$$cov(X,Y) = E[(X - E[X])(Y - E[Y])]$$

$$cor(X,Y) = E[(X - E[X])(Y - E[Y])]/(\sigma_x \sigma_y)$$

FIGURE 14.4 *Scaling and formatting data.* (a) Scaling high content features. Since the feature is so heterogeneous (and in particular, the scales of the feature are so discordant), scaling is required. The formula shows the simplest method for scaling data, converting it to a scale of 0 to 1. (b) Generating correlation and covariance matrices. The processes treat the data in essentially identical ways, except that the correlation matrix will scale the output through the denominator term, as explained in the text.

data. Hierarchical clustering evaluates the differences between the samples through the feature data, so a **distance matrix** is calculated. A covariance matrix demands that the data be scaled appropriately, so rescaling of HCS data needs to be performed first (as discussed above). A correlation matrix accounts for scaling differences, so calculating one from raw data incorporates rescaling implicitly. This is highlighted in Figure 14.4b. The covariance matrix compares feature relationships across samples directly, the correlation matrix performs the same comparison but then scales the magnitude through the product of the standard deviations for each feature (seen as the denominator in the equation). The correlation matrix is literally a matrix of correlation coefficients, and reduces feature relationships to a scale of -1 to 1. Since the covariance matrix is not scaled, the terms can be outside of the range of -1 to 1, and therefore this matrix is generally preferred for PCA and other methods, as it provides more sensitivity to the relationships between the features.

Distance matrices are used in hierarchical clustering, because the method determines relationships by how far away samples are from each other (in fact, many teaching examples use distances between cities as the metric to cluster groups of cities). When considering multiple features for clustering, one can see that having measures at different scales would complicate the concept of "distance" between features. Therefore, a distance matrix also requires the data be rescaled a priori. The steps involved in preparing data for various UL approaches are outlined in Figure 14.5.

14.3.8   Analyze the Data

We introduced several methods for analyzing data earlier in this chapter. Each of them can be useful for characterizing a dataset. The most common questions are whether treatment effects are related in some way, so performing a clustering approach is a good way to start. For questions about what the feature relationships are that define these differences, PCA or FA can provide some insights. PCA and FA can also be used to identify sample groupings, because the results can be mapped back to the samples, such as plotting samples along the first two or three principal components in PCA. So if UL methods are not highly robust, they are at least flexible, and the learning process may focus more on what trends in the samples or features are observed frequently or are most closely related to known relationships (such as classes of treatments) that can be used to infer common properties with novel treatments or samples. UL methods can also be combined. For example, hierarchical clustering is sensitive to overrepresentation of correlated features and one of the key strengths of PCA is to eliminate redundant features. It is possible to cluster samples after establishing the principal components of the feature data.

14.3.9   Perform Quality Assessments

Since most UL methods will classify all of the data, quality metrics are essential to check when the reliability of the analysis has started to drop off. For example, PCA is able to classify a dataset into as many components as there are variables, the value in the approach is based on knowing how many of the components are needed to explain the biology effectively. The number of k-means clusters is not chosen with a

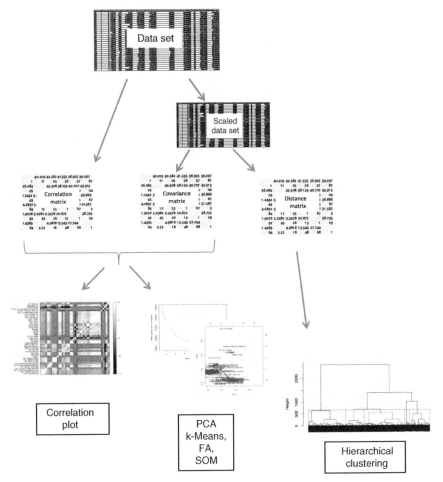

FIGURE 14.5 *Preparation of data for multivariate analyses.* Raw cell-level data for which the features have been reviewed and reduced (as described in Figure 14.3) need to be prepared for multivariate analysis. For methods that reduce dimensionality or define relationships based on similarity, either a correlation or a covariance matrix needs to be generated. These are square matrices that describe the extent of similarity for the features; features are compared in a sequential, pairwise fashion. A covariance matrix is more common in other fields, which tend to compare variables (features) that are already scaled. Since HCS features are typically based on widely differing scales, a correlation matrix is typically used, which explicitly scales the data. Alternatively, the data can be rescaled in a separate step, and the scaled feature data used to generate a covariance matrix. In some cases, further multivariate analysis may not be necessary; a graphical review of feature correlations may be sufficient, shown by the correlation plot in the figure. For methods that separate samples based on the extent to which they are different from other samples, such as most clustering methods, a dissimilarity matrix needs to be generated, again using scaled data.

strong idea of how many are needed (in fact, a common purpose of the procedure is to determine how many clusters exist). In some cases, it may be necessary to pare the results (such as pruning a dendrogram) or rerun an analysis with an optimized set of parameters (such as the number of k-means clusters). For most of these approaches, methods exist for checking on their reliability. Robustness measures for UL methods has been a very active area of research, now that they are finding more uses in emerging areas such as business analytics and "Big Data."

The most well-known example of assessing results in a learning analysis is the use of a Scree plot in PCA. Recall that in PCA, the initial component is defined as that which measures the greatest variability in the data. The second component measures the greatest variability in the data that remain. With each component, additional variance in the data is explained, but each successive component explains less overall variance. At what point is increasing quantification of the features offset by the increasing complexity incurred by adding additional components? A Scree plot is a very common method of making this assessment, an example is shown in Figure 14.6a. As new components are added, each contributes to the description of the variance in the data, but the contributions become smaller. A Scree plot is a visual way of deciding the number of components that materially add to the description of the data. This is usually done by looking at where the slope levels off, meaning that each successive component adds little to the description of the data. In the example, the number of components that could be retained would be between 3 and 5. The ambiguity here exemplifies the fluid nature of exploratory multivariate data analysis, frequently a truly correct answer does not exist. Other exploratory methods have similar measures for assessing rigor. A similar plot to the Scree plot, called the Elbow plot, exists for comparing the sequential increase in the number of k-means cluster for the Elbow Method centroids, in this case using the sum of the squares of the error to evaluate the effect. In k-means and hierarchical clustering, a related measure, the Gap Statistic can be calculated for each branch to determine the number of statistically distinct clusters [13]. Although hierarchical clustering will separate all samples in a dataset, there is a limit to the number of divisions that are supported by the data. To find this limit, each successive segmentation is compared for the variability between the groups versus within the groups. The point at which this balance is maximized is generally considered to be the optimal number of clusters, but again, this is subject to other considerations. The computational methods vary, but plotting the gap statistic produces a peak value (as opposed to flattening out), as shown in Figure 14.6b. Hierarchical clustering can then be revised to reflect the extent to which the data can be clustered. In the example shown in Figure 14.6b, seven clusters can be defined by the data, these clusters are grouped in the dendrogram shown in Figure 14.6c. Highlighting the clustering in this way helps to limit over-interpretation of the data.

14.3.10 Generalize the Results

When a critique of the methodology itself has been completed, the next phase is to assess whether the trends in the data suggest new relationships between the features and/or the samples. Do the features used to classify the samples make sense? Do

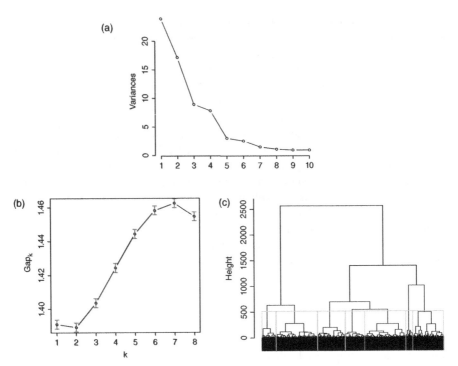

FIGURE 14.6 *Graphical methods for evaluating the strength of multivariate classifications.*
(a) A Scree plot used to evaluate the contributions of components in PCA. The amount of
variance accounted for by each principal component is plotted as a function of the number of
principal components. The appropriate number of principal components to use is indicated by
the point at which the slope of the line flattens, and the subsequent contribution of additional
components becomes minimal. (b) The GAP statistic plotted as a function of number of clusters
for data classified by hierarchical clustering. The optimal number of clusters is where the peak
in the GAP statistic is reached. (c) Using the GAP statistic to define clusters. Since plotting
the GAP statistic indicates that the data supports seven clusters, the point in the vertical scale
where seven clusters have been separated is used to draw boxes around the clusters. The fair
interpretation is that the data has been able to distinguish seven groups of samples, but that
further subclusters are not supported by the data, even though the samples have been further
subdivided by the hierarchical clustering algorithm.

they include the features that relate directly to the experimental conditions, such as
the localization, abundance, or area of the parts of the cell that should have changed?
For features that are correlated, are they redundant features that were not eliminated
earlier, or do they indicate an unappreciated biological connection (such as effect on
an organelle and the cell cycle). After reviewing the features, does the reduced dataset
help make an assessment of the samples or treatment conditions? For a PCA, does
the reduced number of components allow samples to be easily classified (such as in
a 2D or 3D plot)? Do the groupings fit experimental conditions, or do they appear

mixed or even random? This is the point where you may start to plan follow-up or confirmatory experiments. How well does an analysis facilitate this process?

This final step, generalization, is the most critical in terms of being able to take away any form of learning from a high dimensional dataset. The way to test for this is to split the dataset up into subparts, and test whether the results are still meaningful when compared across training and validation phases, either through independent datasets or through separating a single dataset into subsets. The latter approach is called cross-validation, and is a very common approach for protecting against overfitting. Simply put, for cross-validation, the dataset is split into five parts, and the unsupervised learning algorithm is trained on four-fifths of the dataset, and then tested on the remaining one-fifth. The process is repeated five times, each time on a different subset of the total data, and the error rate is averaged over the five train/test iterations. The method sounds somewhat painful, but machine-learning software (such as Weka) can implement cross-validation easily for techniques such as clustering, k-means, and principal components. While 5-, 7- and 10-fold cross-validation are common approaches, it is also fairly common to see a simple test set-training set split in publications. Splitting the dataset into two has some advantages (it is easier to implement and understand and there is no variation in the error rate). However, this comes at a price (all of the data is not used by the model, and the single estimate of the error rate that you get from this method may be significantly biased away from the true error rate). If the dataset is large enough, the distinction between test/training validation and cross-validation becomes less important, or it may even be possible to cross-validate on the training set and then use a separate (typically one-third) split of hold-out data as the test set.

However, even in cases where cross-validation returns an encouraging result, relationships identified by exploratory data analysis methods need to be verified in additional experiments, for example, to confirm the effect of a perturbation on a cellular process using additional reagents for a broader range of pathways or processes. If cells are classified into four or five groups, review images of cells from each group. Some systems are better at being able to locate the cells than others. Recall that each cell in the dataset had a unique identification for the well, field, and cell number. In a completely home-built analytical workflow, you may need to add these columns back into the dataset. Do the classifications truly identify different classes of cells in the images? Can you identify the relevant features used to make these classifications? Exploratory data analysis is alright for fun and for profit, but in the end, traditional biology experiments performed in a hypothesis-testing framework remain the gold standard!

When you have completed the assessment of the study, there are three general paths forward.

- *Retest.* Repeating the study with an expanded series of test conditions should produce the same general patterns in the data if the initial analysis was robust.
- *Validate results through follow-up experiments.* If the results suggest that a set of test conditions are linked to an interesting biological process, following the biology could be the most productive path forward. Characterizing the pathway

involved, seeing if it is recapitulated in related cell lines, and related experiments that support biological finding may be the way to go.

- *Develop a multivariate assay through machine learning.* The critical distinction between unsupervised and supervised learning is that the latter is a statistical test that is used to classify newly tested samples. Do they fit into one of the classes we have defined or not? The process behind building a statistically robust phenotypic assay is discussed in the next chapter.

14.4 CONCLUSIONS

Unsupervised learning brings powerful data analysis tools that were developed for a range of disciplines and have gained a lot of attention recently for being part of the "Big Data" movement. While truly powerful, they are still largely untested and we are not yet at the point of best practices. This is actually true for most methods of analyzing Big Data, but is particularly true for HCS, as there are currently many methods for quantifying cellular features and we are still way off from a consensus or biological framework for many of them. This not to say that UL methods should not be examined, just that the results may be speculative, and it can be a mistake to hold up a dendrogram as the demonstration of a subtle relationship between two samples.

KEY POINTS

1. The methods are readily implemented and easy to run.
2. Processing of the data is the major challenge and can take some time.
3. It is important to take advantage of robustness measures.
4. A hypothesis (even a general one) and a verification/validation plan are essential to actually learn something from the exercise.

FURTHER READING

The introduction here was kept brief, to help a scientist gain an essential appreciation of the breadth of methods and technical issues that are part of unsupervised learning. What happens next is largely driven by need. If good statistical and bioinformatics support is available, then extending ones understanding of these methods (but still at a high level) can help in planning a study. One example would be *Data Mining: Concepts and Techniques (3rd ed.)* by Han, Kamber, and Pei, particularly Chapter 10. Note that even in a high level introduction such as this one, comfort and familiarity with mathematical functions and the programming process are very helpful and could even be considered prerequisites. The computational process that is needed to produce a dendrogram or scatterplot should not be considered as a black box. For more detailed discussions that can be challenging but rigorous, *R and Data Mining* by Yanchang Zhao and *The Elements of Statistical Learning*, by Hastie, Tibshirami, and Friedman, are excellent references (most of these were cited in the text explicitly). Finally, Big Data and R are parts of a very active interdisciplinary community of researchers, programmers, statisticians, and other experts who discuss these methods and share best practices through the web. Tapping into these communities should be a frequent event for anyone who wishes to learn about the emerging best practices and streamlined methods that are emerging.

REFERENCES

1. Tan, P.-N., Steinback, M., and Kumar, V. *Introduction to Data Mining*, Addison-Wesley, 2005.

2. Han, J., Kamber, M., and Pei, J. *Data Mining: Concepts and Techniques*, 3rd edn, Morgan Kaufmann, 2011.

3. Kohonen, T. Image-based assessment of growth and signaling changes in cancer cells mediated by direct cell-cell contact. *Applied Optics*, 1987, **26**: 4910–4918.

4. Tamayo, P. et al. Interpreting patterns of gene expression with self-organizing maps: methods and application to hematopoietic differentiation. *Proceedings of the National Academy of Sciences*, 1999, **96**: 2907–2912.

5. Eisen, M.B. et al. Cluster analysis and display of genome-wide expression patterns. *Proceedings of the National Academy of Sciences*, 1998, **95**: 14863–14868.

6. Hastie, T., Tibshirani, R., and Friedman, J. *The Elements of Statistical Learning: Data Mining, Inference, and Prediction*, 2nd edn. Springer-Verlag, 2008, 763 p.

7. Ester, M. et al. *A density-based algorithm for discovering clusters in large spatial databases with noise.* Proceedings of the Second International Conference on Knowledge Discovery and Data Mining, 1996, pp. 226–231.

8. Slack, M.D. et al. Characterizing heterogeneous cellular responses to perturbations. *Proceedings of the National Academy of Sciences*, 2008, **105**: 19306–19311.

9. Perlman, Z.E. et al. Multidimensional drug profiling by automated microscopy. *Science*, 2004, **306**: 1194–1198.

10. Jackson, A.L. et al. Expression profiling reveals off-target gene regulation by RNAi. *Nature Biotechnology*, 2003, **21**: 635–637.

11. Barteneva, N.S., Fasler-Kan, E., and Vorobjev, I.A. Imaging flow cytometry: coping with heterogeneity in biological systems. *Journal of Histochemistry & Cytochemistry*, 2012, **60**: 723–733.

12. Lapan, P. et al., Image-based assessment of growth and signaling changes in cancer cells mediated by direct cell-cell contact. *PlosOne*, 2009, **4**: e6822.

13. Tibshirani, R., Walther, G., and Hastie, T. Estimating the number of clusters in a data set via the gap statistic. *J. R. Statist. Soc. B*, 2001, **63**: 411–423.

15

SUPERVISED MACHINE LEARNING

Jeff Palmer and Arijit Chakravarty

15.1 INTRODUCTION

In many high content screens, the images collected from automated microscopy platforms are typically analyzed by breaking the image down into individual cell-level measurements. These cell-level measurements are then further processed as ratios or differences to classify the cells into various phenotypic classes or outcomes. This works when a simple rule of thumb can be used to derive a metric that behaves well, and we discussed many of the rules for generating such metrics in Chapter 8. However, there may be situations when simple metrics are insufficient—if the desired phenotypic outcome is complex, or if there are several different closely related outcomes. In these settings, a more robust approach may be beneficial. Machine learning algorithms, the focus of this chapter, provide a solution for these types of problems.

Machine learning algorithms emulate human learning, reasoning, and decision making, using techniques borrowed from artificial intelligence, data mining, and statistics. There are two classes of machine learning algorithms—supervised and unsupervised. In the previous chapter, we discussed unsupervised learning algorithms, where the question was "how many different classes of outcome or cellular phenotype are there in this image?" On the other hand, supervised learning algorithms are best suited for questions of the general form of "how many cells of class X and how many cells of class Y are there in this image?" A popular example of supervised learning outside the field of high content screening is the classification of handwritten digits in an image for automated sorting of postal zip codes. In this example, the input is the pixel intensities in the image, and the output is the classification of a number

An Introduction to High Content Screening: Imaging Technology, Assay Development, and Data Analysis in Biology and Drug Discovery, First Edition. Edited by Steven A. Haney, Douglas Bowman, and Arijit Chakravarty.
© 2015 John Wiley & Sons, Inc. Published 2015 by John Wiley & Sons, Inc.

(0, 1, etc.). To construct a successful classification algorithm, one would need to train the algorithm using a set of images where the true digit classification is known. Given a new input image, the algorithm could then be used to identify the number in the image with a certain degree of certainty (or probability). It is pretty easy to see that, in concept, very similar approaches could be used to teach a supervised machine learning algorithm to differentiate between different cell types in an image, for example.

The goal of the chapter is not to teach the implementation of machine learning algorithms, but to clue the reader in to what these algorithms are, what they are capable of, and when the situation may call for their use.

15.2   FOUNDATIONAL CONCEPTS

For the majority of decision-making problems in machine learning, we must construct a model (or algorithm) that adequately captures and processes the required input and, in the case of supervised learning, also provides robust (accurate and precise) predictions of the output. As we will soon see, there are many such algorithms that are suitable for any one particular task. Given the wide variety of algorithm choices, we will discuss in a later section how best to choose from among these options. (By the way, accuracy and precision sound like about the same thing, but they are actually quite different! Accuracy refers to the lack of bias in a set of measurements or predictions, while precision refers to the lack of variance. See Figure 15.1 for a more intuitive explanation of these terms.)

The input for an algorithm may be regarded as a set of features, for example image-processing-derived measurements of cellular morphology. Take for example a mitotic index assay, where after preprocessing an input image (nuclei detection and segmentation, erosion, dilation, etc.), we have recorded information such as average stain intensity and area of each nuclei detected in the image (see Figure 15.2). Now we have a list of features associated with each cell that we have made measurements for. This list of features (in math terms, a vector of random variables) may also be referred to as the feature space.

In real-life machine learning problems, very often the action starts here, as one of the first steps will often be "feature extraction" (dimensionality reduction), where a long list of potential input features is whittled down into a smaller number of features, based on their ability to effectively and efficiently represent the data and also have strong predictive capabilities. Ideally, the features in this shorter list are also uncorrelated with each other. In our mitotic index assay example, these features may be derived quantities such as area and average intensity of nuclei identified in an image. This was discussed in the previous chapter. As you recall, it may not always be obvious that two features may be highly correlated, so it is best to test these potential relationships directly through a correlation matrix or scatter plots.

Now, the next step becomes tying this shortlist of measured "input" values for each individual cell to a predetermined "outcome" measurement (in our mitotic index example, and no prizes for guessing) this measurement is actually the classification of

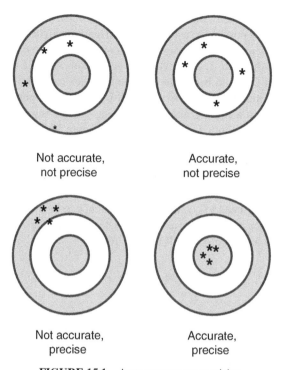

FIGURE 15.1   *Accuracy versus precision.*

each individual cell as mitotic or nonmitotic. To train a machine learning algorithm, these measurements would typically have to be made by a human operator, who scores each cell by hand. The training process will then optimize the algorithm so that it can recapitulate the judgment (for what it is worth) of the human operator in scoring each cell in the same way as the human operator did, using only the measurements that are provided to it as inputs. In the mitotic index assay example, there is only one

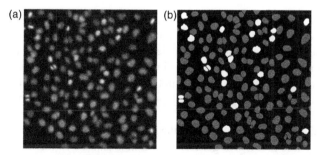

FIGURE 15.2   *Mitotic index assay example.* (a) Original image of nuclei. (b) Identified nuclei after image processing. Bright nuclei signify mitotic nuclei.

outcome variable, and it takes one of the two possible values (mitotic or nonmitotic). However, it is entirely possible to use a supervised learning algorithm to predict more than one outcome simultaneously, and the outcome variable(s) can take more than one value (such as digits: 0, 1, 2) or even be a continuous variable in its own right.

So, for now, let us zoom back into the mitotic index example. A sample image from a typical mitotic index assay is shown in Figure 15.2. Figure 15.2a shows an image of nuclei from a mitotic assay experiment. Nuclei have been classified as mitotic (bright), or nonmitotic (grey), as shown in Figure 15.2b. Now, we will try to build an automated supervised learning algorithm that will be able to take as input a set of metrics derived from images like the one in Figure 15.2a, and be able to predict with a high level of accuracy and precision whether each nuclei is mitotic or nonmitotic.

To do this, we will need to construct a model that can be used to learn from our existing data, and then be applied to new future data to predict the outcome of interest. In its most general form, a model will take the form:

$$Y = f(X, \theta) + \varepsilon.$$

Here, the term $f(X, \theta)$ represents a function that maps the input X to the output Y, and will capture the deterministic component of the relationship between X and Y. The θ term represents a set of unknown parameters associated with this function that need to be estimated (learned). The quantity ε represents a random variable that is independent from the input X and has an expected value of 0. This quantity is meant to capture the random variation and nondeterministic nature of the data and is often called the error term.

So, in this case, let us pretend for the moment, that a model (or machine learning algorithm) can be built that takes as its inputs, say the area and average intensity of individual nuclei, and produces as its output the classification of individual cells as mitotic or nonmitotic.

15.3   CHOOSING A MACHINE LEARNING ALGORITHM

Before we get too far into the details of machine learning algorithms, let us talk first about how we decide how well a machine learning algorithm is performing. This part is actually crucial, as many machine learning algorithms are now available right out of the box, either online (or as downloadable graphical user-interface programs like Weka, which is a favorite of many in the machine learning community). So it is pretty easy to run a machine learning algorithm on a dataset, but surprisingly tricky to know if it was done right.

Why is this? The answer is simple—the better that a model fits a particular dataset, the less likely it is that it will generalize to a new dataset. In the machine learning field, this is known as the bias-variance tradeoff. Bias is the inverse of accuracy, and variance is the inverse of precision, so this is another way of saying that models that maximize accuracy also reduce precision, and vice versa. A model that has been overly well fitted to a given dataset will perform poorly on the next dataset—it has been *overfitted*. In the next section, we will explore this concept in practical terms looking at a couple of common machine learning algorithms.

For now, let us discuss how to assess model fit (accuracy) as well as model generalizability (precision). For model fit, the commonest performance metric is the *error rate*, which as the name implies, is a metric that assesses the prediction error (performance) of a given algorithm. For a discrete outcome in a classification problem, this rate would simply be estimated by the proportion of misclassifications in a test sample where the outcome is known. For more complex outcomes, it would be estimated as some quantified difference/distance between the predicted outcome and true outcome in some test sample (or within the training sample).

For model generalizability, things are a little trickier. The most straightforward way to assess this, and thereby guard against overfitting, is to employ some form of empirical *cross-validation*. In K-fold cross-validation (KF-CV), suppose we have a sufficiently large set of training data that we partition into K equally sized groups. Then we randomly choose one of these K groups to serve as a test set, and train our model on the data in the remaining K-1 groups. We then record the misclassification rate of this fitted model in the test set. We repeat this exercise for each of the K groups and estimate the overall error rate as the average of the misclassification rates computed for each of the K test sets. An often used variant of this approach, termed leave-one-out cross-validation (LOO-CV), is when one observation serves as the test set. Either way, cross-validation provides a quick way to estimate how well a machine learning algorithm will perform on a dataset that it was not fitted to. Another common technique for evaluating model generalizability is bootstrapping, which is discussed in the section on statistical modeling (in Appendix C). The details of bootstrapping are a little different, but in the context of machine learning, it offers many of the same benefits as cross-validation.

In addition to considering future model performance in terms of prediction characteristics, a straightforward assessment of model adequacy and fit is the error or rate of misclassification within the full set of training data itself (training error). This can often quickly lead us to a determination of whether a candidate algorithm is worthwhile to pursue, or better yet, whether or not our data and the way we have constructed our feature space carries enough information to have viable predictive power. Relating back to the discussion on feature space, it is not uncommon in a machine learning application to allow features to be selected based on some assessment of model fit or error rate estimate. This approach is termed feature selection (or variable selection), and can be quite useful in determining which features have strong predictive capabilities (are correlated with the outcome of interest), and which should be considered for inclusion in the model specification.

In addition to choosing a type of algorithm to apply to your data, most machine learning algorithms also have a complexity parameter whose value must be specified by the user. In the next section we will review one of the simplest machine algorithms in common use, the k-nearest neighbors (k-NN) algorithm, in which a prediction for a new data point is the average outcome of the k observations in the training set that have the "closest" feature values. Here, while there are no parameters to be estimated from the training data, we must a priori specify the complexity parameter k. The complexity parameter is often closely related to the notions of accuracy (bias) and precision (variance) of a prediction. Generally speaking, as the model complexity increases, the prediction accuracy also increases while precision typically decreases. As noted earlier, an optimal algorithm will have a high level of accuracy and precision, so it is often helpful to consider this bias-variance tradeoff when selecting a model.

15.3.1 Two Common Supervised Learning Algorithms

With the basic foundational underpinnings of supervised learning now at our disposal, let us turn to a few specific algorithms that can be directly utilized in the context of an HCA experiment.

In the previous section we introduced the conceptually simple k-NN algorithm. This algorithm is nonparametric, in that there are no parameters to be estimated via training. Instead, predictions for new input data (features) are derived directly from data that exists in the training set. An example will help make this more clear, so let us revisit the mitotic index assay example introduced in the previous section. Suppose our feature data (area and intensity) from a training set are as displayed by the points in Figure 15.3. The lighter dots in this figure represent nonmitotic cells and the darker dots represent mitotic cells. From this figure, it is clear that there is good separation between the groups based on the feature values (mitotic cells are more likely to have small area and high intensity), so this data would be a good candidate for constructing a learning algorithm.

Suppose we arbitrarily let $k = 3$ and apply a 3-nearest neighbors (3-NN) algorithm to this training data. There is no estimation taking place here, but based on this algorithm if we were to take any random new cell, our prediction (mitotic or not) would be determined by the squiggly decision boundary within Figure 15.3. Any

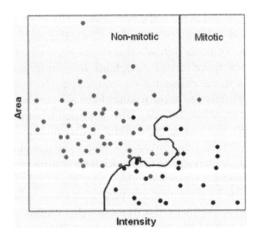

FIGURE 15.3 *Mitotic index assay example.* Solid curve represents 3-NN decision boundary.

new point to the right of the curve would be classified as mitotic and any new point to the left would be nonmitotic. We can easily compute the training error of this algorithm as 5.6% (percentage of training points incorrectly classified), which is very reasonable suggesting first that we can successfully build a learning algorithm for this problem. Other values of k would yield different training error rates. For $k = 1$, the training error is 0%, as every training point essentially "votes" for itself. As k increases though, the training error will typically also increase. Training error does not address the predictive power of an algorithm. In the next section we will look at test error in an effort to determine an optimal choice for the complexity parameter k.

The parameter k may also be viewed as a smoothing parameter, where larger values of k would yield smoother decision boundaries. For $k = 1$, the decision boundary is typically very jagged and will not perform very well in terms of prediction performance. At this point, we can say that a 1-NN algorithm is overfitted. Moving up to 2- or 3-NN algorithms, the training error would be expected to increase, but the test error (under cross-validation) would be expected to decrease. (This is the bias-variance tradeoff in action. More about this later.) Despite its simplicity, the k-NN algorithm has proven to be very effective in many machine learning applications. It is also attractive for its lack of reliance on often unverifiable model assumptions that are associated with more complex models. Finally, this algorithm is not restricted to classification problems, it can work equally well in the context of a continuous outcome.

Now, let us move on to a different machine learning algorithm, one that is in fact so simple that it might come as a surprise to hear it mentioned in this context. This is the parametric logistic regression model. A logistic regression model is basically a very steep "dose-response curve" that switches immediately from 0 to 1. (See Appendix C for a more detailed treatment.) The parameters of a logistic regression model are commonly estimated from training data using maximum likelihood, and most data mining and statistical software applications have built-in functions for doing this. For the same mitotic index example, the decision boundary for a logistic model that includes both area and intensity as features is displayed as a solid line in Figure 15.4.

The training error for this model is 11.3%, which is slightly higher than that for the 3-NN algorithm. A complexity parameter for logistic regression (or most regression algorithms for that matter) can be thought of as p, the number of features used in the model (with more complex models having higher p). Suppose we instead fit the logistic regression model to only one of the two features in this example. The decision boundaries for these two models are displayed in Figure 15.4 as vertical and horizontal dashed lines (vertical line for the model with intensity only). As we reduce the complexity of the model, the training error is significantly increased (to 18.3% for both models). In terms of feature selection, these results preliminarily indicate that an optimal model should include both features, but we will confirm this when we look at test error in the next section (note that there may even exist additional unobserved features that would decrease the training error, and possibly the test error, even further).

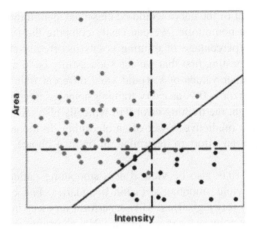

FIGURE 15.4   *Mitotic index assay example.* Logistic regression boundaries for model including both features (solid line) and models including only a single feature (dashed lines).

We briefly mentioned earlier that training error can be useful for determining whether it is feasible to apply a machine learning algorithm to your problem, or even to assist in feature selection for a particular model. As mentioned previously, at the heart of machine learning is the bias-variance tradeoff, so we will look to balance the *training error* against the *cross-validation error* (variance or lack of generalizability).

Now, let us examine the two models/algorithms introduced in the previous section and assess their performance based on various error rates in a test set. Figure 15.5 shows the training error, fivefold cross-validation error, and LOO-CV error both the k-NN algorithm (Figure 15.5a) and the logistic regression model (Figure 15.5b). For k-NN, the error rates are reported for each k from 1 to 20. For the logistic regression figure, error rates are reported for p = 2 (intercept and intensity) and p = 3 (intercept, intensity, and area). The error rates for the logistic regression model with intercept and area were similar to those for the intercept and intensity model.

For this dataset, the error rates for the k-NN algorithm are consistently lower than those for the logistic regression model. In terms of choosing the complexity parameter *k* for the k-NN algorithm, we will balance the training error against the cross-validation error. After k = 3, the cross-validation error rates appear to stabilize, so a reasonable choice would be k = 3, the "arbitrary" choice from the previous section.

15.3.2   More Supervised Learning Algorithms

In the previous section we introduced two of the simplest supervised learning algorithms; k-NN and logistic regression. Now, let us explore the extensions of these algorithms, to more powerful approaches that build on their basic principles. (Once again, the focus in this section is on explaining the concepts behind common machine learning algorithms. The interested reader would benefit from a more detailed treatment provided in Machine Learning by Ian Witten, a book that accompanies Weka,

Error rates for kNN algorithm

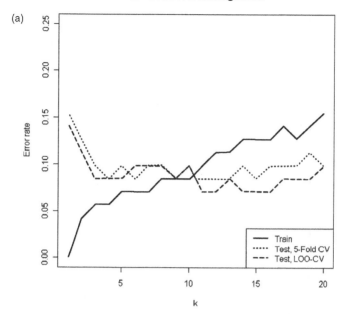

(a)

Error rates for logistic regression

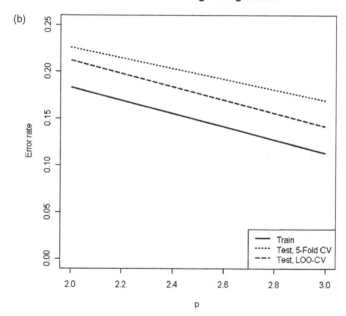

(b)

FIGURE 15.5 *(a) Error rates for k-NN algorithm. (b) Error rates for logistic regression.*

a free graphically driven machine learning software package. All of the algorithms described here are available as part of pull-down menus in Weka).

Logistic regression is perhaps the most common classification algorithm used in machine learning. However, for situations where the outcome is continuous (regression problem), there are several additional algorithms that we may consider. The simplest method (closely related to logistic regression) is standard linear regression (see Appendix C for a more detailed treatment). The application of linear regression typically requires that the outcome data (as well as the features) satisfy some minimal distributional assumptions (e.g., that the outcome variable of interest is normally distributed). The most common approach to training a linear regression model is to employ the method of least squares, where the parameters are estimated by minimizing the sum of the squared distances between the true outcomes and the predicted outcomes on some training set (residual sum of squares, or RSS). For regression problems, we can no longer characterize their performance in terms of a misclassification rate. Training error is quantified using the quantity RSS, and likewise, test error can be quantified by looking at the same distance metric (sum of squared distances) in a test set. As a slight variant on this approach, the fitting of a linear regression model can also be done by using the least median squares approach, which is more robust to outliers.

The k-NN algorithm is conceptually related to a larger class of classification algorithms called prototype methods. With prototype algorithms, a set of prototype observations (not necessarily from a training set) are constructed in the feature space and labeled with a desired classification. Prediction would then entail finding the closest (Euclidean distance) prototype to a new candidate input observation and labeling the candidate with the outcome (classification) of the closest prototype. As a cell enters mitosis, the DNA condenses, leading to a smaller, but brighter chromosomal DNA stain for mitotic cells as opposed to the signal from uncondensed interphase nuclei. Thus, the simplest prototype method for the mitotic index assay example would be to choose two prototype points, one for each outcome (nonmitotic or mitotic). A prototype point for nonmitotic could be (intensity = low, area = high), and for mitotic, (intensity = high, area = low). A prediction for a new input observation (cell intensity and area) would then be the label of the closer of the two prototypes. This concept can be extended to a situation where there are multiple prototypes per outcome/class. Several methods have been proposed to assist in choosing these representative prototypes, one of which is k-means clustering, where observations from the training set are clustered together based on some similarity metric. The prototype points would typically be chosen as the "centers" of the derived clusters. We can now see that the 1-NN classifier is an extreme case of a prototype classifier, where every observation in the training set in fact serves as a prototype. Perhaps unsurprisingly, this is also why the 1-NN classifier performs poorly under cross-validation, as this algorithm does not really "learn" anything, it simply treats each new data point as a brand new prototype (kind of like an amnesiac version of machine learning, if you will).

Logistic regression and the very closely related method of linear discriminant analysis (LDA) are often considered to be adequate for solving a large array of classification problems. As the name implies, LDA (and logistic regression) assumes that the boundaries among classes or outcomes are linear (refer to Figure 15.4). As we saw previously for the mitotic index assay example, the optimal model (3-NN) among those considered is in fact not linear. Several extensions of LDA exist that overcome this issue and allow us to model the input data with nonlinear, or less smooth, decision boundaries. In some cases there may be too many input variables (or equivalently, the number of dimension of the feature space may be too high) for traditional LDA classification methods to work efficiently. Certain extensions of LDA can be used to reduce the number of input variables by effectively eliminating those that have little relationship with the outcome variable, and giving more weight to those features that have more powerful predictive characteristics.

As an alternative to reducing the number of input features based on importance, another class of supervised learning algorithms called support vector machines (SVM) implicitly increases the number of input features by calculating new derived features. At its heart, an SVM algorithm will automatically determine a set of nonlinear transformations and combinations of the input data (feature space) that yield a good linear fit. With respect to the original feature space, the best SVM model fit will be able to capture nonlinear trends in the data, and in such cases will often outperform the best alternative linear classifier. Now this might sound like an invitation to overfitting, as the initial set of features has now spawned a whole new set of derived features based on transformations and combinations of the input data. To get around this, SVM algorithms implicitly penalize (or regularize) overly complex models to ensure parsimony.

Let us now turn to a different class of algorithms called classification and regression trees (CART). Tree-based methods in supervised learning are so called due to the ability to represent a model fit with a hierarchical decision tree. A tree is built by partitioning the feature space one input feature at a time based on some decision criterion. Refer to Figure 15.6a below for a classification tree fit to the training data from the mitotic index assay example.

The first (best overall) split of the data (into "intensity high" and "intensity low") is represented by the vertical reference line. The next best split is for "intensity low," where the split is made between "area high" and "area low." This split creates the regions **R1** and **R3** labeled in the figure. The final split is for "intensity high," where the feature space is split into "area high" (region **R2**) and "area low" (region **R4**). Regions **R1** and **R2** are classified as nonmitotic, and regions **R3** and **R4** are classified as mitotic. An alternative decision tree representation of this same model fit is displayed in Figure 15.6b. For the prediction of a new input observation using a decision tree, one would start at the top of the tree and follow the heuristic rules following each node of the tree until you reach a terminal node (diamond in Figure 15.6) at the bottom. Prediction using the method in Figure 15.6a would simply entail finding the region the candidate observation belongs to, and assigning to it the label associated with that region.

For this example, the training error is 8.4%, which lies between that for the 3-NN classifier and logistic regression algorithms that we explored earlier. For a continuous outcome (regression trees), the same logic described above would apply,

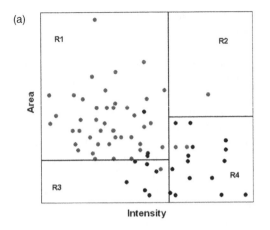

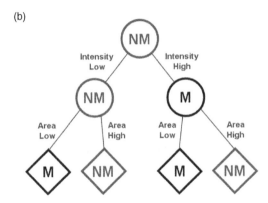

FIGURE 15.6   *Mitotic index assay example.* (a) Decision boundaries for a classification tree. (b) Hierarchical decision tree associated with the same model fit. M = Mitotic, NM = Nonmitotic.

however different cost criteria would need to be used to find the optimal split points. CART have gained significant popularity over the last decade or so due primarily to their interpretability. In this space, a method that is very common these days, and that often performs well on most datasets, is Random Forest, which consists of a randomly generated collection of classification trees. One drawback of these approaches, however, is that the model fits are necessarily linear and may not be adequate for input data with subtle nonlinearities.

Neural networks are another class of powerful supervised learning algorithms that have received plenty of attention over the years, and are often considered to be one of the foundational tools used in artificial intelligence. Indeed, the methodology has been named after its original intent of trying to model the complex neuron–synapse relationship in the brain. Similar in flavor to the SVM, neural networks introduce a new (unobserved) set of variables that are derived from the input features. The relationship between these derived features and the outcome or response is then modeled. Like the SVM, the form of these derived features are adaptively determined by learning from the data in a training set. Unlike the SVM however, the number of derived features is fixed in advance. There are other technical differences between these two approaches which are beyond the scope of this book. Interestingly, when you write out the equations, standard linear regression and logistic regression are actually special cases of neural network algorithms! A closely related network-based class of supervised learning algorithms are Bayes Nets, which differ subtly from Neural Nets in the way their weights are calculated during the training process.

15.4   WHEN DO YOU NEED MACHINE LEARNING, AND HOW DO YOU USE IT?

The take-home from this chapter is simple; when you are trying to construct a metric where from many different input features (or with many different possibilities for how to combine the input features), it is possible to overfit the metric to the test dataset. This overfitting will result in the metric performing well on the test dataset but poorly on subsequent datasets. Supervised machine learning approaches guard against this kind of overfitting with cross-validation, where a subset is used for training, and a different subset is held out as a "test set."

The key with machine learning algorithms, then, is to recognize when your metric-based approach may actually be overfitting the data, and when to engage a formal machine learning algorithm. Although there is a really wide range of machine learning algorithms, most practitioners of machine learning have one or two of their own favorites depending on the class of the problem. For example, for most problems in high content analysis, the outcome is typically categorical and the input is continuous. For this category of machine learning problems, Logistic Regression, SVMs, and Random Forests are popular (although some people also use Bayes Nets or Neural Nets).

For other classes of problems, the options are a bit different. If your output should happen to be continuous (with continuous inputs), Linear Regression may be the best option, although SVMs, Bayes Nets, and Neural Nets will work as well. If your inputs are categorical, typically tree-based methods are used (such as Random Forests).

Pick one or two of your favorite methods and stick to them—as you read further you will find that there are nuances to each method, and the better you get to know the method, the more tuned in you will be to these nuances. Although there is a large body of literature comparing different machine learning methods for certain classes of problems, there are no systematic surveys in the high content space, and in fact the answer is likely to be different for each particular subclass of problem. There is a theorem in machine learning called the No Free Lunch theorem, which (loosely paraphrased) states that for every machine learning algorithm, there exists a dataset for which it will outperform all other algorithms. So the point being, pick one or two algorithms and stick to them—the good news is that performance differences in machine learning algorithms are usually small. Once you have identified your machine learning algorithm (or two) of choice, filter your dataset to reduce the number of correlated features, and apply the algorithm, using appropriate cross-validation techniques to guard against overfitting.

Over the last 5 years, machine learning approaches have been applied in a variety of contexts for high content analysis, for instance with phenotypic analysis of siRNA screens. A number of software packages exist for implementing such machine learning screens, starting from Weka (which is the tool of choice for many in the machine learning community but which does not integrate directly with images) to tools that are optimized for high content analysis (such as Enhanced Cell Classifier, Cell Profiler, Cell Organizer, and Pattern Unmixer, all of which are freeware, or off-the-shelf commercial options such as Perkin Elmer's PhenoLogic plug-in). One can expect the range of options to keep growing in this space!

Hopefully after reading this chapter, you have a sense of the broad outlines of what machine learning is and what it can do for you. If you are interested in developing your own expertise in the subject further, your next step may be to download Weka and to experiment with the datasets that come with it, reading the accompanying book (Data Mining by Witten, Frank, and Hall) in parallel. A deeper exposition of the concepts behind machine learning can be found in the Hastie and Tibshirani book. Alternatively, you may know someone whose day job is machine learning or data mining! In this case, hopefully, this chapter provides a basis for a shared understanding of how to apply machine learning to your high content problems.

15.5   CONCLUSIONS

Machine learning, or supervised learning, are a set of approaches that add the capacity to discriminate between clusters in a statistically defined algorithm. It is distinct from unsupervised learning in that it is prospective, having the ability to apply the algorithm to new samples rather than describe patterns in the original dataset.

KEY POINTS

1. Always use supervised learning approaches when building "rulesets" from Big Data.
2. Cross-validate (many options exist, and they are all reasonable) to guard against overfitting.
3. Find a few algorithms that work best for you (such as Bayes Nets, Random Forest, Logistic Regression, or Neural Nets) and stick to them.
4. Do not get bogged down in choosing an algorithm!

FURTHER READING

Witten, I., Frank, E., and Hall, M. *Data Mining: Practical Machine Learning Tools and Techniques*, 3rd edn. Morgan Kaufmann, 2011.

Hastie, T., Tibshirani, R., and Friedman, J. *The Elements of Statistical Learning: Data Mining, Inference, and Prediction*, 2nd edn. Springer Series in Statistics, 2009.

Jones, T.R.et al. Scoring diverse cellular morphologies in image-based screens with iterative feedback and machine learning. *Proceedings of the National Academy of Sciences of the United States of America*, 2009, **106**(6): 1826–1831.

Zhang, B. and Pham, T.D. Phenotype recognition with combined features and random subspace classifier ensemble. *BMC Bioinformatics*, 2011, **12**: 128.

Misselwitz B. et al. Enhanced CellClassifier: a multi-class classification tool for microscopy images. *BMC Bioinformatics*, 2010, **11**:30.

Eliceiri, K.W. et al. Biological imaging software tools. *Nature Methods*, 2012, **9**:697–710.

Shariff, A. et al. Automated image analysis for high-content screening and analysis. *Journal of Biomolecular Screening*, 2010, **15**(7):726–734.

APPENDIX A

WEBSITES AND ADDITIONAL INFORMATION ON INSTRUMENTS, REAGENTS, AND INSTRUCTION

STEVEN A. HANEY, DOUGLAS BOWMAN, ARIJIT CHAKRAVARTY, ANTHONY DAVIES, AND CAROLINE SHAMU

General microscopy information and demonstrations		
Nikon Microscopes	http://www.microscopyu.com/index.html	General microscopy tutorials, including fluorescence, brightfield and live-cell imaging
	http://micro.magnet.fsu.edu/primer/index.html	Descriptions of general microscopy methods
Olympus Microscopes	http://www.olympusmicro.com/index.html	Descriptions and examples of general microscopy methods
Instrument vendors		
Thermo/ Cellomics	http://www.thermoscientific.com/ecomm/servlet/productscatalog_11152_L10629_82095_-1_4	Cellomics ArrayScan and CellInsight imagers
GE	http://www.gelifesciences.com/webapp/wcs/stores/servlet/catalog/en/GELifeSciences-us/products/AlternativeProductStructure_12997/	InCell 2000 and InCell 6000 imagers

(continued)

An Introduction to High Content Screening: Imaging Technology, Assay Development, and Data Analysis in Biology and Drug Discovery, First Edition. Edited by Steven A. Haney, Douglas Bowman, and Arijit Chakravarty.
© 2015 John Wiley & Sons, Inc. Published 2015 by John Wiley & Sons, Inc.

Molecular Devices	http://www.moleculardevices.com/ Products/Instruments/ High-Content- Screening.html	ImageXpress Micro and Ultra imagers
Perkin Elmer	http://www.perkinelmer.com/ Catalog/Category/ID/ High%20Content% 20Screening%20Systems	Opera, Operetta imagers
Yokogawa	http://www.yokogawa.com/ scanner/index.htm	Cell Voyager imagers
TTP Labtech	http://ttplabtech.com/ cell-imaging/acumen/	Acumen explorer, line scanning cytometer

Image analysis

Perkin Elmer	http://www.perkinelmer.com/ pages/020/cellularimaging/ products/acapella.xhtml	Acapella is obtained through Perkin Elmer and is generally used with the Opera series of imagers
Molecular Devices	http://www.moleculardevices.com/ Products/Software/High- Content-Analysis.html	MetaXpress image analysis and AcuityXpress data processing software
Thermo/Cellomics	http://www.thermoscientific.com/ ecomm/servlet/ productscatalog_11152_ L10623_82094_-1_4	Cellomics BioApplications image analysis algorithms
Broad Institute	http://www.cellprofiler.org/	CellProfiler image analysis and CellProfiler Analyst data analysis open-source software
LOCI (Univ Wisconsin), Max Plank (Dresden, DE), others	http://fiji.sc/	FIJI image analysis and related applications including BioFormats, KNIME and other open source tools

Fluorescence optimization applets

Life Technolo- gies/Molecular Probes	http://www.invitrogen.com/ site/us/en/home/ support/Research- Tools/Fluorescence- SpectraViewer.html	
Semrock Filters	http://searchlight.semrock.com/	

APPENDIX B

A FEW WORDS ABOUT ONE LETTER: USING R TO QUICKLY ANALYZE HCS DATA

STEVEN A. HANEY

B.1 INTRODUCTION

The R statistical programming software was introduced in the text as an alternative to commercial data analysis packages. There are several reasons for considering using it for general data analysis and visualization.

- It is free and open source. R can be downloaded and started in just a few minutes from the R-project website (www.r-project.org). It runs on Windows, Macintosh, and Linux systems. While we work with commercial software routinely, working in an open-source environment ensures transparency, which is essential to advancing the science of HCS and makes for a better learning experience.

- Despite being a command-line language (you type commands at a prompt), it has an enormous number of built-in routines for analyzing and visualizing data. In literally three commands, you can be looking at a Q–Q plot of samples in your experiment.

- It is the language of choice for bioinformaticians. Chances are, your data will be analyzed in R by someone else. Having a working knowledge of R is a fantastic way to communicate more directly with a data analyst, and can take some of the work off of him or her, especially the general "can you tell me how the data looks" questions in deciding whether you will need to rerun the experiment.

An Introduction to High Content Screening: Imaging Technology, Assay Development, and Data Analysis in Biology and Drug Discovery, First Edition. Edited by Steven A. Haney, Douglas Bowman, and Arijit Chakravarty.
© 2015 John Wiley & Sons, Inc. Published 2015 by John Wiley & Sons, Inc.

- Many packages for analyzing bioinformatic data exist and are used to analyze large-scale datasets. One important family of these is the Bioconductor set of data analysis packages (www.bioconductor.org), and include packages for the analysis of transcriptional profiling, metabolomics, RNAi screens, and image-based experiments. Just because you are not using R now does not mean that you will not be doing so in the future.

So, despite many reasons for using R yourself, they must be weighed against the barrier of actually downloading it, using it, and getting something in return for doing so before the time you can afford to spare on such an effort expires. The aim of this appendix it to guide through few of the things that R comes prepared to do, and to do so with only a few commands.

B.2   SETTING UP R

B.2.1   Locate the R Website and Installation Server

There are a collection of servers around the world that maintain the current versions of R for each platform (Microsoft Windows, Macintosh OS, and Linux). As noted above, the website is www.r-project.org. Following the link for downloading R will take you to a page of server addresses, or mirrors, that comprise the Comprehensive R Archive Network, or CRAN. There are over 80 mirrors to choose from, picking one takes you to a page that asks for the computer platform you will be working under (common sense would suggest that you select a server that is close to you, but do not let us fence you in). If this is all new to you, you probably want the precompiled binary (ready to use) versions written for the Mac OS or Windows.

B.2.2   Download R and Install

Clicking on the link for Windows or Macintosh will start the transfer of the base version of R, and an additional number of packages. Packages are supplemental software routines contributed by the R community, and are tailored to specific areas of research, including Psychology, Epidemiology, and Astrophysics. A number of the most useful packages have been incorporated into the basic installation of R. Additional packages can be installed from the mirror servers. For these examples, we will only work with the base form of R and a couple of routines from packages that are bundled with the basic installation, so installing the "base" package is all that you will need. If you decide to move on to Bioconductor, you will download those packages yourself, and as you gain more familiarity with R, you may want to work with packages that include specific statistical tests or extend the graphics capabilities. Although R is a command-line language, some standard file operations with traditional menus (Save, Print, etc.) are available. Installing packages is easily accomplished through the menu. In addition, there are some interfaces available that add additional menu operations and can help organize work in R. A popular example

```
R version 2.14.1 (2011-12-22)
Copyright (C) 2011 The R Foundation for Statistical Computing
ISBN 3-900051-07-0
Platform: x86_64-pc-mingw32/x64 (64-bit)

R is free software and comes with ABSOLUTELY NO WARRANTY.
You are welcome to redistribute it under certain conditions.
Type 'license()' or 'licence()' for distribution details.

  Natural language support but running in an English locale

R is a collaborative project with many contributors.
Type 'contributors()' for more information and
'citation()' on how to cite R or R packages in publications.

Type 'demo()' for some demos, 'help()' for on-line help, or
'help.start()' for an HTML browser interface to help.
Type 'q()' to quit R.

>  |
```

FIGURE B.1 *R window.* The console space for entering commands and review tabular output in R. Commands are entered at the prompt at the bottom.

is RStudio that is also freely available and open source (www.rstudio.com). We will work with the basic R interface with these examples, but anyone wishing to continue with R will want to look into these options.

B.2.3 Prepare the Data File

The cell level data file will be a bit unwieldy, as we discussed in Chapters 8 and 10. It will have many columns and very many rows. Image analysis algorithms are proactive, they will capture data that may be useful for some experiments, although in most cases, the additional feature data will not be used. If we know that we will be only using one or two features as metrics, we can eliminate all other columns and save the table with a new name. Identifying and eliminating redundant and highly correlated columns was discussed in Chapter 14. These can be done to reduce the size of the data file, and it will need to be saved as a .csv file in the working directory for R. In cases where R is being used to evaluate feature columns for potential redundancies, they can be retained in the file and the file can be loaded into R. For the examples described here, the data is saved to a folder called "rdata" on the C: drive.

B.2.4 Launch R and Load Your Data

Launching R is as easy as double clicking on the icon on the desktop or application folder, depending on how you chose to install it. When you do so, you will get a window that looks pretty close to Figure B.1. The header will update as revisions are uploaded, but it is ready to go, as shown by the ">" symbol at the bottom,

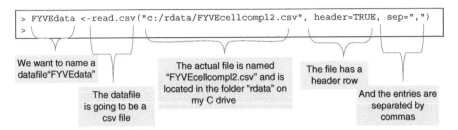

FIGURE B.2 *Anatomy of a command in R.* The command defines an object called FYVEdata, denoted by the arrow. It will be a table that is read from a CSV file in the folder rdata. Additional specifications are shown in the figure.

which prompt for a command from us. R will use a working directory for saving information generated during a session. R will work with your computer and find a folder designated by your operating system, which will be the default folder for any document you save to your computer. R will use this default initially. To find this path, type getwd() at the command prompt (>) and see what comes back. In Windows, it should be something like: "C:/Users/HCS User/Documents." You can set up your files in this folder. You can also designate another folder to find and store data; you will need to call this path explicitly when loading or saving a file. As noted above, these examples will load data from C:/rdata. An example is shown in Figure B.2. The complete command is broken down into parts. A couple of conventions:

- R is case sensitive: Fyvedata and FYVEdata are two different things.
- R is most useful when data is written into working files or objects. We designate these objects in the command line at the left of the arrow (<-). From here we can call the object (a table of data, this case) or parts of the object (such as a set of rows, a particular column, or even a single data point) from this object using commands we will introduce shortly. Calling an operation directly (entering just the command to the right of the arrow) will post the file or results directly on screen, which can be thousands of lines long, and also makes it impossible to call specific rows or columns for subsequent operations.
- Most aspects of R are the same in all platforms, but one thing that is not is the use of forward and backward slashes. The backward slashes used in these examples are for Windows systems, forward slashes are used in Mac and Linux. Calling a new window is another inconsistency. Calling a new graphic window is needed to compare plots (such as density plots from two different wells). Typing windows() will call a new window in the Windows platform, typing quartz() or X11() will open a new window in Mac OS or Unix/Linux, respectively.

The first thing we need to do is load the dataset we want to analyze, and give it a name to refer to it by. We do so twice in Figure B.3, the first time incorrectly, to highlight that mistakes happen and are easily corrected, all you typically need to do is just try again. The first time we tried this, we did not call the file name correctly,

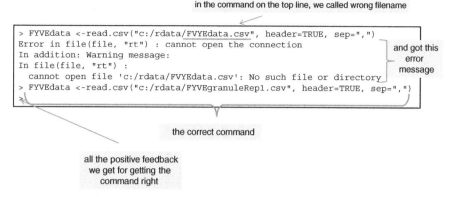

in the command on the top line, we called wrong filename

```
> FYVEdata <-read.csv("c:/rdata/FVYEdata.csv", header=TRUE, sep=",")
Error in file(file, "rt") : cannot open the connection
In addition: Warning message:
In file(file, "rt") :
  cannot open file 'c:/rdata/FVYEdata.csv': No such file or directory
> FYVEdata <-read.csv("c:/rdata/FYVEgranuleRep1.csv", header=TRUE, sep=",")
>
```

and got this error message

the correct command

all the positive feedback
we get for getting the
command right

FIGURE B.3 *Working with the R interface.* Calling a command incorrectly will produce several lines of feedback that can generally be traced back to the problem, but is not always literally interpretable. Calling a command correctly will produce output to the console window (when called without assigning the output to an object), produce a graph in a new window, or, when assigning the output as an object, will produce only a new command prompt.

so it returned a warning saying that it could not find the file we asked for. When we got it right, all we got was a somewhat underwhelming command prompt. Feel free to let out a small "woohoo" on your own.

The first few rows and columns of the original file are shown in Table 8.2. You can confirm that the correct file was loaded by checking the first few rows in R, using the command head (FYVEdata). As noted above, you could also call the entire file by typing its name directly, in this case typing FYVEdata at the prompt and hitting return. This will scroll the entire file. Using the head() function is much safer, even if you are pretty sure that the object is a small file or output. You can also check the end of the file by typing tail(FYVEdata). Some common commands are listed in Table B.1.

B.3 ANALYZING DATA IN R

If you have made it this far, you are doing very well, and you have made it past almost all of the problems that could derail your efforts to take charge of your data. The next steps are much easier to execute, and tap into the exceptional power of using R. We will go through some quick exercises that will enable you to look at data as we did in the analytical chapters in the main text.

B.3.1 Summary Statistics

The first thing we talked about in Chapter 9 was the utility of the summary statistics. We will utilize the puncta formation assay in this discussion. The experiment consists

TABLE B.1 Common Commands in R

	Function examples	Syntax explanation, notes
		Basic file functions
Import a file	workingfile <- read.csv("c:/rdata/myoriginalfile.csv", header=TRUE, sep=";")	Loads data into R that can be analyzed as the object "workingfile," which was copied from "myoriginalfile.csv," and was found in folder "rdata" on the c drive. The first row contains the column headings and data entries are separated by commas.
Create a new window	windows(), for Windows platforms, quartz() for Macintosh OS and x11() for Linux	Opens a new window in the R working environment, useful if two graphs want to be compared. Otherwise, a new plot will overwrite the initial plot. Note that operations specific to R are platform independent, but operations that interact with the computer can be platform specific. Opening a new window and saving a plot as a pdf file are examples.
Check that the file was imported or analyzed correctly	head(workingfile) tail(workingfile) head(workingfile, n=10)	Default number of rows is 6, "n=" specifies how many rows should be shown
Open a supplemental library or package	library(lattice) library(bioclite)	Call additional functions that are not part of the basic R package; must be called explicitly if one of the functions is to be used, even if the package is part of the basic R installation (such as "lattice")

Select specific rows from a data file	fyveB11 <-subset(featuredata, well=="B11") fyveB11 <-featuredata[featuredata, well=="B11",]	These options produce the same object. The latter uses square brackets to denote selected data in a [row, column] convention. The row term selects all rows where the value in column "well" is B11, the column term selects all columns, as none are specified.
Summarizing data for many rows	table(workingfile$well)	Tabulates the number of rows with each well name and writes a table that summarizes well names and the number of rows (cells) per well
	workingfiletable <-table(workingfile$well)	Designates an object (workingfiletable) with the table data, allows the table to be saved or new columns to be added
Save data or figures	write.table(workingfile, file="c:/rdata/workingfile2.csv", append=FALSE, sep="", rownames=FALSE)	The example as written would rewrite the table loaded as "workingfile" into a new file in the same folder called "workingfile2"; however, any object that results from an operation in R can be saved as a file. rownames=FALSE is needed to prevent a shift in rownames for the saved table.
	(saving a plot that has been drawn on a new window in R)	OK, while there are command-line processes in place for saving graphical output, the practical situation is that one of two things will be done: the graph will be saved as a file through the menu bar (or "right clicking" on the figure), or the graph will be copied (again, through right clicking), and pasted into another application window.

(continued)

255

TABLE B.1 *(Continued)*

	Function examples	Syntax explanation, notes
		Summary statistics functions
Summary statistic values for a dataset	workingfilesumstats <- summary (workingfile)	Setting the summary stats as an object (workingfilesumstats) means that the output will not deluge the screen with up to 384 rows and many columns; instead the output can be called by row or column, or saved as a file
	mean(workingfile$columnname)	Using $ to specify a column limits the analysis to that column only
	median(workingfile)	If no column name is specified, the indicated analysis for each column in the dataset "workingfile"
		Additional notes:
		All summary statistic options are available: median(), range(), min(), max(), summary(), quartile()
		Results from these operations can be called to the screen by typing the object name (workingfilesumstats, in this case), checked (by using the head() or tail() functions), or saved to a file (using write.table(), as shown above)
		Plotting data
General plotting	plot(workingfile$columnname1, workingfile$columnname2)	A general plotting function, plots an x–y plot of the two named columns; the function is very general, and can plot scatterplots (as called here), line plots and Q–Q plots (see below); plots can be saved, but can also be formatted extensively, including plot titles, legends, etc.

256

Q–Q plots	`qqnorm(workingfile$columnname)` `qqplot(workingfile$columnname, distribution=norm)`	The two terms are equivalent, but using the second allows for nonnormal distributions to be used, including Poisson, lognormal, and exponential distributions
	`qqline(workingfile$columnname)`	Supplementary command that is needed to draw the comparison line on the plot; overlays the line on the original plot
Histograms	`hist(workingfile$columnname, breaks=20)`	Creates a histogram that divides the data into 20 bins, plots in current (active) graphics window

Density plots	`workingfilecolnamedensity`
	`<- density (workingfile$columnname)`
	`plot(workingfilecolnamedensity)`	

Other plots	Bar plot Box plot Violin plot	See online documentation for additional information on these plot types

Lattice plots

Opening the lattice plot functions	`library(lattice)`	Technically, this is a file function and not a lattice-specific function, but before beginning any lattice graphics operation, the set of lattice functions needs to be loaded; each time you venture into a fundamentally new area of R functions, one or more packages may need to be called
Basic lattice plots	`plot(workingfile[,2:12])`	Plots a set of X–Y plots for columns 2 through 12 and all rows of workingfile; skipping column 1 would be done because (for most HCS data) it contains the well numbers; this will return an 11 × 11 array of plots, where each column is plotted against all others in a pairwise array; see Figure 10.3
Q–Q plots for a feature, grouped by well number	`qqmath(~columnname2|columnname1, workingfile, distribution=qnorm)`	Calculates Q–Q plots for all wells as grouped by columnname1, for the data in columnname2; the vertical bar (usually found above the enter/return key on most keyboards) designates the separation of values in columnname2 by groups defined by columnname1

(a)

```
> FYVEgranuleSumstats <-aggregate(FYVEdata$granuleintegarea,
list(FYVEdata$well), summary)
```

(b)

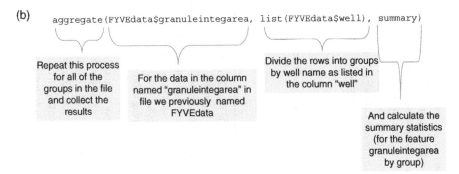

FIGURE B.4 *Anatomy of the command used to generate the summary statistics for the first plate of the FYVE granule assay.* The data from the CSV file was previously read into R and named FYVEdata. The command as entered into R is shown in (a). The command is parsed in (b).

of a dose titration experiment of multiple compound treatments. We will look at the summary statistics for features related to the formation of puncta, such as number of puncta, average area of puncta, and integrated (total) area of puncta. Also, it would be most helpful if we had these data for each well. This was a dose-response assay with increasing dose along each row, so checking the mean values per row is where we want to start. The command we will use to generate the summary statistics is shown in Figure B.4. OK, we said that we could get you analyzing your data in R in a few minutes, and we hit you with the command you see in Figure B.4. This command is complex, actually the most complex example we will use in this introduction, but we break down the function in the figure as well, and hopefully the logic can be parsed. However, the upside of all of this was we were able to do a lot of work very quickly. Looking at the output of this command (back in Chapter 9), consider what we were able to do. From the default spreadsheet, and without any significant restructuring of the data, we were able to generate six statistical measures for one feature for each well in the experiment. The command would be the same if we had thousands of cells per well, and if we had a 384-well plate (but working across multiple plates is probably where we would want to reach out for help). We are now starting to focus on the data and less so on R To check the results, we could save the table to a file and open it in another application, such as Excel™ or Spotfire™, load the entire table of summary statistics into the console window, or perform a quick check, using the head() command. This is easy and does not load an excessive amount of information into the console. An example is shown in Figure B.5.

```
> head(FYVEgranuleSumstats)
  Group.1 x.Min. x.1st Qu. x.Median x.Mean x.3rd Qu. x.Max.
1     B02   2.00     20.00    30.50  32.12     39.00  93.00
2     B03   1.00     25.00    32.00  40.20     54.25 183.00
3     B04   7.00     27.00    38.00  42.28     48.00 126.00
4     B05   2.00     23.00    36.00  41.56     49.00 162.00
5     B06   1.00     23.00    35.00  36.93     41.50 112.00
6     B07   8.00     26.00    36.00  39.30     48.50  82.00
>
```

FIGURE B.5 *Output sent to the R console.* The head (the first 6 rows) of the summary statistics of the FYVE assay that was called as an object using the head() command.

B.3.2 Drawing Q–Q Plots, Histograms and Density Plots

In the text, we noted that Q–Q plots are a very informative way of looking at the distribution of your data. Drawing these plots is almost as easy as getting the summary statistics. Histograms and density plots are also very effective, so we will discuss them here as well. For this tutorial, we have emphasized methods that minimize reformatting of the data file itself, and instead use commands in R to extract the data we are interested in. For the Q–Q plots we have presented in this book, the data was selected and analyzed as follows.

- The rows corresponding to a specific well were selected and written to a new file (or object, in the terminology of R).
- The Q–Q plot was made for a feature (column) that we are interested in.
- The Q–Q line was drawn over the data.
- During initial data analysis, the qqplot() and qqline() functions are typically called without assigning the results to a new object (using the <- arrow). This is because the output is sent to a new window rather than the command screen, and we can open additional windows if we want to compare results side-by-side. Plotting results to part of a window can be done when you want to format several plots for a figure. This is not difficult, but is just marginally beyond the scope of this very basic introduction to R.

The process and the results are shown in Figure B.6. It is also possible to plot all of the data for a feature without selecting an individual well. Note that the Q–Q plot is a general method for comparing data against a model distribution. Although usually used to compare against a normal distribution, other distribution types can be used, examples are presented in Table B.1.

A histogram can be plotted with a single command, the command includes a specific argument telling R how many bins (divisions) you would like. An example is shown in Figure B.6 as well. A distribution plot presents data in much same way as a histogram, but the data is preprocessed before plotting. A density distribution is calculated, and then plotted, using two commands, as is also shown in Figure B.6. If

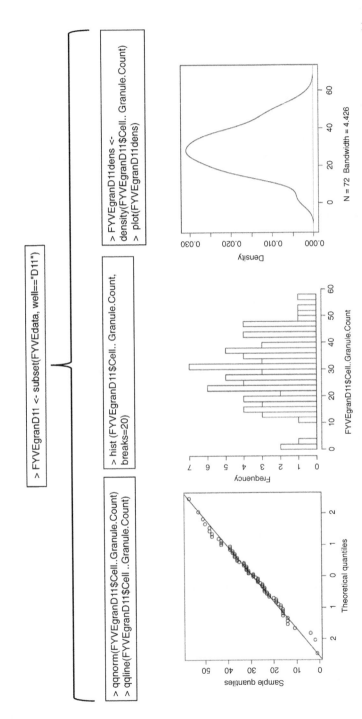

FIGURE B.6 *Producing statistical graphs in R.* To look at data for a single well, the rows corresponding to that well are written as a new object using the command at the top (well D11 in this example). From this object, options available for plotting the data include the Q–Q plot, a histogram or a density plot, as shown in the figure. The Q–Q plot draws the data points in the first command, the line is overlayed with the second command, qqline(). For the density plot, a density distribution is calculated from the data, and the distribution is then plotted by the second command.

you want to compare the plots on screen, such as from positive and negative control wells, you will need to open a new window before drawing the density plot.

B.3.3   Arrays of Graphs

While performing many of these analyses is going to be new to most readers, looking at arrays of data is less foreign. Transcriptional profiling and some other genomics methods plot output from self-organizing maps (Chapter 14) in arrays of plots, most frequently in Spotfire™. There is a package that it available in R that can plot data in this way, it is one of the supplemental packages that is downloaded with the basic installation of R, but as a supplemental package, still needs to be called (loaded) explicitly before the routines can be used. The package is named lattice, and it allows you to look at data for every well in a plate with just a couple of commands. An example is provided in Figure B.7. After loading the lattice package (by entering the term library(lattice) at the command prompt, we have been able to generate histograms of the granule counts per cell for every well in the plate using a single command. Like the aggregate() command used to generate the summary statistics, this one is complex, but based on examples presented in the previous figures, the pattern is starting to become clear. The results are also presented in the figure. In the histograms, the impact of compound treatments are shown by the progression of large numbers of cells with few granules (large bars on the left side of the plots) to increasing numbers of cells with greater numbers of granules per cell, bars to the right of the lowest bar. Highly skewed distributions can leave much of the data plotted to the left side of the plots, even for wells showing strong responses, because a few extreme values will affect the binning of all of the samples. However, clear trends can be observed, and are similar to the alternative representation of compound effects shown in Figure B.7b.

B.3.4   Exporting and Saving Results

Although we discussed how R is a command-line language, it is not completely hostile to the concept of a graphical user interface (GUI), the use of pull-down menus and double clicking on folders. For figures, R has several options in the menu for saving, including PDF, EPS, BMP, and TIFF formats. Saving images can be done at the command line as well, which is important if you will write routines or functions that will automatically save one or more images. For tables, such as summary statistics, writing the object to a CSV file using a command-line instruction is the best way. Examples of this are given in Table B.1.

B.4   WHERE TO GO NEXT

So we got you using R in just a few minutes! What to do next? OK, in order to achieve our goal, we had to put the cart before the horse. We fed a few specific lines of code into the console and got some output that helped us look at one of our

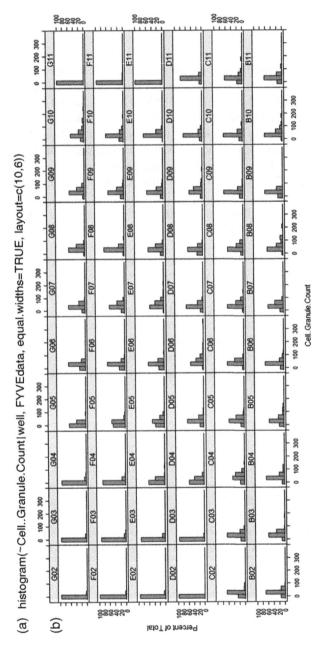

(a) histogram(~Cell..Granule.Count|well, FYVEdata, equal.widths=TRUE, layout=c(10,6))

FIGURE B.7 *Lattice (or arrayed) plots in R.* (a) After loading the supplementary package lattice() using the library() command, an array of plots can be called using the histogram() command (note the distinction between this and the hist() command for drawing a single histogram in the base **R** package, as shown in Figure B.6). The tilde (~) and the vertical slash (|, usually above the return key on a standard keyboard) are used to define the variable to be plotted and the category to be used to array the plots. Plotting by well is most common for an initial assessment, but any categorical information that is included in the data (such as dose or gene/RNAi reagent) can be used. The following information in this command, FYVEdata, is the name of the object. The last two parts are part of the many optional specifications that can be added to a command. equal.widths=TRUE fixes the widths of the bars in the histogram (the default value of FALSE will adjust the widths to the number of data points in the bin, so that highly skewed distributions will have very wide bins as they extend beyond the median) and the layout of the array is intended to preserve the organization of the inner 60 wells of a microtiter plate. A complete microtiter plate would have the dimensions of (12, 8). The array of plots are shown in (b). Note that the default plotting order is from first to last groups in the data object, but drawn from bottom to top, so the order of the wells in this figure reversed relative to the actual microtiter plate.

experiments. Our goal in this exercise was to link the process of R to output that matters scientifically. Clearly the next step is to get the horse back in front of the cart. Learning the basics of R is no different than what we did in the examples, but you need a good source for presenting all of the tools in R, as well as managing files and improving the visual quality of the graphs.

As a very robust open-source programming environment, many blogs and discussion groups are available online (but reviewing past postings and the FAQ information is essential for interacting with them effectively). In addition, some excellent books are available, as noted in the Further Reading section. Options are available for instruction that is geared toward absolute beginners and those who want to focus on programming, statistics, graphic, or data analytics. Some will be exhaustive tome-like references, others will take a very conversational approach.

FURTHER READING

Several books that utilize R extensively for applications such as data analysis and analytics were mentioned in earlier chapters. The sources discussed here are more focused on R itself. *R in Action* by Rober Kabacoff (Manning Publications) is very popular, as it is an extension of an online resource that is a good first source for getting your bearings in R called Quick-R (www.statmethods.net). *R Cookbook* by Paul Teetor (O'Reilly Cookbooks) leverages a popular approach to teaching programming: detailed solutions to dozens of common problems, essentially setting up a set of case studies that one can read through to find an example that matches the problem of the moment. *The Art of R Programming: A Tour of Statistical Software Design* by Norman Matloff (No Starch Press), is—as advertised—focused on programming for those who want to develop more detailed functions. These are just a beginning, to help find a book that is written at the level you are looking for, sample chapters are typically available.

APPENDIX C

HYPOTHESIS TESTING FOR HIGH CONTENT DATA: A REFRESHER

LIN GUEY AND ARIJIT CHAKRAVARTY

C.1 INTRODUCTION

The connection between biology and statistics is firm, but the statistical analysis of biological data is often reflexive (repeating past approaches to new data), and occasionally, incorrectly applied. This is largely because most genuine statistical training that biologists have had is typically years old. The problem is not so much that statistics changes frequently (although within the context of "Big Data" and analytics, it is getting a bit of a makeover), but that it is frequently compartmentalized to a few concepts, such as a p-value of 0.05 as being all that needs to be checked before drawing a conclusion. However, the need for a refresher is in order, and one that is specific for HCS is particularly relevant. As noted earlier in this book, HCS generates high volumes of data, whereas statistical tests have been developed for a limited number of variables; also recall that the data is frequently not normally distributed. For the digital pathology and pharmacodynamics discussion in Chapter 13, the use of multivariate measures and quantitative texture features are new for most investigators. This appendix reviews concepts invoked earlier in this book, as a point of reference and to lay out the conceptual bases of most statistical evaluations. Specifically, this appendix will focus on inferential statistics, the process of determining whether an authentic difference exists between two samples.

An Introduction to High Content Screening: Imaging Technology, Assay Development, and Data Analysis in Biology and Drug Discovery, First Edition. Edited by Steven A. Haney, Douglas Bowman, and Arijit Chakravarty.
© 2015 John Wiley & Sons, Inc. Published 2015 by John Wiley & Sons, Inc.

C.2   DEFINING SIMPLE HYPOTHESIS TESTING

Inferential statistics are when we go beyond describing or summarizing the data and start making conclusions about the experiment: "did we get a hit?," or "what is the IC50." Inferential tests work on sample estimates, which are used to estimate the "true" or underlying population parameters. For example, the sample mean is not the true population mean, but rather an estimator of the population mean (because you did not measure all the cells for a given treatment condition, you measured some of them and use the sample group to infer what the effect on the entire population would be). It can be shown mathematically that for extremely large samples, the sample mean is an unbiased estimator of the population mean. The goal of a hypothesis-driven statistical test is to make inferences regarding population parameters, using these sample estimators. This section will give an overview of the fundamentals of statistical hypothesis testing as well as introduce common statistical tests for the comparison of two groups.

C.2.1   The Standard Error and the Confidence Interval

All sample estimators, due to the nature of their estimation from empirical data, are subject to sampling variability. Hence, without an estimate of the standard error, an estimate of the variability of an estimator between multiple samples, we have no idea how precise the sample estimator is in a given dataset. One naive approach to assess precision would be to repeat an experiment multiple times. Clearly this is a time-consuming and unsophisticated method of estimating the standard error. Fortunately, probability and mathematical statistics theory have derived well-established properties such that the standard error can be estimated from a single dataset or experiment. Note the conceptual distinction between the standard deviation, a measure of variability of *observations* in a dataset, and the standard error, a measure of variability of an estimator (such as the mean) from the true mean of all the instances ever.

Consider as an example the mean of a cytological feature for a specified well. From statistical theory, the standard error (SE) of a mean is calculated as $s/\sqrt{n}$, where s is the standard deviation of the cytological feature and n is the number of observations or data points. Now, armed with the standard error of the mean, we can make statements regarding the precision of an estimator and construct a range of plausible values of the true population mean. That is, instead of using the sample mean as our only estimate of the population mean, we can derive a confidence interval that reflects a range of values within which will likely contain the true population mean. A wide confidence interval indicates that there is substantial variability surrounding the estimator and a narrow confidence interval indicates that the estimator is quite precise. A confidence interval can be constructed according to differing levels of certainty; the most commonly reported confidence interval is the 95% confidence interval. The interpretation of a 95% confidence interval is that if the experiment were repeated multiple times and confidence intervals were calculated for each experiment, we expect the true population mean to fall within 95% of the confidence intervals. Note that the confidence interval does not indicate that there is a

95% chance that a *particular* confidence interval contains the population mean. This interpretation is incorrect because the population mean is a fixed parameter and is either contained within the confidence interval, or not. It cannot be contained inside an interval 95% of the time. This is an important distinction. Assuming a normal distribution, the generalized 95% confidence interval formula is $\hat{x} \pm 1.96\,[SE(\hat{x})]$, where $\hat{x}$ is the sample estimator and $SE(\hat{x})$ is the standard error of the estimator.

Although the 95% confidence interval is the most frequently reported confidence interval, the specified degree of certainty for a confidence interval can be changed to be more conservative or more liberal. A more conservative confidence interval of 99%, for example, would multiply the standard error by 2.58 since 99% of the area of a normal curve lies within 2.58 standard deviations from the mean.

C.2.2 The Null Hypothesis and Fundamentals of Statistical Hypothesis Testing

Before we introduce several basic statistical significance tests for the comparison of two groups, let us dig into the principles behind statistical inference. Significance tests allow us to formally examine data with respect to a scientific hypothesis. It reflects the degree to which we can be certain that a population parameter, as estimated by a sample estimator, follows a scientific conjecture. To perform a statistical test, the scientific conjecture needs to be specified in terms of a null hypothesis (denoted as H_0) and an alternative hypothesis (denoted as H_A or H_1). The null and alternative hypotheses are statements regarding the behavior of the true population parameter and are generally specified to represent opposite conclusions. The null hypothesis typically hypothesizes a null effect (e.g., no difference between groups, no association between treatment and outcome, etc.). This is in contrast to the alternative hypothesis, which typically hypothesizes that there is a nonnull effect (e.g., there exists a difference between groups or association between treatment and outcome, etc.). The null hypothesis is the hypothesis that we would like to reject through statistical inference. In other words, our scientific conjecture posits that the alternative hypothesis is more likely to be true but we cannot accept the alternative hypothesis without rejection of the null hypothesis.

Once a hypothesis has been formulated, the most appropriate statistical test can then be chosen. This statistical test allows us to decide whether the empirical data is more consistent with the null hypothesis in which case the null hypothesis cannot be rejected. Statistical theory informs us about the theoretical distribution of the test statistic, assuming the null hypothesis is true. This distribution can then be used to determine a critical value such that the null hypothesis is rejected if the test statistic is greater than (or equal to) the critical value and the null hypothesis is not rejected if the test statistic is less than the critical value (Figure C.1).

The two types of errors encountered in significance testing are the type I error and type II error. A type I error is analogous to a false positive, and a type II error is analogous to a false negative. The type I error rate for a given statistical test is denoted as α, and is often referred to the level of significance. The type II error rate is denoted as β, and determines the power of a statistical test (calculated as $1 - \beta$).

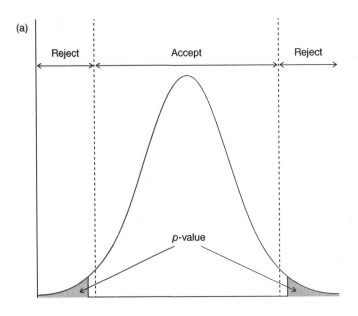

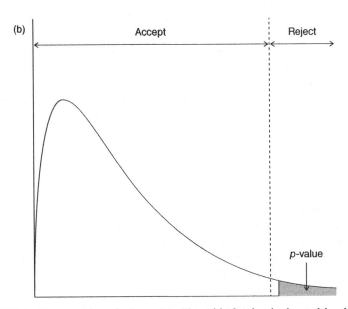

FIGURE C.1 *Statistical hypothesis testing.* The critical value is denoted by the dashed vertical line and partitions the null distribution into rejection and acceptance regions. Test statistics that fall within these regions either refute or support the null hypothesis being evaluated. In this example, the test statistic is denoted by the solid vertical line within the rejection region. The *p*-value is the probability of obtaining a test statistic at least as large as the one observed and is calculated as the area under the curve greater than the observed test statistic (shaded in gray). These values are shown for a two-tailed hypothesis for a null distribution that follows (a) a normal distribution and (b) a chi-square distribution.

The power of a statistical test is dependent upon the sample size (number of data points), size of the effect under the alternative hypothesis, and type I error rate (which is usually defined ahead of time, most commonly at the 0.05 level). It is important to understand that there is a see-saw relationship between the two error rates such that they cannot be simultaneously minimized for a given dataset.

In addition to comparing the observed test statistic to a critical value to determine whether the data favor the null or alternative hypothesis, the p-value (abbreviated for probability value) provides a complementary measure of the degree to which the data support either hypothesis. The p-value is the probability of obtaining a value at least at extreme as the test statistic under the null hypothesis (Figure C.1). The more extreme the test statistic, the smaller the p-value and the more "significant" the result or the less likely the data favor the null hypothesis (or to put it colloquially, the less likely the data is to have arisen purely by chance). A p-value of 0.001, for example, implies that we only 0.1% of experiments are expected to yield a test statistic as large as the one observed in the dataset, suggesting that the null hypothesis is likely to be false.

C.2.3 Inferential Statistics and High Content Data

We can already see how the conversations between a biologist and a statistician start to change with the analysis of high content data. For well-level data analysis, the discussions will be similar to other conversations regarding assays with replicates. However, if cell-level comparisons are to be made, the impact on inferential statistics can be significant. Consider the impact on the standard error. If hundreds of cells are measured, the sample number will play a sizable role on the calculation of the standard error. What this means is that there will be little concern about the sample data and the true population mean (a 99% confidence interval can be achieved routinely with such large sample numbers). Likewise, even very minor changes to cells through a perturbation will report a statistically significant change. The challenge is to identify the biologically relevant effects from among the statistically significant measurements.

C.3 SIMPLE STATISTICAL TESTS TO COMPARE TWO GROUPS

Now that the fundamentals of statistical hypothesis testing have been described, we introduce a few of the most frequently used statistical tests for the comparison of a continuous measurement between two groups. Guidelines are also provided regarding which test is most appropriate given the data and biological question being asked. A common utility of these tests, as an example, is to determine if a cytological feature is statistically different between two treatments (e.g., treatment vs. control). The tests introduced in this chapter all assume that the observations (samples) are independent, or are not correlated, which is a reasonable assumption for the majority of high throughput or RNAi screening experiments. Correlated observations can arise in an experiment when a comparison is made between the same wells pre- and posttreatment. Since the comparison is made using the same wells, the observations are clearly not independent and require differing statistical tests that incorporate this information. Such tests are beyond the scope of this chapter and the interested reader is suggested to read Box et al (2005) or Sokal and Rohlf (1995).

C.3.1 The t-test

The t-test is by far the most popular and well-known statistical test to compare two independent samples of observations for a continuous measurement. It specifically assesses whether the means of two groups are statistically different. The null hypothesis states that the means of the two groups are the same and the alternative hypothesis states that the two means are different. In mathematical notation, this is expressed as $H_0 : \mu_1 = \mu_2$ versus $H_A : \mu_1 \neq \mu_2$ where μ_1 and μ_2 are the true means of groups 1 and 2 in the population. Note that when the alternative hypothesis is specified irrespective of direction (i.e., $H_A : \mu_1 > \mu_2$ or $\mu_1 < \mu_2$), the test is referred to as a two-tailed test and the p-value is calculated using both upper and lower tails of the null distribution, this is shown graphically in Figure C.1a. Conversely, when only one of the two directions is specified, the test is known as a one-tailed test. This is the same situation as in Figure C.1a, but only one of the directions is considered. Because a one-tailed test only uses one tail to calculate the p-value, its p-value is always half that of the equivalent two-tailed test. However, unless one has an a priori hypothesis regarding the direction of the alternative hypothesis, a two-tailed test should always be used. A one-tailed test should never be used simply to obtain a smaller p-value.

There are two forms of the t-test depending on whether the variances between the groups are equal are unequal. The t-test assuming equal variances is:

$$t = \frac{\bar{x}_1 - \bar{x}_2}{\sqrt{\dfrac{(n_1 - 1)\, s_1^2 + (n_2 - 1)\, s_2^2}{n_1 + n_2 - 2}}\sqrt{\dfrac{1}{n_1} + \dfrac{1}{n_2}}}$$

where $\bar{x}_i$ is the sample mean for groups $i = 1$ and 2, s_i^2 is the sample standard deviation within each group, and n_i is the sample size or number of observations within each group.

This hairy-looking expression makes a lot more sense when it is parsed out term by term. The numerator is simply the difference between sample means and the denominator is simply the standard error of that difference. The denominator can be further parsed into two terms: the term on the left-hand side of the denominator is the pooled standard deviation across the two groups and the term on the right-hand side is a sample size weight to convert the pooled standard deviation to the standard error. Thus the test statistic is simply the mean difference divided by its standard error.

The test statistic, referred to as Student's t-test, follows a t-distribution with $n_1 + n_2 - 2$ degrees of freedom under the null hypothesis. The degrees of freedom is an additional parameter necessary to fully describe distributions such as the t-distribution, which represents the number of independent values on which the statistic is based (calculated as the total number of observations subtracted by the estimated number of parameters). For the t-test, for example, this then is $n_1 + n_2$ (the total number of observations) minus two (the number of estimated sample means). A t-distribution has a symmetrical bell-shaped curve similar to a normal distribution except that it has a higher kurtosis. As the number of observations (and therefore the degrees of freedom) increases, the t-distribution closely approximates a standard normal distribution. The t-distribution is subsequently used to calculate the p-value after which the null hypothesis is either rejected or not rejected depending on whether the p-value is less than or equal to the significance level.

The major underlying assumption of this t-test is that the variances across groups are equal (homogeneity of variances or homoscedasticity). Figure C.2a shows an example of two symmetric distributions belonging to two hypothetical treatments

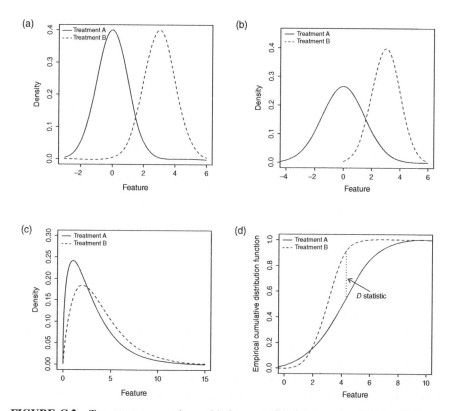

FIGURE C.2 *Two group comparisons.* (a) An example of data appropriate for Student's t-test. The feature distributions of treatments a and b are both normally distributed and have similar variances. (b) An example of data appropriate for Welch's t-test. The feature distributions of Treatment A and Treatment B are both normally distributed but do not have equal variances. (c) An example of skewed data appropriate for Wilcoxon's rank sum test. Note that transformations may be helpful in making the distributions more symmetrical. (d) Example of K–S statistic. Two empirical cumulative distribution functions belonging to Treatments A and B are plotted and the maximal discrepancy between the two lines is calculated (D statistic).

groups with equal variances, which would be appropriate for the Student's t-test. If the two groups exhibit unequal variances, the unequal variances t-test should be used.

$$t = \frac{\bar{x}_1 - \bar{x}_2}{\sqrt{\dfrac{s_1^2}{n_1} + \dfrac{s_2^2}{n_2}}}.$$

This test statistic, referred to as Welch's t-test, also follows a t-distribution, but its degrees of freedom is given by another hairy-looking expression known as the Sattertwaite approximation.

$$\frac{(s_1^2/n_1 + s_2^2/n_2)^2}{s_1^4/(n_1^3 - n_1^2) + s_2^4/(n_2^3 - n_2^2)}.$$

So that looks really ugly, you say. If you look closely at it, Sattertwaite approximation functions as a kind of "fudge factor" that penalizes samples for being unequally sized. When the two samples are weighted equally, the $n^3 - n^2$ terms are then identical, so the term is at its smallest. The larger the difference between the two sample sizes, the larger the denominator gets relative to the numerator, and the fewer degrees of freedom you are left with!

Figure C.2b shows an example of two treatment groups with symmetric distributions but unequal variances, which would be appropriately analyzed with Welch's t-test. Unequal variances can be inspected visually through plots like histograms and boxplots, and evaluated formally with a statistical test comparing between-group variances, such as Levene's test for equality of variances. Because Welch's t-test is more conservative than Student's t-test, it is preferable to use Student's t-test if the variances across group appear equal.

The t-test is the most appropriate statistical test when the research question is to determine whether there is a locational shift between two groups that are normally distributed. The t-test is fairly robust to the normality assumption and can be used for symmetrical distributions or distributions with modest skew. Because the t-test is highly sensitive to outliers, the t-test is not suitable for distributions that are highly skewed and a nonparametric alternative such as the Wilcoxon rank sum should be considered. However, because the t-test has more statistical power than the Wilcoxon rank sum test when the data are normally distributed, it is preferable to use the t-test if the group distributions are not highly skewed. Lastly, when reporting results from a t-test, it is useful to report the effect size in addition to the p-value, so that the reader can assess the magnitude of effect. Effect sizes such as fold change and standardized mean difference can tremendously help the interpretation of a p-value.

C.3.2 Wilcoxon Rank Sum Test

The Wilcoxon rank sum test (also called Mann–Whitney U or Wilcoxon—Mann–Whitney test), is a nonparametric, or distribution free, alternative to the t-test to determine whether observations from one group tend to be larger than observations from another group. Highly skewed distributions, such as the distributions shown in Figure C.2c, are inappropriate for nonrobust tests such as the t-test. The Wilcoxon rank sum test, which makes no assumptions regarding the behavior of the data, is the appropriate t-test alternative for such highly skewed data. The null hypothesis states that the probability distributions of the two groups are equivalent and the alternative hypothesis states that the rank sum of one group is greater than the other group. That is, the null hypothesis postulates that the two groups are from the same distribution, but the observations have been randomly assigned to the two groups and thus should be equally distributed among the groups. The alternative hypothesis, conversely, states that observations from one distribution are greater than the other distribution. The test statistic, commonly denoted as U, is calculated as follows.

1. Combine observations from both groups and rank all observations in ascending order. To assign ranks, the ith smallest observation should be assigned rank i and in case of ties, each tied observation should be assigned the average rank.
2. Sum the ranks of each group separately. Let R_A and R_B denote the summed ranks for groups A and B, respectively.
3. Calculate $U_A = R_A - \frac{n_A(n_A+1)}{2}$ and $U_B = R_B - \frac{n_B(n_B+1)}{2}$ where n_A and n_B are the number of observations (sample size) for groups A and B, respectively.
4. Calculate the test statistic $U = \max(U_A, U_B)$.

To determine the significance of the test, the observed test statistic U is compared to its theoretical distribution under the null hypothesis, which for small samples ($\sim n < 20$) has been explicitly tabulated in many statistical textbooks and software packages. For large samples ($\sim n > 20$), the null distribution of U approximates a normal distribution with mean $n_A n_B/2$ and variance $n_A n_B(n_A + n_B + 1)/12$.

It is a common misconception that the Wilcoxon rank sum test is a test comparing medians across two treatment groups. The Wilcoxon rank sum test is probably perceived to be a medians test as it is the nonparametric alternative to the t-test; however, while the t-test assesses whether two group means are equivalent, the Wilcoxon rank sum test does not. Although rare in practice, situations can occur where the medians between two groups are equivalent but the groups have significantly different rank sums. Note also that the alternative hypothesis does not state that there may be arbitrary differences at any point throughout the distributions, but that one distribution has larger observations than the other. To detect any discrepancy between two distributions, the Kolmogorov–Smirnov test, described in the next section, is appropriate.

C.3.3 Kolmogorov–Smirnov Test

The Kolmogorov–Smirnov test (or K–S test) is another nonparametric statistical test to determine whether the distributions from two groups are equal. Unlike the previous tests, it is a general statistical test that makes no constraint in how two distributions differ. The null hypothesis is the same as the Wilcoxon rank sum's null hypothesis, namely, that the two distributions are equivalent or drawn from the same population. However, the alternative hypothesis specifies that there is a substantial deviation at any point between the two distributions. This is conceptually different than the alternative hypothesis of the Wilcoxon rank sum test, which states that one distribution is (stochastically) greater than the other. The K–S alternative hypothesis makes no assumption regarding how the two distributions deviate from one another and thus, is more sensitive to changes in shape parameters such as kurtosis. Also, because the K–S test is a nonparametric test similar to the Wilcoxon rank sum test, it makes no assumption on how the data should behave and thus can be used for any dataset. The test statistic, D, is calculated as the greatest discrepancy between the two **cumulative distribution functions** estimated empirically from the data (Figure C.2d). This is mathematically notated as $D = \max[F_1(x) - F_2(x)]$ where $F_1(x)$ and $F_2(x)$ are the cumulative distribution functions for groups 1 and 2, respectively. The empirical cumulative distribution functions are estimated as the proportion of observations less than or equal to each possible observation. Under the null hypothesis, it can be shown that D follows the Kolmogorov distribution, from which statistical inferences can be made.

Although it is possible to perform the K–S test for a small dataset (e.g., well- or field-level data with less than 20 observations), it is difficult to obtain an accurate representation of the distribution from a small number of observations. This is obvious for anyone who has tried to plot a histogram with only a few observations. Therefore, cell-level data, which can yield hundreds or thousands of observations from a single well, is the most appropriate for the K–S test. However, one needs to interpret cell-level data with caution as the statistical power to detect smaller differences increases dramatically with the usage of cell-level data compared to well- or field-level data. A "significant" p-value from cell-level data may translate to a trivial or nonmeaningful biological difference. As a general note, it is important to report the size of the effect (such as fold change or standardized mean difference) in addition to the p-value when analyzing cell-level data. It is important to report D in addition to the p-value so that one can gauge the size of the discrepancy between the two distributions. In addition, one can compare control wells using the K–S test and obtain $D_{control}$, the maximal discrepancy between the control wells, and constrain the test statistic D such that D has to be larger than the $D_{control}$ in order to be considered statistically significant.

C.3.4 Multiple Testing

The problem of multiple testing arises when a statistical test is repeated numerous times in a given dataset due to an inflation of the type I error (false positive) rate. For example, in high throughput screens, one may be interested in assessing the mean difference between several small molecule compounds (or RNAi treatments) and a negative control. However, the more statistical tests we perform, the more likely a type I error (rejection of the null hypothesis when the null hypothesis true in the population) is committed. For example, if 100 t-tests were performed comparing 100 small molecule compounds to a negative control, then just by chance, 5 tests would be "statistically significant" given a significance level of 5%, even if there are no real differences in the dataset. Similarly, 1 out of 100 tests would be declared significant just by chance given a significance level of 1%. This is because a significance level is specified for a single statistical test, not for a series of statistical tests. To overcome the problem of obtaining spurious positive results due to multiple testing, a number of approaches have been developed.

One simple yet crude multiple testing correction is to lower the significance level to a fixed acceptable rate. The Bonferroni correction method assumes that the overall type I error rate is at most α, the given significance level. Therefore, if the overall type I error rate is at most α and there are m independent statistical tests, then the Bonferroni correction specifies the significance level for each individual test should be lowered to α/m. A statistical test is declared significant only if its p-value is less than the Bonferroni-corrected significance level of α/m. Alternatively, the observed p-value may be multiplied by m (note this quantity may be greater than 1 which is impossible for probabilities) and then compared to the desired significance level α. The Bonferroni correction is attractive for its simplicity and is the most commonly used multiple testing correction for this reason, but is also the most conservative, as it assumes all tests are independent, when in practice there frequently are correlations. Recall that there is a see-saw relationship between the type I error and type II error rates such that if the type I error rate is decreased, the type II error rate is consequently increased. Thus, a Bonferroni-corrected significance level leads to a dramatic decrease in statistical power making it difficult to identify true positives.

A more sensitive alternative to the Bonferroni correction is the false discovery rate (FDR) method, commonly used in population genetic and gene expression studies. The FDR is the expected proportion of false positives among the statistically significant results. For example, if 100 tests were declared significant with an FDR of 5%, then 5 of those tests declared significant can be expected to be false positives. The FDR should not be confused with the false positive rate. The FDR is proportion of false positives among results declared significant, whereas the false positive rate is the proportion of false positives among results declared nonsignificant. To control the FDR, a step-wise procedure was proposed in which the m observed p-values are rank ordered and each successive p-value, $p_{(i)}$, is compared to the quantity $q\frac{i}{m}$ where q is the desired FDR and i is the p-value ranking (1 for minimum and m for maximum ranking). The largest p-value ranking with a p-value less than its corresponding $q\frac{i}{m}$ quantity is recorded and all statistical tests with smaller p-value rankings are declared significant. This step-up procedure has been shown to control the FDR to at most q, the desired FDR. A simple mathematical arrangement shows that $q_i = p_{(i)}\frac{m}{i}$, coined the q-value, which can be interpreted as the FDR adjusted p-value. The q-value estimates the minimum FDR if a particular p-value is declared significant. Because the q-value estimates the proportion of false positives among significant results, a more lenient threshold than the significance level (e.g., q-value of 0.10) is often used.

Lastly, a more computationally intensive multiple correction method is permutation testing, which is one of the only multiple correction methods that allows the statistical tests to be correlated. Briefly, the data is randomized in a manner consistent with the null hypothesis. A test statistic is calculated across all tests for each permutation, or randomization of the data, and the maximum test statistic (or alternatively minimum p-value) is recorded and contributes an observation to the empirical null distribution. Typically, one performs at least 1000 permutations, from which the empirical null distribution of the test statistic is obtained. The permutation adjusted p-value is then calculated by comparing each observed statistical test to the permuted null distribution.

C.4   STATISTICAL TESTS ON GROUPS OF SAMPLES

C.4.1   The Foundation of Analysis of Variance

Analysis of variance (ANOVA) is a series of statistical models designed to partition the observed variance into separate components. The simplest form of ANOVA is the one-way ANOVA, which compares the means across several treatment groups and is a direct extension of the t-test. Whereas the t-test is limited to assessing whether the means differ between two groups, the one-way ANOVA generalizes this test to compare the means across as many groups as specified. The null hypothesis states that the means across all groups are equal ($H_0 : \mu_1 = \mu_2 = ... = \mu_k$ for k groups) and the alternative hypothesis states that at least one of the means is different (H_A: at least some $\mu_i \neq \mu_j$ for any $i \neq j$). It is named the "one-way" ANOVA because it models only one factor, defined as a set of treatment groupings. For example, the goal of an experiment may be to determine whether cell viability differs among different transfection reagents. In this experiment, the factor is the type of transfection reagent and cell viability among transfection reagents can be compared using the one-way ANOVA (note that cell viability most likely needs to undergo the square-root transformation to follow a normal distribution as it is a count variable). The one-way ANOVA model is formulated as

$$y_{ij} = \mu + \tau_i + \varepsilon_{ij}$$

where y_{ij} is jth observation in the ith group, μ is the grand mean or mean across all observations, τ_i is the deviation of the ith treatment mean from the grand mean (or the "treatment effect"), and the ε_{ij}s are random errors.

Table C.1 shows the total variability partitioned into mutually exclusive sources in a one-way ANOVA. A measure of total variability in a dataset can be specified as $\sum_{i=1}^{k} \sum_{j=1}^{n_{i}} \left(y_{ij} - \bar{\bar{y}} \right)^2$, which squares the deviation between each observation and the sample grand mean ($\bar{\bar{y}}$, the sample mean across all observations) and sums this over all observations. This quantity is referred to as the total sum of squares (SS_{total}), since it is calculated as the sum of squares of grand mean deviations over the entire sample set. An easier conceptualization of the SS_{total} is that it is equivalent to the total sample variance multiplied by its degrees of freedom, $N - 1$ where N is the total number of observations across all treatment groups. That is, if the SS_{total} were divided its degrees of freedom ($N - 1$), this would yield the total sample variance, a conventional measure of total variability. The SS_{total} can be partitioned into exactly two

TABLE C.1   One-way Analysis of Variance (ANOVA)

Source of variability	Sum of squares (SS)	Degrees of freedom (df)	Mean square (MS)	F (test statistic)
Treatment	$SS_{treatment} = \sum_{i=1}^{k} n_i(\bar{y}_i - \bar{\bar{y}})^2$	$k - 1$	$MS_{treatment} = SS_{treatment}/(k - 1)$	$F = MS_{treatment}/MS_{error}$
Error	$SS_{error} = \sum_{i=1}^{k} \sum_{i=1}^{n_{ii}} (y_{ij} - \bar{y}_i)^2$	$N - k$	$MS_{error} = SS_{error}/(N - k)$	
Total	$SS_{total} = \sum_{i=1}^{k} \sum_{i=1}^{n_{ii}} (y_{ij} - \bar{\bar{y}})^2$	$N - 1$		

Notation: $\bar{\bar{y}}$ is the grand sample mean (or mean across all observations), $\bar{y}_i$ is the mean of the ith group, y_{ij} is the jth observation within the ith group, n_i is the number of observations for the ith group, k is the number of groups ($i = 1,2,...k$), and N is the total number of observations across all groups.

quantities: (1) the variability of each treatment mean to the grand mean (known as the treatment sum of squares, $SS_{treatment}$, or between-group sum of squares) and (2) the variability of each observation to its group mean (known as the error sum of squares, SS_{error}, or within-group sum of squares). The SS_{total} is exactly equal to $SS_{treatment}$ + SS_{error}. The $SS_{treatment}$ sums the squares of the departure of each treatment mean with respect to the grand mean across all treatments. If the null hypothesis were true, all treatment means should be similar to the grand mean and the $SS_{treatment}$ should be small. Conversely, if the null hypothesis were false, at least one treatment mean should differ substantially from the grand mean yielding a large $SS_{treatment}$. However, the $SS_{treatment}$ clearly increases as the number of treatment groups increases therefore the $SS_{treatment}$ must be adjusted by the number of treatments (k). Dividing $SS_{treatment}$ by its degrees of freedom ($k - 1$) yields the mean square treatment ($MS_{treatment}$), a measure of treatment variability adjusted by the number of treatments. The $MS_{treatment}$ seems like an appropriate test statistic to determine whether the data favor the null or alternative hypothesis, but still needs to be corrected for the amount of variability observed within the treatment groups. A pooled estimate of the common within-group variance, assuming the variances across treatment groups are equal, is the mean square error (MS_{error}), calculated as the $SS_{error}/(N - k)$ where $N - k$ is the degrees of freedom associated with SS_{error}. The one-way ANOVA test statistic is then $F = MS_{treatment}/MS_{error}$, which follows an F distribution with $k - 1$ and $N - k$ degrees of freedom under the null hypothesis. Similar to the sum of squares decomposition, the degrees of freedom can be partitioned such that the total degrees of freedom ($N - 1$) equals the treatment degrees of freedom ($k - 1$) plus the error degrees of freedom ($N - k$).

C.4.2   Assumptions and Alternative One-way ANOVAs

Like all parametric statistical tests, the one-way ANOVA F-test relies on a set of underlying assumptions regarding the data. The first assumption is that the observations within each group are normally distributed with mean μ_i and variance σ^2

for the ith group. Consequently, the random errors, ε_{ij}, are assumed to be normally distributed with mean 0 and variance σ^2. This assumption is commonly assessed by examining whether estimates of the random errors, known as residuals, follow a normal distribution. Residuals defined as the difference between the observed value and predicted value ($e_{ij} = y_{ij} - \bar{y}_i$) are useful for identifying outliers. Normality can be evaluated with a Q–Q plot of the residuals, or formally assessed with statistical tests such as the K–S or Lilliefors test of normality. The second assumption of the one-way ANOVA is that the variances are similar across treatment groups. This can be assessed visually by plotting the residuals by the predicted values (Figure C.3).

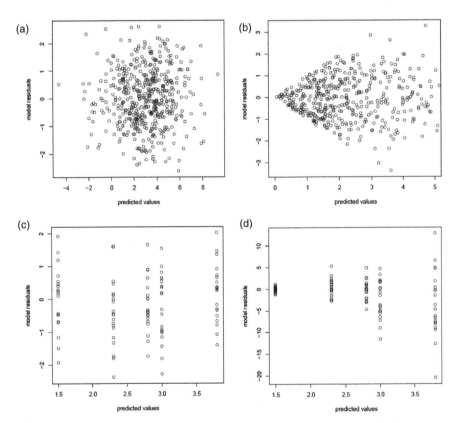

FIGURE C.3 *Assessment of equality of variances assumption for ANOVA and linear regression.* A plot of the model residuals and the predicted values is useful for evaluating whether the equality of variances assumption is satisfied in the data. (a) and (b) Residuals from a linear regression are plotted against their predicted values. There is no obvious trend in (a) suggesting that the homogeneity of variances assumption is satisfied. However, the funnel shape in (b) indicates that the equality of variances assumptions has been violated, suggesting that the data need to be transformed, a nonparametric alternative should be used. (c) and (d) Residuals from an ANOVA are plotted against the treatment means. While there is no obvious pattern in (c), there is clearly an increase in variability as the treatment means increases in (d), suggesting a violation of the equality of variance assumption.

The spread of the residuals should remain constant as the predicted value increases if the homogeneity of variances' assumption is satisfied.

If the one-way ANOVA assumptions seem unreasonable given the data, an alternative ANOVA should be considered. For data that exhibit unequal variances, Welch's ANOVA, which explicitly allows for the treatment groups to have different variances, should be utilized. Welch's ANOVA is a direct extension of Welch's unequal variances t-test. For data that do not resemble a normal distribution, the Kruskal–Wallis ANOVA should be used. The Kruskal–Wallis ANOVA is a nonparametric alternative of a one-way ANOVA and is a direct extension of the Wilcoxon ranked sum test, which compares whether the ranked sums across multiple groups differ significantly. The details of such ANOVAs are beyond the scope of the chapter, but can be found in classic statistical textbooks including Sokal and Rohlf (1995) and Tamhane and Dunlop (1999).

C.4.3 *Post Hoc* Multiple Comparisons of Means

The ANOVA F-test differs from the previous two group comparison tests in that it is an overall test. That is, if the null hypothesis is rejected by the F-test, this indicates that all treatment means are not the same. This, however, does not pinpoint exactly which treatment mean is statistically different from another treatment mean. It could be that all treatment means are statistically different from one another, or only a single pair of treatment means is different. To understand the source for a given significant F-test, more concise hypotheses need to be tested through multiple tests such as pairwise comparison tests. Pairwise comparisons of means can identify exactly which two treatments are statistically different. The number of pairwise comparisons for k treatment groups is $k(k - 1)/2$. For example, the total number of pairwise comparisons given four treatment groups is six. However, as previously discussed, an increase in the number of statistical test inflates the overall false positive rate, increasing the likelihood of incorrectly declaring a result significant.

The most conservative and crude methods of controlling for an inflation in the overall false positive rate are the Bonferroni and Dunn–Sidak corrections, which lower the significance level to a rate that accounts for the number of multiple tests performed. Individual pairwise comparisons are performed with a t-test and the p-values of these t-tests are compared to a Bonferroni-corrected (or Dunn–Sidak adjusted) significance level. For example, if an ANOVA F-test suggested that there was an overall difference between three treatments (e.g., control, compound A, compound B), three additional t-tests would be performed comparing each treatment with the other. Considering an overall significance level of 0.05 across the three pairwise comparisons, the Bonferroni-corrected significance level would be 0.05/3 or 0.0167. Thus, a t-test would only be declared significant if its p-value were less than 0.0167. More sophisticated methods have been developed that are less conservative and have greater statistical power. The Tukey–Kramer method is a popular pairwise comparison method that uses the most extreme observed discrepancy between treatment means ($\bar{y}_{max} - \bar{y}_{min}$) to determine the null distribution, called the Studentized range distribution. The Studentized range distribution controls for multiple testing by setting the overall false positive rate to the specified significance level. Usual t-test statistics are calculated for each pairwise comparison and are compared to the Studentized range distribution to determine whether the pairwise test is statistically different.

Pairwise comparisons are sometimes not of interest. There may be specific hypotheses, typically defined a priori, that may not be relevant for pairwise comparisons. For example, we may be interested in assessing whether a control group differs from two combined treatment groups, say compounds A and B, with respect to some cytological feature. Pairwise comparisons could tell us that the control group is significantly different from compound A and compound B, and that the two compound groups are not statistically different. From this, we could infer that the mean of the control group significantly differs from the pooled compound groups. However, there are statistical methods, named contrasts, that allow us to formally test this. Contrasts are a set of weights assigned to each group that mathematically specify a focused comparison. The only constraint placed on contrasts is that they must sum to 0 across all groups. For example, to continue with the above example, a contrast specifying a comparison between the control group and pooled compound groups is $(-2,1,1)$, where -2 is the weight assigned to the control group and 1 is the weight assigned to each compound group. This contrast states that the mean in the control group is equal to the average mean of the combined compound groups under the null hypothesis $(-2\mu_{control} + \mu_A + \mu_B = 0 \rightarrow 2\mu_{control} = \mu_A + \mu_B \rightarrow \mu_{control} = (\mu_A + \mu_B)/2)$. Contrasts are especially useful because they can specify any comparison among the groups, including pairwise comparisons. For example, $(0, -1,1)$ is a contrast comparing compound groups A and B. The contrast assigned a weight of 0 to the control group, thus excluding it from the analysis. Test statistics are formed with contrasts by summing the sample group means weighted by their contrast and dividing this sum by its standard error. This test statistic follows a t-distribution under the null hypothesis with $N - k$ degrees of freedom. More sophisticated methods, such as Scheffe's method, have been developed to correct for simultaneously performing multiple contrast tests. Scheffe's method is an overall test, similar to the one-way ANOVA that assesses whether at least one of the specified contrasts statistically deviates from the null hypothesis.

C.5   INTRODUCTION TO REGRESSION MODELS

Regression analysis is a widely used statistical technique aimed at explaining and quantifying the relationship between a set of explanatory variables to a dependent variable. It is especially useful for experiments with possible confounding factors. An experiment should ideally be designed in such a way that it minimizes the contributions of possible confounding factors that may bias results so that the biological hypotheses can be effectively evaluated. However, if confounding factors are present in an experiment, they can be entered into a regression model to "adjust" for its potential bias. In regression analysis, an empirical model is built from the data that estimates the amount of change in the dependent variable for a given change in an explanatory variable, holding all other explanatory variables constant. The dependent variable is specified in the model as the measurement that is affected due to changes in the explanatory variables. The explanatory variables are a set of variables entered into the model to try to "explain" the dependent variable. Explanatory variables are additionally referred to as predictors, covariates, and independent variables. As an example of dependent and explanatory variables, an experiment may wish to examine the influence of various concentrations of a compound on a cytological feature. In this example, the cytological feature is clearly the dependent variable and the compound concentration level is the explanatory variable. In this section, we introduce two types of regression models: linear regression, which is arguably the most commonly used regression model that models a continuous dependent variable, and logistic regression, which models a dichotomous dependent variable.

C.5.1 Linear Regression

Linear regression models assume a linear relationship between the explanatory variables and the dependent variable, which is assumed to follow a normal distribution. The simplest form of linear regression is a model in which a dependent variable is regressed on a single explanatory variable (called "simple linear regression" for obvious reasons). This model is mathematically notated as $y_i = \alpha + \beta x_i + \varepsilon_i$, where y_i and x_i are the dependent and explanatory variables, respectively, for the ith observation and the ε_is are the random errors. α and β are called regression coefficients (or model coefficients) and represent the intercept and slope of the regression line. Consider as an example that we are interested in modeling the relationship between some cytological feature, such as total intensity which we define as the dependent variable, to another cytological feature, such as area which we define as the explanatory variable. A plot of these two features shows that they are clearly correlated and that a line can be fitted through the scatterplot (Figure C.4). The slope of this line represents the magnitude of association between the variables, or more precisely, the amount of predicted change observed in total intensity for one unit change in area. The line is specifically fit with the ordinary least squares (OLS) method, which estimates the intercept and slope such that they minimize the deviations between the line and the observed dependent variable $(y_i - (\alpha + \beta x_i))$. More specifically, OLS takes the squares of these deviations and sums them across all observations to obtain an overall goodness-of-fit measure (mathematically notated as $\sum_{i=1}^{n} [y_i - (\alpha + \beta x_i)]^2$). This quantity should clearly be small in order for the regression line to fit well and when minimized, should yield the "best" intercept and slope estimates. These estimates are commonly referred to as OLS estimates and are denoted with a "^" sign as $\hat{\alpha}$ and $\hat{\beta}$.

Simple linear regression is closely related to correlation analysis, as apparent from Figure C.4. Correlations measure the strength and direction of the relationship between two measurements. The most familiar type of correlation is the Pearson correlation coefficient (r), which is appropriate for two normally distributed measurements that appear to have a linear relationship. It is calculated as the covariance between the two measurements divided by the product of their respective standard

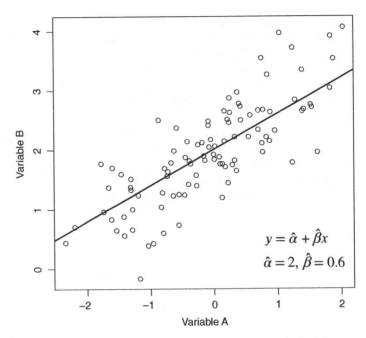

FIGURE C.4 *Simple linear regression.* A scatterplot of two cytological features and their estimated linear regression line.

deviations. The Pearson correlation coefficient, r, is equivalent to the standardized OLS slope estimate. That is, if both variables were z-score standardized to have a mean of 0 and a variance of 1, the OLS slope estimate would be equivalent to the Pearson correlation coefficient. Rather than standardizing the two variables, the slope estimate can be standardized with the following transformation: $r = \hat{\beta}\frac{s_x}{s_y}$, where s_x and s_y are the standard deviations of the explanatory variable and dependent variable, respectively. This standardized regression coefficient is equivalent to the Pearson correlation coefficient. In simple linear regression, the correlation coefficient is also the square root of the coefficient of determination, r^2, which represents the proportion of variation explained or accounted for by the regression model. Thus, an r of 0.84 translates to an r^2 of 0.7, implying that 70% of the variation in the dependent variable is explained by the fitted regression model. The coefficient of determination is often used as a goodness-of-fit measure, to indicate how "well" the model fit the data. As a side note, the Pearson correlation coefficient is not appropriate if two variables exhibit a nonlinear relationship. For these scenarios, a nonparametric correlation alternative should be considered such as Spearman's rank correlation, which measures the relationship between rankings of the variables.

Statistical inferences can be made regarding the estimated slope coefficient for a simple linear regression model. The null hypothesis states that the explanatory and dependent variables are independent or that there is no relationship between the two variables ($H_0 : \hat{\beta} = 0$) and the alternative hypothesis states that the two variables are not independent ($H_0 : \hat{\beta} \neq 0$). The test statistic is formulated as the OLS estimate divided by its standard error, which follows a t-distribution with $N - 2$ degrees of freedom under the null hypothesis. The slope estimate is interpreted as the expected amount of change in the dependent variable for one unit change in the explanatory variable. Thus, the larger the slope estimate, the stronger the association between the two variables. Positive slope estimates indicate that the two variables are positively correlated (i.e., as one variable increases, so does the other) and negative slope estimates indicate that there is a negative correlation between the two variables (i.e., as the explanatory variable increases, the dependent variable decreases).

Assumptions of the linear regression model are similar to those of the one-way ANOVA. Specifically, linear regression models assume that the random errors, ε_i, are assumed to be normally distributed with mean 0 and variance σ^2. The normality assumption can be assessed by model residuals, defined as the difference between the observed value and its predicted value $e_i = y_i - \hat{y}_i$. Conceptually, residuals show how well the model was able to predict the dependent variable. A very large (or small) residual indicates that the model was unable to "explain" that observation well, suggesting that the observation may be an outlier. Note that although linear regression assumes the dependent variable (and thus the random errors) to be normally distributed, it makes no assumption regarding the explanatory variables. Thus, one could model a dichotomous (yes/no) explanatory variable, a high skewed explanatory variable, and so on. Although there are no assumptions placed on the distributions of the explanatory variable, it should be checked whether their relationship with respect to the dependent variable is linear. Lastly, linear regression assumes equality of variances across different levels of the explanatory variable similar to ANOVA. This can be inspected visually by plotting the residuals e_i by the predicted values $\hat{y}_i$. The plot should not reveal any obvious relationship between the residuals and the predicted values, otherwise there is evidence of nonconstant variance (Figure C.3).

C.5.2 Multiple Linear Regression

The simple linear regression model can be extended to a multiple linear regression model, which models additional explanatory variables. These models are referred to as multivariable models as they contain more than one explanatory variable. In multiple linear regression, the model is formulated such that the dependent variable is predicted by a set of variables:

$$y_i = \alpha + \beta_1 x_{1i} + \beta_2 x_{2i} + ...\beta_k x_{ki} + \varepsilon_i$$

where β_j denotes the regression coefficient for the jth explanatory variable and x_{ji} denotes the ith observation for the jth explanatory variable. Estimation of the regression coefficients and statistical inference is made with the same methodology introduced for simple linear regression. However, the interpretation of the estimated regression coefficient is slightly different in a multivariable model. The estimated regression coefficient indicates the expected amount of change in the dependent variable for one unit change in the explanatory variable, holding other variables constant. In other words, the regression coefficient indicates the relationship between the explanatory variable and dependent variable for fixed levels of the other explanatory variables. It can be considered as the effect between the two variables, "adjusting" for the other covariates. Sometimes there may be confounders in an experiment that one was unable to control in the experimental design. These confounders can be entered in a multivariable model as additional covariates to "control for" or "adjust for" their effect on the outcome.

Explanatory variables in a multivariable model should not be highly correlated, otherwise multicollinearity problems may arise. Multicollinearity is present when explanatory variables are not linearly independent, or can be computed from one another. For example, total intensity, mean intensity, and area are not linearly independent since mean intensity is calculated as total intensity divided by area. Thus, only two of these features should be included in a multivariable model, since the third feature can be calculated from the other features. Multicollinearity can lead to biased regression coefficients as well as inflated standard errors.

C.5.3   Logistic Regression

Linear regression models are appropriate for normally distributed dependent variables. They are not appropriate for categorical or discrete outcomes, as they would predict observations that are outside the range of the categorical outcome. For example, if we were interested in modeling the probability of a dichotomous outcome (such as the proportion of cells undergoing mitosis), we would need to transform this probability so that it resembles a normal distribution. Without the transformation, a linear regression model would predict values greater than one and smaller than zero. To model the relationship between a set of explanatory variables and a dichotomous outcome, we need to use the logistic regression model. Logistic regression models are a type of generalized linear models, which are a set of models that have been generalized from the linear regression model through some transformation of the dependent variables. In logistic regression, the dichotomous variable is first converted to proportions (e.g., proportion of cells undergoing mitosis and proportion of cells not undergoing mitosis) which are then transformed using the logit transformation, defined as $\log(\frac{p}{1-p})$ where p is the probability of the dichotomous outcome. This transformed outcome variable can now be regressed among the explanatory variables. Thus, the logistic regression model relates the explanatory variables to the dependent variable through the logit link. Specifically, the logistic regression model formulation is

$$\log\left(\frac{p}{1-p}\right) = \alpha + \beta_1 x_{1i} + \beta_2 x_{2i} + \dots \beta_k x_{ki}.$$

This model always predicts probabilities (p) between one and zero. Interpretation of the regression coefficients is different from linear regression models, as one can infer from the formulation of the model. The exponents of the regression coefficients (e^{β}) represent odds ratios, defined as the ratio of odds of the outcome for different levels of the explanatory variable. The odds ratio (OR) is also a common measure of association between two dichotomous measurements. An OR greater than 1 indicates that there is an increased likelihood of the outcome for increasing values of the explanatory variables and an OR less than 1 indicates that there is a decreased likelihood of the outcome for increasing values of the explanatory variable. An OR of 1 indicates no association between the variables. Similar to the linear regression model, the null hypothesis states that there is no association (i.e., $OR_j = 1$) versus the alternative hypothesis which states there exists some association (i.e., $OR_j \neq 1$). In addition, logistic regression makes no assumptions regarding the distribution of the explanatory variables. In fact, logistic regression does not impose as many assumptions as linear regression (such as the normality and constant variance assumptions). The only real limitation of logistic regression is that the outcome is assumed to be dichotomous (e.g., yes/no).

C.6   CONCLUSIONS

Data analysis is an integral step in any high content screening experiment. This supplement introduced the fundamentals of data analysis and highlighted commonly used univariate statistical tests. Prior to any statistical or data analysis, the data should always be inspected visually to reveal whether any experimental artifacts or outliers should be excluded from the dataset. Systematic inter- and intraplate variability can be attenuated with normalization procedures. Because the majority of statistical tests rely on the assumption that the data follow a normal distribution, it is important to assess normality to decide whether a parametric or nonparametric statistical test should be utilized. Whereas parametric tests assume the data to follow some theoretical distribution, nonparametric test make no assumptions regarding the distribution of the data. However, in general, parametric tests are more powerful than nonparametric tests and thus data transformations should be explored to try to make the data resemble a normal distribution.

Statistical tests to compare a continuous cytological measurement among two treatment groups include the t-test, Wilcoxon rank sum test, and the K–S test. The t-test compares the means among two treatments and is a parametric test that assumes the data to be normally distributed. There are two alternatives of the t-test: Student's t-test which assumes the treatment groups to have equal variances, and Welch's t-test which allow the variances among treatment groups to be unequal. The Wilcoxon rank sum is a nonparametric alternative to the t-test that is most suitable for highly skewed data. It assesses whether observations from one treatment group tend to be larger than the other group. The K–S test is the most general two-group comparison, and assesses whether there exists any difference at any given point between the two treatment distributions. The K–S test similarly is a nonparametric test and can be applied to any dataset.

The ANOVA and regression analysis were additionally introduced in this chapter. ANOVA is a direct extension of the t-test that compares whether more than two treatments have equivalent means. ANOVA is an overall test that indicates whether any of the treatment means differ and thus, does not pinpoint exactly which treatment mean is statistical different. Following a statistically significant ANOVA F-test, pairwise comparisons are often performed to identify the treatment(s) that produced the statistically significant overall test. In addition, hypothesis driven contrasts can be tested that can compare any combination of treatments. Regression analysis aims to quantify the relationship between a set of explanatory variables to a dependent variable. Linear regression models a continuous dependent variable, which is assumed to follow a normal distribution, among a set of explanatory variables. Standardized regression coefficients in multiple linear regression models can be interpreted as partial correlation coefficients, or correlation coefficients adjusting for other covariates. Logistic regression, in contrast, models a dichotomous dependent variable and the exponents of the regression coefficients can be interpreted as odds ratios.

KEY CONCEPTS

Inferential statistics is rooted in the concept of the sample set as data used to estimate the total population. The measures generally quantify the robustness of the sample as a reflection of the total population.

When performing large numbers of hypothesis tests, it is critical to perform multiple comparison testing. When a 100 t-tests are performed, 5 of them will (by definition) be significant purely by chance at a significance level of 0.05. Three methods for multiple comparison testing were discussed in this chapter—Bonferroni correction, the FDR, and permutation testing.

FURTHER READING

Box, G.E.P., Hunter, W.G., and Hunter. J.S. *Statistics for Experimenters: Design, Innovation, and Discovery.* 2nd edn. Wiley-Interscience, Hoboken, NJ, 2005.

Brideau, C et al. Improved statistical methods for hit selection in high-throughput screening. *Journal of Biomolecular Screening*, 2003, **8**: 634–647.

Kuehl, R.O. *Design of Experiments: Statistical Principles of Research Design and Analysis.* Duxbury Press, Pacific Grove, CA, 1999.

Sokal, R.R. and Rohlf, F.J. *Biometry: The Principles and Practice of Statistics in Biological Research.* 3rd edn. W.H. Freeman, New York, 1995.

Tamhane, A.C. and Dunlop, D.C. *Statistics and Data Analysis: From Elementary to Intermediate.* Prentice Hall, Upper Saddle River, NJ, 1999.

Zar, J.H. *Biostatistical Analysis.* 4th edn. Prentice Hall, Upper Saddle River, NJ, 1999.

GLOSSARY

Term	Chapter introduced	Definition
Absolute IC50	Chapter 8	The 50% inhibitory concentration when the data conform to a standard dose–response curve and the signal ranges from the minimum to the maximum range of the control wells.
Adaptive threshold	Chapter 4	Identification of positive signal based on the difference in intensity relative to local background, for example, intensity must be 100 grey values above the local background. This is generally more robust and less prone to variability as compared to a fixed threshold.
Agglomerative clustering	Chapter 10	Clustering method based on finding the two samples nearest each other, grouping them as a cluster, and finding the next nearest sample, adding that member. Distance measures used in dendrograms are based on the distance between a cluster and a new member.
Bandpass	Chapter 2	The range of wavelengths that will be transmitted through a filter.

(continued)

An Introduction to High Content Screening: Imaging Technology, Assay Development, and Data Analysis in Biology and Drug Discovery, First Edition. Edited by Steven A. Haney, Douglas Bowman, and Arijit Chakravarty.
© 2015 John Wiley & Sons, Inc. Published 2015 by John Wiley & Sons, Inc.

Term	Chapter introduced	Definition
B-score	Chapter 9	A method of normalizing values in a microplate by accounting for systematic changes based on location, particularly along the outer wells of a microtiter plate. Uses median values instead of means.
Centroid	Chapter 14	A point of reference for authentic data points to be measured against in a multivariate (multifeature) data set in k-means clustering. Several centroids will be added to the data set for the purpose of grouping the data points as clusters defined by being more related (closer) to one centroid than any of the other centroids.
Chemical genetics (or genomics)	Chapter 1	Similar to HTS, but at a smaller scale and seeking to find chemicals that perturb specific cellular processes for the purpose of using these compounds to characterize the biology and mechanism of this process, rather than to begin the process of developing a drug.
Coefficient of variance (CV, or %CV)	Chapter 8	Standard deviation divided by the mean, a method for comparing and integrating standard deviations across a dose range.
Convoluted image	Chapter 4	A confocal image that has not completely removed light from outside the plane of focus.
Correlation matrix	Chapter 10	The extent of similarity between a set of feature measurements with each other when the measurements are scaled closely with each other. Used to group similar features in procedures such as Principal Components Analysis and Factor Analysis.
Covariance matrix	Chapter 10	Similar to a correlation matrix, but used when the feature data are not closely scaled (such as intensity measurements and p-values). Used in the same way as a correlation matrix, but the data are not considered as robust.
Deconvolution	Chapter 3	An interpretative image processing algorithm that removes light that is out of focus (such as light that is outside of the focal plane) and is used to sharpen the image.
Difference matrix	Chapter 10	A measure of distance or difference between a variable with all other variables in a table of feature measurements. Used for clustering samples or features in a dataset.
Distribution-free (Statistical test)	Chapter 10	A statistical test or measure that does not assume the data values fall into a specific pattern or distribution in order for the test or measure to be valid.

Term	Chapter introduced	Definition
Divisive clustering	Chapter 10	Clustering method based on looking at dividing a group of samples into two most clear-cut subgroups, dividing each of these groups by the same criteria, until all samples in all subgroups have been separated. The distance between the groups once separated defines the extent of separation, as depicted in a dendrogram.
Dose–response relationship	Chapter 8	The change in activity (e.g., inhibition of an enzyme) as a function of compound dose. Ideal behavior is a sigmoidal curve with a Hill slope of 1 and clear plateauing of the signal change at the lowest and highest doses. Deviations from these characteristics are indicative of complex effects of the compound on the assay metric.
Euclidean distance	Chapter 10	Roughly, geometric. The distance between two samples when plotted as a graph using the feature data to position the point along n number of axes.
Feature	Chapter 7	Something that can be measured by an image analysis algorithm from an image. Some features are direct measurements, such as length of a nucleus along the major axis, others are interpreted by comparing fluorescence channels of the same field, such as speckles in one channel per cell, as measured by proximity to the nucleus, which was measured in a different channel.
Feature data	Chapter 8	Measurements made by the image analysis algorithm; quantitative properties of identified objects.
FFPE (formalin-fixed, paraffin embedded)	Chapter 13	Preservation of a tissue sample that fixes and supports the tissue for processing by IHC, tissue microarrays or other methods.
Fixed threshold	Chapter 4	Identification of positive signal based on fixed intensity threshold, for example, every pixel with an intensity value greater than 1000 is considered signal. This can be fast and work well on images with high signal-to-noise ratio, but is prone to errors with on images with varying background or with the relative fluorescence changes from image to image.
Fluorophore	Chapter 2	A molecule that absorbs light within a range of wavelengths and emits most of the absorbed energy as light at a longer, lower energy (red-shifted) range of wavelengths.

(continued)

Term	Chapter introduced	Definition
Heteoskedasticity	Chapter 10	When the variance in a distribution changes between samples. Homoskedasticity is assumed for some statistical tests, so heteroskedasticity will invalidate presumptions for these tests.
High content screening (HCS)	Chapter 1	A term that was coined to contrast the rich data acquired through automated, image-based screening with the high-volume, single-measurement data generated from high throughput screening (HTS).
High throughput screening (HTS)	Chapter 1	A highly automated screening process that seeks to find drug candidates through their effect on a single cellular or biochemical effect. Greater than a million compounds can be screened in a single HTS campaign, although smaller campaigns are common.
Homoskedasticity	Chapter 10	Equal variance across samples in an experiment. This is presumed for many statistical tests, especially those that anticipate a normal distribution (such as the t-test).
IHC (immunohisto-chemistry)	Chapter 13	Labeling of tissue samples with visible (non-fluorescent) antibody-based dyes, typically with a colorimetric counterstain.
Lattice	Chapter 10	An array of graphs (as called from the statistical package R), that plot a table of data in pairwise combinations.
LIMS (laboratory information management system)	Chapter 13	A software-based system for storing information related to laboratory operations. In the context of WSI and digital pathology, these systems store images and related sample data that include sample processing data (antibody, dilution) and treatment data (compound name, dosage, time point).
MAD	Chapter 8	Median absolute deviation, similar to the standard deviation, but based on dispersion around the median rather than the mean, and therefore a more robust measure of error, due to being less sensitive to outliers.
Mean	Chapter 8	The average value, the total value of all of the samples divided by the number of samples.
Median	Chapter 8	The value of the 50th percentile, or middle value for samples that have been rank ordered.
Metadata	Chapter 8	Information about an experiment, such as treatment conditions for each well (e.g. dose, compound name).
Metric	Chapter 8	Relevant feature or value derived from a set of features that describe the relevant biological phenotype.

Term	Chapter introduced	Definition
Microplate	Chapter 3	A synonym for microtiter plate.
Microtiter plate	Chapter 3	A multiwell plate for running experiments with multiple samples simultaneously. Wells are arranged in a 2×3 array, most commonly for 96 wells (8×12 array) or 384 wells (16×24 wells), but 6-, 12-, 24-, 48-, and 1536-well formats exist.
Multiwell plate	Chapter 3	Essentially a synonym for microtiter plate, but more common for plates with lower well numbers (6–48 wells).
Non-normal Distribution	Chapter 10	A group of values that does not conform to a normal distribution pattern. Such a distribution cannot be accurately evaluated by many common statistical tests, including Student's t-test.
Nonparametric	Chapter 10	A distribution that cannot be characterized by parameters such as the mean and standard deviation. Such distributions cannot be modeled (and therefore evaluated) by these parameters, and must therefore be compared by methods that do not explicitly rely on such patterns or shapes.
Normal distribution	Chapter 8	The most common type of parametric distribution, and the form that the most common statistical tests assume (or have been proven) to be true. A distribution of values where the data that fall away from the mean value do so equally often on either side of the mean, and according to proximity from the mean. The resulting distribution is the classic "bell-shaped" distribution, and can be accurately described by the parameters of the distribution, such as the mean and the standard deviation.
Overfitting	Chapter 10	When a formula or analysis used to characterize data is too complex, explaining both the signal and the noise in a dataset.
Parametric	Chapter 10	A distribution that conforms to a set of parameters, such as the mean and the standard deviation. Distributions conforming to such patterns can be compared using the parameters characterizing the distributions.
Q–Q plot	Chapter 10	Quartile–quartile plot; a comparison of measured data to theoretical data, if it were to conform to a specific parametric distribution (usually a normal distribution). Deviations from the fit line indicate a lack of fit to the theoretical distribution.

(*continued*)

Term	Chapter introduced	Definition
Quantile	Chapter 10	Specific percentile points for rank-ordered data, dividing the measures into four quadrants, the 25th, 50th, 75th, and 100th percentiles correspond to the 1st through 4th quartiles, respectively.
Quintile	Chapter 10	Identical to quartiles, but explicitly includes the 0th percentile as the first quintile. The quartile values correspond to the remaining quintile values.
Range	Chapter 8	The min and the max values.
Rank-ordered data	Chapter 10	A set of data values, such as feature values for one or more groups of samples, that are arranged in rank order. In some statistical tests, the data values are not compared, but the rank ordered membership in one group or another are compared.
Resolution	Tutorial	The extent to which cells or fine cellular structures can be distinguished, dependent on magnification and the numerical aperture of the objective.
Relative IC50	Chapter 8	The 50% inhibitory concentration that is deduced from the data when they either do not conform to a standard dose–response curve (e.g., not clearly plateauing at the highest concentrations), or when the signal plateaus at a range less than what is observed in the positive control samples.
Robust	Chapter 8	Resistant to minor errors or deviations from expected (idealized) patterns. As examples, robust statistical methods are those that do not rely on measurements that are most sensitive to outliers (such as the mean of a sample), and alternatively use parameters that are less sensitive to such anomalies (such as the median).
Segmentation	Chapter 4	Identification of objects based on thresholding and other image processing steps, for example, identification of nuclei based on DNA marker.
Signal-to-noise	Tutorial	The extent to which authentic fluorescence can be distinguished over the readings of background pixels in the image. Defines whether objects can be detected, but is not the same thing as an assay window, where some of the detected objects in the negative control well may be "positive", for example, localized to the nucleus in a translocation assay.

Term	Chapter introduced	Definition
Site acceptance test (SAT)	Chapter 5	A set of tests a manufacturer performs upon installation of an HCS system to confirm operation of all the components, from the microscope to image analysis workstations. A user can also include a specific assay of interest to add to the vendor's SAT. For example, if the instrument was purchased to perform a specific assay, this assay can be performed upon installation, and the results can be compared to some predetermined results.
Skew	Chapter 10	When a distribution shows an enhanced or exaggerated extension of values on one side (from the median or mean). In some cases, a skewed distribution can be converted to a normal distribution through an algebraic transformation, such as taking to log of the original distribution when the skew is on the right, or toward the higher values, when the data are plotted as a histogram or density plot.
Skew	Chapter 10	A shift from a normal distribution in one or both of the tails of the distribution, such as a large number of outliers.
Standard deviation	Chapter 8	A number that describes the variability of a set of values around its mean value. For a normal distribution, the spread of values defined by the mean $\pm$ the standard deviation represents where the majority of the values lie (68.2%, in an ideal normal distribution).
Standard operating procedure (SOP)	Chapter 11	A detailed document describing all aspects of the process (wet lab, reagent sources, image acquisition, image analysis, etc.) required to perform an assay. The goal of the SOP is to achieve better uniformity and reproducibility across multiple users, multiple sites, and multiple experimental runs.
Summary statistics	Chapter 8	A set of commonly used statistical measures, typically used for an initial inspection of sample groups/treatment conditions in an experiment. Inspection can include determining which groups showed an effect or which ones may be comprised of data that are not normally distributed. Typically four or more measurements of the following, for each treatment group: min/0th percentile, 25th percentile/1st quartile, mean, median/50th percentile/2nd quartile, 75th percentile/3rd quartile, max/100th percentile/4th quartile/5th quintile, range (min and max), and standard deviation (assumes data are normally distributed).

(continued)

Term	Chapter introduced	Definition
Truncation	Chapter 8	Summarizing data in a way that loses information, such as using percent of cells positive and losing information about the intensity, shape, area, and granularity of the region measured.
V-factor	Chapter 8	Similar to an Z-prime score, but incorporates the standard deviations of all of the doses in a dose–response curve instead of just those for the positive and negative controls.
Widefield microscopy	Chapter 2	The most common type of microscopy, that views an image through an objective with a wide focal plane, capturing a relatively thick image with reasonably consistent focus.
Working distance	Chapter 3	The distance from the objective that can be viewed in focus by the objective. Long or extra-long working distance objectives are required for capturing cells away from the well bottom, such as in matrices or tissue samples or spheroids.
WSI (whole slide imaging)	Chapter 13	Technology that enables the acquisition or digitization of an entire sample on a slide to a digital image. These images can be viewed on a computer workstation much like one would view on an optical microscope.
Z-score	Chapter 8	A common measure of robustness for a screening assay, based on the size of the standard deviations and difference between the means for a set of positive and negative controls.
Z stack	Chapter 3	A set of images from the same field, but at progressively farther distances from the well bottom that can be used to generate a 3D image of the cells in the field.

TUTORIAL

DOUGLAS BOWMAN AND STEVEN A. HANEY

T.1 INTRODUCTION

With a steady stream of software and other highly technical products being released each month, one of the best ways for a potential user to become familiar with new offerings is through the use of one or more tutorials. Tutorials change the problem from "how do I understand all the things this product has to offer" to "how does this new product help me get something done." A single tutorial will not address all aspects of the new product, but will show how it can be used to solve a specific problem. Multiple tutorials will highlight different capabilities. Importantly, tutorials are generally better at discussing the preprocessing of data that are needed get started.

The following is a virtual tutorial, designed to allow the reader to work through two simple HCS assays, starting from acquisition on an HCS imaging system to initial plate-based quality assurance, image analysis, data analysis, and finally to quality control. A genuine tutorial will require loading software to analyze images, but this virtual version is designed to discuss the material presented in this book, in the context of a "what you need to do to go from the laboratory bench to the group meeting."

Genuine tutorials are available, some using open-source tools, so it is possible to work through all the steps of image analysis and data processing without stepping into a laboratory or by purchasing equipment or software (assuming you have access to a computer and an internet connection). Images are available from

An Introduction to High Content Screening: Imaging Technology, Assay Development, and Data Analysis in Biology and Drug Discovery, First Edition. Edited by Steven A. Haney, Douglas Bowman, and Arijit Chakravarty.
© 2015 John Wiley & Sons, Inc. Published 2015 by John Wiley & Sons, Inc.

several sources, the most relevant source for high content image analysis is the Broad Bioimage Benchmarking Collection (www.broadinstitute.org/bbbc/). The collection covers many diverse experiments, generally a single plate, and includes complete image sets, a description of the experiment and results, and complete instructions for processing and analyzing the images. The instructions use CellProfiler for the image analysis, an open source image analysis application, but any application can be used. CellProfiler may be downloaded through the Broad Institute (www.cellprofiler.com). An alternative open source image analysis application is FIJI (http://fiji.sc/Fiji), an updated version of ImageJ, the classic open source image analysis application.

This tutorial will first introduce two common biological assays used in a drug discovery program. These assays will be described in enough detail for the reader to understand the biological event, but the important aspect is to understand what biological phenomenon (phenotype) is important and how to use the high content imaging system to capture and measure the phenotype.

We will attempt to model the process of image acquisition for HCS. One of the changes in the way concepts are discussed in this section is that elements of image analysis are used in image acquisition. Whereas we have dealt with concepts such as camera exposure settings and object segmentation separately in the main text, in practice, segmentation efficiency is frequently used to assist in setting the exposure parameters. An iterative approach that includes adjustments of exposure parameters with evaluation of segmentation accuracy helps insure that *post hoc* image analysis will be robust. One may also utilize image analysis during image acquisition, for example, if software allows for adaptive acquisition to collect a predefined number of cells per well. In this case, nuclei counting is done at acquisition to estimate cell count. Taking a set number of fields can result in some wells having few cells to quantify, so setting a minimum number of cells per well enables the instrument to collect additional fields for a sparse well.

T.2   THE ASSAYS

The two biological assays are the Mitotic Index assay, an imaging assay designed to measure the number of cells within a population that are undergoing mitosis, and the Forkhead Translocation assay, an imaging assay designed to measure the extent of translocation of a protein from the cytoplasm to the nucleus. At several points during this discussion, critical decisions for each assay will be compared side-by-side to emphasize places where considerations that are unique for one situation will impact these decisions. A bit more about the assays:

(i) **The Mitotic Index assay.** This simple two-color assay measures the percentage of cells undergoing mitosis, within a population of cells in a well. The first label is needed to label the entire population of cells (nuclei, specifically), and the second label is required to identify the fraction of this population of cells that are in mitosis. A DNA stain such as DAPI (a fluorescent stain that binds to DNA), Hoescht, or DRAQ5 is used to label the cell nucleus and a mitotic marker such as phospho-histone H3 (Ser 28) is used to mark cells in mitosis. Most biologists are familiar with the analysis of DNA content to monitor the cell cycle through flow cytometry. In these cases, data are collected on many thousands of cells for each sample (often more than 40,000 cells) and the resulting histograms are robust measures of the population. Since HCS is typically a well-based format (and 96-wells or greater, at that), collecting a sufficient number of cells is one of the challenges for adapting DNA content to HCS. Key decisions in this experiment are faithful recognition of nuclei through all phases of cell division (where nuclei in metaphase through anaphase will be very different from those in G1), and collection of enough cells per well to produce an accurate measure of their growth state. A platemap showing the layout of the treatments and example images of the DNA and pHH3 staining is shown in Figure T.1.

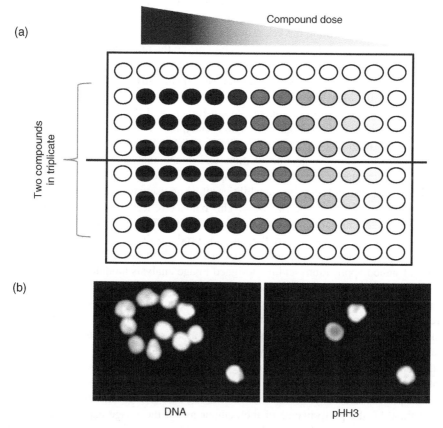

FIGURE T.1 *Mitotic Index assay.* (a) Plate configuration showing dose titration of two compounds in triplicate. (b) Example images showing DNA marker to label all nuclei and pHH3 marker to label mitotic nuclei.

(ii) **The Forkhead Translocation assay.** FKHR, a mammalian transcription factor, is normally localized in the cytoplasm, but translocates to the nucleus upon inactivation of the PI3K/AKT pathway. The FKHR Redistribution Assay (BioImage, part of Thermo Scientific) utilizes a green fluorescent protein (GFP) fusion protein to fluorescently label FKHR. A DNA stain such as DAPI, Hoescht, or DRAQ5 is used to label the cell nucleus. As shown in Figure T.2, the GFP is primarily localized to the cytoplasm in the control sample and primarily localized to the nucleus in the treated sample.

T.3 THE IMAGER AND IMAGE ANALYSIS SOFTWARE FOR IMAGE ACQUISITION

Although the intent of this book has been to be system-independent, it is necessary to illustrate these assays using an imaging platform and image analysis software. For this tutorial, the image acquisition phase and some of the image analysis will use the MetaXpress HCS platform (Molecular Devices, Sunnyvale, CA) to walk through the process: from defining the key acquisition parameters, to acquiring images with sufficient resolution and signal/noise, to configuring the image analysis algorithm to quantify cellular activity by automatically measuring a variety of image and cellular features. Images for these assay examples and quantitative image processing was performed using MetaXpress on the ImageXpress Micro HCS System. Some figures use images generated on the Opera (Perkin-Elmer) and the Thermo Insight NXT (Thermo-Cellomics, Pittsburgh, PA). The MetaXpress software is used in both the image acquisition and image analysis phases as MetaXpress includes components for instrument control, image acquisition, and image analysis. The image analysis capabilities include a mix of "canned" application modules designed for specific biological assays, and scripting capabilities allowing users to customize image analysis algorithms for their specific assay. This balance reflects the range of users and applications that are anticipated by the manufacturer. This is common to all of the major instrument vendors, where they provide products that will be deployed in academic Cell Biology Departments and industrial high throughput screening (HTS) cores. Canned applications and predesigned image analysis modules that anticipate the most common HCS assays are an integral part of image-based assay biology. Specific iterations are detailed for each platform but are expected to give similar results (determination of the percentage of cells in S phase should be relatively consistent across all platforms that use a canned assay module). Canned applications are validated on a few cell lines (typically HeLa, U2OS, and a couple of others), so difficulty may be encountered if working with nonvalidated cell lines. Another aspect of canned applications is that they are frequently used during image acquisition. Up to this point in the book, we have worked through the model that image acquisition is defined by the performance of the cellular assay, the image capture and image processing to generate the data; this is followed by a data analysis phase. Canned applications can link steps during image acquisition with their impact on the results (typically performed on control wells), before the images are acquired for the entire

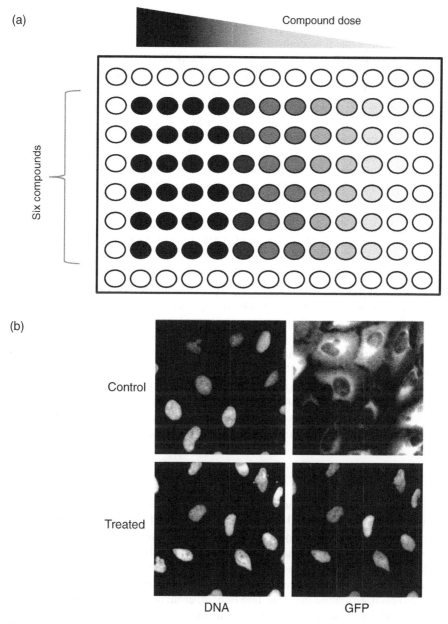

FIGURE T.2 *Translocation assay.* (a) Plate configuration showing dose titration of six compounds. (b) Example images showing primarily cytoplasm localization of GFP signal in control and nuclear localization in compound treatment.

plate. In this way, the algorithm can be adjusted to optimize the results and then run to produce the final results. For acquisition, a cell counting algorithm could be utilized to assure that a predefined number of cells are acquired per well. There are many cases where this is fine, but it does open the potential for problems with some of the samples to be missed (as discussed at several points in the main text). Review of images and data should still be a part of the complete experiment.

This tutorial will primarily expose users to the canned application modules (specifically the Mitotic Index, Cell Scoring, and Translocation). Many users will be running basic assays and have no need to run more sophisticated script-based analyses. Many of the other high content imaging systems available include a variety of tools similar to these two software products. Although the specific tools are different, the concepts introduced in this tutorial should be easy to translate to the other software packages.

T.4   LESSON 1: GETTING STARTED IN IMAGE ACQUISITION

The first lesson will work through the setup of the instrument parameters for image acquisition while stressing the importance of "good data in = better data out".

The quality of an HCS assay depends on a number of factors. The common saying "bad data in = bad data out" (there are several versions of this) applies to HCS assays as much as any other quantitative method. It is important to optimize each stage of an assay. The tutorial will focus on optimizing the instrument parameters and image analysis parameters, but equally important is the "wet lab" optimization as described in earlier chapters. In fact, the image visualization and image analysis measurements are often important in determining the optimal wet lab protocol. This process is often an iterative process among wet lab protocol optimization, image acquisition, and image analysis. Two key acquisition parameters are **resolution** and **signal/noise**.

T.4.1   Resolution

The images should have enough resolution, so the image analysis algorithm can accurately define the required cellular compartment and can determine the cellular features needed for each assay. Resolution on HCS systems is typically defined by the magnification (microscope objective) and binning (digital camera) as described in Chapter 3. As you increase magnification, you increase the ability to measure small objects but with a reduced field of view, which results in fewer cells in a single image. This may require the collection of a larger number of sites per well in order to collect a large number of cells. Also, if the magnification is too low, you may lose the ability to detect individual cells or to differentiate between the nucleus and the cytoplasm. Binning increases signal at the expense of resolution, so if signal detection is essential, binning can help, but if texture or granularity is more important, binning may compromise the analysis. So in these two examples, the Mitotic Index assay is essentially a binary endpoint (even if the specific endpoint is "intensity per nuclei," a real number, the quantification will ultimately be "% positive cells"). The challenge here is to include enough cells to determine whether a treatment has had an effect, so

FIGURE T.3   *Magnification and camera binning are key instrument parameters to define resolution.*

fine details are less important than definitive decisions on whether a cell is in mitosis or not.

As for the Forkhead assay, robustness is affected by segmentation as much as tabulating the total numbers of cells. So maintaining clear definitions between cellular boundaries is a key factor. High resolution is a strong consideration during assay development. Increasing resolution is one thing that separates this assay from the Mitotic Index assay.

Having said that, it is important to note that good data can often be extracted from lower resolution images. We return to subtle issues that were raised at various points in this book. Magnification sounds like a simple decision; more is better. Although high magnification images are more visually appealing, during data analysis, it is frequently found that lower magnification does not result in poorer performance. In fact, lower magnification results in higher number of cells acquired and reduced scan times on the instrument. There are occasions where resolution is important. For example, higher resolution is required if the biological phenotypes are small objects, such as markers for DNA damage that exhibit small spots in the nucleus, and in experiments where texture is relevant (e.g., smoothness vs. graininess of the cytoplasm). "On-the-fly" image analysis (image analysis performed during image acquisition) is one of the strengths of canned algorithms, as they assess the impact of image acquisition parameters in real time, through adjusting options available in the instrument controls (an example is shown in Figure T.3). Key parameters for image resolution are listed in Table T.1.

In the simple example shown in Figure T.4, cells were acquired with both a 4× and 10× objective. If the desired output was a simple nuclei count (proliferation assay, live-dead assay), both resolutions result in desired segmentation accuracy. The added benefit of the 4× is 2.5 times as many cells per field.

T.4.2   Signal-to-Noise

The images should have enough signal (over background), so the image analysis algorithm can differentiate fluorescence from background. The higher the signal/noise ratio, the more likely a software algorithm can automatically detect the signal of interest over the background, and better dynamic range to differentiate between high and low responders. Parameters important for these assays are listed in Table T.2.

Signal-to-noise (Signal/noise, or S/N) is determined by a number of factors, including fluorescence markers (antibody concentrations), fluorescence background levels,

TABLE T.1   Key Image Acquisition Parameters to Define Resolution

Resolution	
Mitotic index assay	FKHR translocation assay
Images should have sufficient resolution to:	Images should have sufficient resolution to:
• Accurately segment nuclei • Accurately determine which nuclei are phospho histone H3 positive • Be as low magnification as possible-accumulating data on many cells is essential to robust data • 5–10× objectives, with binning if needed • Higher NA objectives not normally needed (phospho histone H3 staining is typically strong)	• Accurately segment nuclei • Accurate identify the cytoplasmic boundary • Since the assay will be measuring the fluorescence intensity within a cellular compartment, you want to collect a large enough area (enough pixels) to have an accurate measurement of the cytoplasmic signal • 10–20× objectives, without binning unless necessary

and instrument acquisition parameters (primarily exposure time). We will focus on the instrument parameters. These will also be set through the instrument software's acquisition dialog, for example, Figure T.5.

In general, the user should set the exposure time for each fluorescence channel so that the full dynamic range of the digital camera is utilized. In many cases, this is

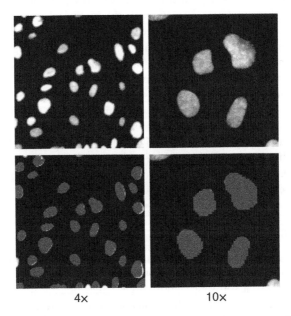

4×   10×

FIGURE T.4 *Sample images and segmentation overlays for images using 4× and 10× objectives.*

TABLE T.2   Key Image Acquisition Parameters to Define Signal/Noise

Signal/noise ratio	
Mitotic index assay	FKHR translocation assay
Images should have sufficient S/N to:	Images should have sufficient S/N to:
• Accurately identify nuclei • Accurately identify the cells that are positive for mitotic marker (ignoring lower intensity, non-mitotic background cells) • Binning can be very helpful to increasing signal intensity of mitotic cells and decrease scan times	• Accurately identify nuclei • Accurately identify GFP-positive signal over background. • Binning may help, but check the impact on segmentation

an exposure time long enough so the brightest signal attained is 4095 (most systems utilize 12-bit CCD cameras). Many systems also include an AutoExpose function, where the software will determine the appropriate exposure time to collect a specific maximum intensity. Systems such as confocal-based scanners, which do not have CCD cameras, will have similar parameters that define S/N. These include gain, scan time, and scan averaging.

A disadvantage of long exposure times is increased plate acquisition times. In the case of detecting nuclei, most image analysis algorithms are robust enough to detect individual nuclei with only a few hundred gray values over background, so you can use a lower exposure time. This will result in a faster acquisition time. In practice, there is a balance between S/N and scan speed. In the example of Figure T.6, nuclei were acquired at four different exposure times that varied from 2.5 milliseconds to 200 milliseconds. The 2.5 milliseconds exposure time results in insufficient S/N in the image to accurately segment individual nuclei, while the other exposure times are sufficient. At very low exposure times, the identification of nuclei degrades, but it is also seen that good segmentation is stable over a wide range of exposure times.

Determining the appropriate instrument parameters (magnification, binning, and exposure times) involves a series of iterations that include acquiring images on the instrument with different settings, running the image analysis algorithm, then checking the accuracy of the image segmentation.

FIGURE T.5   *Exposure time is key instrument parameter to define signal/noise.*

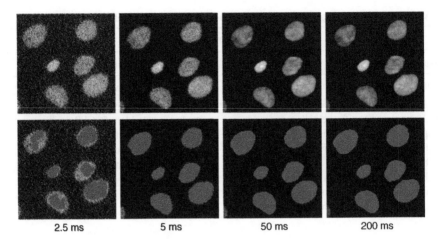

2.5 ms	5 ms	50 ms	200 ms

FIGURE T.6   *Image and segmentation quality.* DNA images were acquired at four different exposure times resulting in different signal/noise values. There is insufficient signal/noise to accurately segment individual nuclei at 2.5 milliseconds, but all other exposure times are sufficient to segment individual nuclei.

One of the benefits of discussing this in the context of a tutorial is that acquiring images blends instrumentation with image analysis itself, whereas we separated these discussions in the main text of the book when we were dealing with concepts. Nevertheless, from this effort, first you will acquire a complete set of images using the initial criteria, listed in Table T.3, and you will have a set of images that can be used in Lesson 2.

T.5   LESSON 2: IMAGE ANALYSIS

This lesson will start with a set of images and work through the steps for image segmentation and analysis. The lesson will describe the use of a "canned" application

TABLE T.3   Initial Criteria for Image Acquisition

Acquisition of images	
Mitotic index assay	FKHR translocation assay
• Select 10× objective • Select exposure times for both channels, so that you are using a large percentage of the dynamic range of the detector • Binning can help reduce plate acquisition times, but is not necessary	• Select 20× objective • Select exposure times for both channels so that you are using a large percentage of the dynamic range of the detector • Segmentation will be more accurate if data are collected without binning, but it is still an option if signal intensity is low

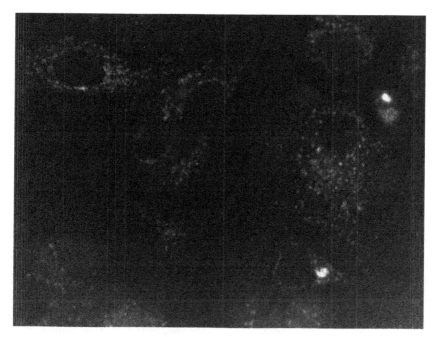

FIGURE T.7   *Sample image exhibiting background variation across field of view.* Puncta in the upper left region are easier to identify because the background is lower than in the rest of the image.

module approach. At the end of lesson 2, the user will have a set of features for all images.

T.5.1   An Introduction to Practical Image Analysis

It is important to understand the basics of image analysis algorithms as we will use the image analysis tools to ultimately optimize the instrument parameters. This is explained in more detail in Chapter 4 and Appendix C, but as a brief refresher.

The basic steps of image analysis are:

- **Find signal of interest over background.** This involves identifying any pixels in the image that are above background intensity. Most segmentation algorithms use an "adaptive" threshold. The user enters the intensity of the signal above background. This allows for better segmentation over backgrounds that may vary across a field of view. An example of an image with strong variation across the background is shown in Figure T.7.
- **Find pixels that are above this intensity threshold and group these signals into objects.** In the case of the DNA marker, these would be all the bright objects that are the general size and shape of nuclei. The algorithm then creates an "object mask" that is used to identify the nuclear region. The object mask

(a)

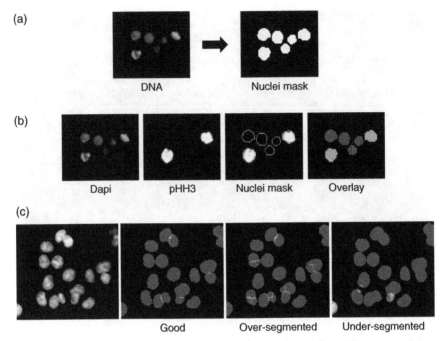

FIGURE T.8 *Basic steps of image analysis.* (a) An initial intensity threshold finds pixels above background and identifies nuclei based on size and shape parameters. (b) A object mask is created and superimposed over the pHH3 image. Nuclei are classified as mitotic or non-mitotic based on an intensity threshold for the pHH3 image. (c) Algorithm parameters are critical for accurate object segmentation. Sample overlays show good, over-segmented, and under-segmented nuclei.

is a binary representation of the identified objects generated by the algorithm. One can think of this as a set of regions that identify each object (e.g., nuclei). The relationship is shown in Figure T.8a.

- **This object or nuclei "mask" can be transferred to any of the other image channels for localization or compartmental analysis.** In the case of the Mitotic Index assay, the software would first identify the nuclei mask, and then transfer the nuclei mask to the mitotic marker image and ask the question: is the mitotic signal positive where the nuclei mask exists, and if so, mark this as a mitotic cell (Figure T.8b). Otherwise, it is a non-mitotic or negative cell.

A cautionary note

- **Be attentive to the extent of over-segmentation and under-segmentation of nuclei.** A common problem with nuclei detection algorithms is inaccurate segmentation, that is, either over-segmentation (splitting of a single nucleus) or under-segmentation (merging of multiple nuclei to a single nucleus). This can be due to the actual algorithm performance, density of cells (high seeding density), or incorrect algorithm parameters, shown in Figure T.8c. In addition to thresholds for any single algorithm, there are multiple algorithms available such as watershedding that treat object definitions with a uniform definition of object intensity or adaptive methods that evaluate the shape of the putative object in addition to an intensity threshold.

As described throughout the tutorial and book, an important step to any assay development process is to look at the images. In this case, it is easy to spot segmentation errors by viewing the segmentation overlay.

T.5.2 Analyzing the Mitotic Index Assay

T.5.2.1 Define Key Measurements Now that we have a basic idea of the image analysis, we want to use the appropriate tools to run the analysis for the Mitotic Index assay. For MetaXpress, select the Mitotic Index Application Module. In this module, there are four simple sets of parameters (see Figure T.9a):

- **Nuclei size:** approximate size of nuclei. The user can use one of the region tools to draw lines across both small and large nuclei to estimate the values to be entered into the dialog. These parameters are very important to minimize the under- or over-segmentation of individual nuclei.
- **Nuclei intensity above background:** intensity of the DNA signal above the background. For example, if the nuclei intensity value is 800, and the background intensity value is 200, the parameter value would be 600. With MetaXpress, the intensity value is displayed in the status bar at the bottom of the application window. The user would first measure intensities in the background area, and then in the nuclei. The difference of these intensities can be entered as the "Intensity above local background."
- **Mitotic marker:** intensity above background for the second channel. Using same process, identify the difference between a positive mitotic cell and a negative cell.

The module has a variety of measurements that can be used for data analysis (Figure T.9b). There are a number of nuclei measurements such as area, intensity, and count. There are also measurements for the mitotic marker such as count, area, and % mitotic. Many of these features may not be relevant to a particular assay or desired outcome. The data analysis section of the tutorial will describe the process for visualizing the different features to determine which metric best represents the biological response.

T.5.2.2 Optimize Image Analysis Parameters Remember, the important aspect of this assay is to accurately identify all nuclei using the DNA marker, and then based

(a)

(b)

FIGURE T.9   *Mitotic Index Assay Application Module.* (a) MetaXpress dialog. (b) Summary feature list.

on the nuclei segmentation, identify those nuclei that are also positive for the mitotic marker.

- Select a representative well to display an image set.
- Using the mouse cursor, determine average diameter of a nuclei, and set minimum and maximum ranges for nuclei size.
- Using the mouse cursor, determine intensity of nuclei and surrounding background. Set intensity above background to the difference of these values, for example 200, and run analysis.
- View segmentation overlay. In MetaXpress, the segmentation overlay is a visual representation of how the image analysis algorithm has identified the different objects in a specific image. For the Mitotic Index application module, mitotic cells are bright grey and non-mitotic cells are dark grey in Figure T.10. Objects that have been classified are typically color-coded (since the book figures are printed as black and white, color overlays are displayed as shaded objects). This is a very important tool to use while fine-tuning the parameters to optimize the accuracy of segmentation. As described below, it is important to run these tests on a variety of wells that include positive and negative controls. There is a button on the image window that turns the segmentation overlay on and off.
- If it is difficult to identify nuclei, you may need to lower the "Intensity above local background" setting to capture the dimmer nuclei. If this does not work,

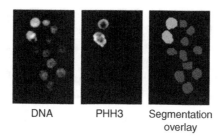

DNA   PHH3   Segmentation
                      overlay

FIGURE T.10   *Testing algorithm performance.*

you may need to return to image acquisition and increase the exposure time. As noted, the analysis and acquisition are often an iterative process—these tests can be done at the time of acquisition to confirm that acquisition settings will be sufficient for image analysis. It is often the case that setting this to a very low value (25–50) results in good segmentation while not picking up any background noise.

- Once the nuclei segmentation is optimized, adjust the threshold for the mitotic marker so the mitotic cells are accurately identified. You can estimate the value by using the mouse to determine the intensity of a mitotic cell relative to a nonmitotic cell. If it is misclassifying some of the mitotic cells, you may need to adjust the intensity threshold for image analysis or increase the exposure time (at acquisition) to increase dynamic range and better separate the mitotic cells from the nonmitotic cells. Continue to iteratively adjust threshold until mitotic cells are accurately identified.
- Select a different well and repeat until this is optimized for conditions across plate.
- Save settings.
- You are ready to run the analysis for the entire plate. The Mitotic Index Module will output a large number of parameters including nuclei count, mitotic count, and % mitotic.

T.5.3   Analysis of the Nuclear Translocation Assay

T.5.3.1   Identify Biologically Relevant Parameters for the Quantification   The Translocation assay involves the localization change of the biological marker between the nucleus and cytoplasm of each cell. Therefore, the image analysis algorithm must be able to identify these compartments. The nucleus is identified by the standard DNA stain such as DAPI, similar to the Mitotic Index assay. The cytoplasm is typically identified by one of the following methods:

- **Use an additional fluorescence marker to identify the cytoplasm.** This is less desirable because of the additional costs (wet lab, reagents, time, and acquisition channel).

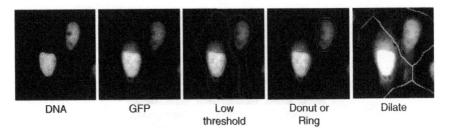

| DNA | GFP | Low threshold | Donut or Ring | Dilate |

FIGURE T.11 *Different methods for defining cytoplasm compartment.* There are a number of different methods for defining cytoplasm compartment: low threshold, donut/ring, and dilate. Setting a very low intensity threshold will often enable the algorithm to define the entire cytoplasm. An alternative is to create a ring region based on growing the nuclear mask. Finally, a simple dilation of the nuclear mask until a neighboring nuclear mask is reached will create a rough estimate of each cell cytoplasm.

- **Use the existing translocation marker, and set a very low intensity threshold for the image analysis.** In this case, use the FKHR image. The mask would be set by segmenting the FKHR protein that remains in the cytoplasm, even in conditions where it has substantially relocalized to the nucleus.
- **Creating a small perinuclear region that represents the cytoplasm.** This is typically called the ring or donut method. The algorithm will generate this ring based solely on the nuclei segmentation.
- **Estimating the cytoplasm by assigning each pixel to its closest nuclei.** This can be done by a number of methods including an "ultimate dilate" processing step, where each nuclear region is continuously dilated until it touches a neighboring region. Depending on the algorithm, the resulting cell boundary may or may not accurately delineate the true boundary. This method can lead to errors if wells contain sparsely seeded cells, leading to substantial regions of the background being counted as cell cytoplasm.

Examples of these are shown in Figure T.11.

T.5.3.2 Implementing the Parameters We will be using Translocation-Enhanced Application Module in MetaXpress.

Configuring the Translocation assay is a little more complicated than the Mitotic Index assay depending on the specific algorithm used. In this tutorial, the donut method will be exemplified. This creates a small perinuclear region that is used to estimate the cytoplasm intensity. The algorithm will first identify all nuclei using the DNA marker; and then based on the nuclei segmentation, it will create the ring region for the translocation marker image.

The module has three sets of parameters:

- **Nuclei segmentation.** The parameters are a little different than the parameters in the Mitotic Index module. The user enters the approximate size, area, and intensity of nuclei.

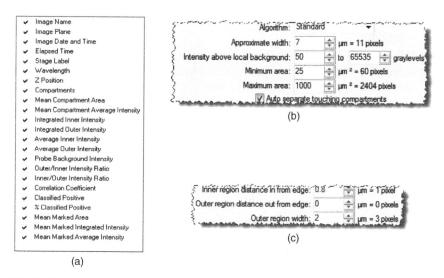

Image Name
Image Plane
Image Date and Time
Elapsed Time
Stage Label
Wavelength
Z Position
Compartments
Mean Compartment Area
Mean Compartment Average Intensity
Integrated Inner Intensity
Integrated Outer Intensity
Average Inner Intensity
Average Outer Intensity
Probe Background Intensity
Outer/Inner Intensity Ratio
Inner/Outer Intensity Ratio
Correlation Coefficient
Classified Positive
% Classified Positive
Mean Marked Area
Mean Marked Integrated Intensity
Mean Marked Average Intensity

(a)

Algorithm: Standard

Approximate width: 7   μm = 11 pixels
Intensity above local background: 50   to 65535   graylevels
Minimum area: 25   μm² = 60 pixels
Maximum area: 1000   μm² = 2404 pixels
Auto separate touching compartments

(b)

Inner region distance in from edge: 0.8   μm = 1 pixel
Outer region distance out from edge: 0   μm = 0 pixels
Outer region width: 2   μm = 3 pixels

(c)

FIGURE T.12   *Translocation-Enhanced Application Module.* (a) Summary feature list. (b) Nuclear segmentation parameters. (c) Ring region parameters.

- **Donut/ring parameters.** The nuclei mask (inner region) and donut mask (outer region) can be defined relative to the identified nuclei. This region is overlaid with the GFP to estimate the set of features in the cytoplasm.
- **Positive/negative classification.** One can define a value to differentiate between a positive cell or negative cell based on the ratio or intensities in the inner versus outer compartment regions. In this case, it is the nucleus versus cytoplasm.

The algorithm generates a set of features based on these compartments. In addition to the nuclei features described in the Mitotic Index assay, features related to the translocation marker are also generated (shown in Figure T.12a). Features include object count, areas and intensity for each of the compartments (nuclei, cytoplasm) as well as a cell classification. The algorithm classifies each cell as positive or negative based on the ratio of inner intensity (nuclear intensity) and outer intensity (cytoplasm intensity).

T.5.3.3   Optimize Image Analysis Parameters   In this assay, it is important to accurately identify both the nuclear compartment and the estimated cytoplasmic compartment.

- Go to the image analysis dialog. For the ring method, you would select the Translocation-Enhanced module. Select a representative well to display an image set. There are two sets of parameters to define: nuclei segmentation and ring parameters (Figure T.12b).

- The nuclei parameters are a little different than the parameters in the Mitotic Index assay.
- Using the mouse cursor, determine average diameter of a nuclei, and enter this as approximate width. Set intensity above background to a low value (50). You should also define minimum and maximum areas for nuclei. This algorithm also has the ability to exclude very bright objects such as mitotic cells or fluorescent artifacts by defining an upper value for Intensity above local background.
- Now define regions for measurement. This module has the ability to shrink the nuclear region in addition to defining the actual ring (Figure T.12.c). A good starting point is to shrink the nuclei region by 1 pixel and define outer region width to 3 pixels.
- Run analysis by selecting Test Run.
- View segmentation overlay and adjust nuclei segmentation parameters to accurately identify nuclear compartment and to accurately define ring region. Depending on the cellular morphology, the outer ring width should not be set too high, where the ring is outside the cellular boundary. You can adjust the nuclei parameters (Figure T.12b), to change the bright grey compartment and the region parameters (Figure T.12c), to vary the size of the ring region (darker grey) (Figure T.13).
- Select a different well and repeat until this is optimized for conditions across plate.
- Save settings.
- You are ready to run the analysis for the entire plate. The Translocation module will output a large number of parameters including compartment area, compartment intensity, inner and outer compartment intensity, as well as the inner/outer and outer/inner ratios.

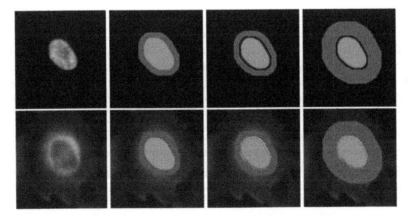

FIGURE T.13 *Estimate cytoplasm with ring region.* Adjusting the inner region distance, outer region distance, and outer region width results in different estimates for nuclear and cytoplasm compartments.

	Total Nuclei (MitoticIndex)	Mitotic Nuclei (MitoticIndex)	Interphase Nuclei (MitoticIndex)	% Mitotic Nuclei (MitoticIndex)	% Interphase Nuclei (MitoticIndex)	All Nuclei Total Area (MitoticIndex)	All Nuclei Mean Area (MitoticIndex)	All Nuclei W1 Integrated Intensity	All Nuclei W1 Average Intensity	All Nuclei W2 Integrated Intensity	All Nuclei Average Intensity
B01											
B02	55.44	39.33	16.11	69.02	30.98	8037.42	146.31	7623408.44	1530.25	4022826.44	83
B03	42.78	29.22	13.56	64.70	35.30	6259.42	140.37	5838612.11	1497.57	3095560.56	79
B04	45.67	28.44	17.22	59.86	40.14	6557.11	150.34	5811917.44	1439.79	2953399.00	74
B05	36.78	23.33	13.44	51.54	48.46	5361.92	149.68	4889298.78	1452.80	2524281.44	71
B06	34.89	18.33	16.56	49.77	50.23	4958.09	138.61	4935435.00	1618.56	2312235.44	79
B07	28.78	15.67	13.11	51.63	48.37	3457.45	128.93	3096343.11	1477.78	1669795.67	74
B08	35.33	18.22	17.11	49.37	50.63	4485.30	120.91	4166132.78	1540.19	1839031.89	74
B09	40.00	16.44	23.56	36.23	63.77	4878.40	126.10	4533249.00	1522.67	1655800.11	53
B10	81.11	32.67	48.44	39.66	60.34	11269.29	139.92	11703886.00	1727.12	4128286.44	61
B11	174.33	10.89	163.44	6.36	93.64	21045.32	121.45	18073075.00	1430.97	3180700.33	25
B12											
C01											
C02	42.78	24.22	18.56	55.58	44.42	6200.25	149.54	5493413.67	1413.71	2851903.56	78
C03	49.11	29.00	20.11	55.84	44.16	6884.01	138.76	5943256.33	1423.68	3064474.67	73
C04	50.33	28.78	21.56	53.89	46.11	7328.33	151.60	6271473.22	1396.62	3049185.33	64
C05	51.89	28.56	23.33	53.23	46.77	7321.49	151.60	6482963.00	1479.45	2985196.36	66
C06	33.89	13.78	20.11	41.15	58.85	4046.91	133.21	3385370.67	1380.95	1557446.22	62
C07	31.78	12.89	18.89	39.32	60.68	3674.33	123.09	2983132.33	1348.00	1556648.33	66
C08	32.00	12.67	19.33	36.36	63.64	3651.78	115.50	3109658.11	1374.15	1307902.78	55
C09	52.00	22.00	30.00	40.67	59.33	6902.50	135.29	6086444.22	1443.32	2357181.67	55

FIGURE T.14   *Sample output feature values from Mitotic Index assay.*

You should now have an entire set of data for both assays. A sample set of the Mitotic Index assay results are shown in Figure T.14. There are additional parameters that are not shown in the screenshot.

T.5.4   Quality Control Algorithm Performance

During the configuration of the image analysis algorithm parameters, the segmentation overlays were viewed on a number of wells to spot-check the segmentation accuracy across different conditions. Although this is an important step, it is also important to spot-check additional wells after the analysis is complete. In MetaXpress, one would view the montage and grid of numerical results, and select individual wells to quickly spot-check the compartment regions.

T.5.4.1   *Mitotic Index Assay*   Example experiment includes three compounds in triplicate. In this example, three views are displayed in Figure T.15.

- The plate montage is viewed as a color-encoded image with actual analysis results (% mitotic) are viewed in lower right corner of each well image.
- Color-encoded image of subset of single well along with segmentation results.
- Data results displayed in heat map colored table. It is easy to assess the relative compound potency differences between the two compounds that are each in triplicate.

One also notices that well F02 has a lower mitotic index compared to other wells in column 2. One can then quickly display the images as well as segmentation overlays from this well to assess why there is a difference. In this case, this well is the top compound dose which results in high cell death and low cell count per well.

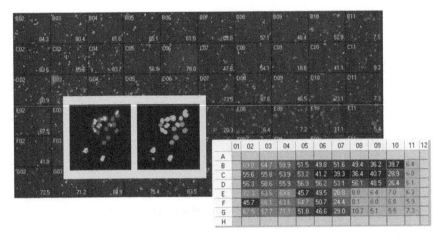

FIGURE T.15 *QC algorithm performance.* Different views can help with assessing algorithm performance and experimental results. A plate montage gives the user easy access to quickly view images throughout the plate. Segmentation overlays are used to assess segmentation accuracy. Data table heat maps are utilized to assess compound effects and potential data outliers.

T.5.4.2 Translocation Assay In this example (Figure T.16a), three similar views are displayed.

- Montage of entire plate, but only the translocation image is displayed.
- Segmentation overlay of single well images. Nuclei segmentation and donut/ cytoplasmic compartment.
- Heat map of data table.

The positive controls are located in B11, C11, and D11. The negative controls are located in E11, F11, and G11 (Figure T.16b). One could check any outliers at this stage. For example, well G10 appears slightly lower—perhaps the image was out of focus, the segmentation failed, or maybe there was some debris in the field of view.

Instead of viewing the nuclear-to-cytoplasmic ratio, one could view average nuclei count per image (Figure T.16c). Since we do not expect the compound treatment to result in any cell death, this view allows us to assess seeding density and if there is any cell death. Note that analysis of the data requires review of the actual data. Heat maps typically take the data range as the color range, but if the data range is narrow (such as in Figure T.16c), the coloring of data will make the differences appear more dramatic than they actually are.

T.6 LESSON 3: DATA ANALYSIS

This lesson will start with an entire plate's worth of features and will work through visualization of segmentation overlays and numerical data.

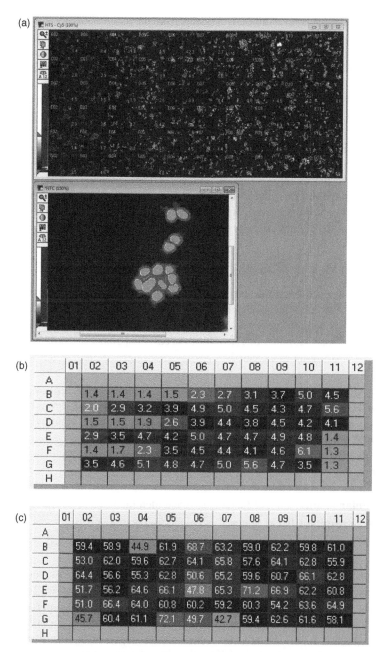

FIGURE T.16 *QC algorithm performance.* Different views can help with assessing algorithm performance and experimental results. (a) A plate montage gives the user easy access to quickly view images throughout the plate. Segmentation overlays are used to assess segmentation accuracy. (b) Data table heat maps are utilized to assess compound effects and potential data outliers. Inner intensity/outer intensity ratio is displayed. (c) Nuclei count per image is displayed.

Now that we have a set of numerical results from the image analysis algorithm, the next step is data analysis. This tutorial is not intended to provide a detailed analysis of the data, but to provide some basic steps and views to assess the feature list and experimental data. The lesson will therefore use the (virtual) demonstration data to generate IC_{50}s with different analysis metrics (e.g., % mitotic, % translocation).

Because data analysis tools vary widely depending on the manufacturer of your instrument, we will export the numerical data to Microsoft Excel for the purpose of this tutorial. Excel is sufficient to generate dose–response curves with add-ins such as XLFit (IDBS)

As we did during initial acquisition of the data, it is important to spot-check the algorithm performance. This is typically done by selecting random wells on the plate, running the image analysis, and checking the segmentation overlay. You also visualize the numerical data in each well, and if any outliers are found, then spot-check the image analysis for that specific site.

T.6.1 Initial View of the Results

In MetaXpress, once you select a plate, a table is displayed with the numerical data for each well. You can select each measurement and visualize via either the numerical data table or basic graph functions. A heat map view is also useful to quickly assess cell number or dose–response effects. This is an easy first glance at the data to confirm the assay is performing as expected.

T.6.1.1 Mitotic Index Assay This experiment contains three replicates of two compounds that are titrated left to right: high dose to low dose. From these heat maps, it is easy to observe the different potencies of the two compounds (as total nuclei or % mitotic nuclei, in Figure T.17a or b, respectively).

There are typically three sets of data that can be exported from the image analysis algorithm.

- **Cellular results:** a list of features for each identified object or cell
- **Image results:** a list of features, averaged for each image
- **Well results:** a list of features, averaged for all sites within a well

For this dataset, the plate map is displayed: a dose titration of two compounds in triplicate (Figure T.17c). Column 11 is control, and only inner 60 wells are used. An initial, simple view is a graph of the raw values, % mitotic on per well basis (Figure T.17d). Since these are replicates, the next step would be to average each replicate (Figure T.18a). Finally, since the columns represent a dose titration, and ultimately we want to generate an IC50 value for potency comparison, XLFit can be used (Figure T.18b).

T.6.1.2 Translocation Assay Similar views can be generated with translocation data, total nuclei (Figure T.19a) or inner/outer intensity ratio (Figure T.19b). This

55.4	42.8	45.7	36.8	34.9	28.8	35.3	40.0	81.1	174.3
42.8	49.1	50.3	51.9	33.9	31.8	32.0	52.0	91.6	147.4
52.8	50.3	46.8	43.0	38.8	44.9	50.0	66.3	101.8	177.6
38.0	41.7	38.4	23.8	47.9	110.3	184.7	210.7	214.0	171.9
21.0	33.2	30.6	32.2	46.4	90.7	192.2	198.4	209.7	182.0
29.6	28.3	41.8	39.2	36.4	79.9	172.8	187.2	201.6	179.6

(a)

69.0	64.7	59.9	51.5	49.8	51.6	49.4	36.2	39.7	6.4
55.6	55.8	53.9	53.2	41.2	39.3	36.4	40.7	28.9	6.8
56.3	58.6	55.9	56.9	56.2	53.1	56.1	48.5	26.4	6.1
72.3	63.6	69.6	45.7	49.5	20.8	8.8	6.4	7.0	6.3
45.7	66.1	63.6	64.7	50.7	24.4	8.1	8.0	6.8	5.9
67.5	67.7	71.1	51.8	46.6	29.0	10.7	5.1	5.5	7.3

(b)

	1	2	3	4	5	6	7	8	9	10	11	12
A												
CPD1, r1		10	3.3333	1.1111	0.3704	0.1235	0.0412	0.0137	0.0046	0.0015	0	
CPD1, r2		10	3.3333	1.1111	0.3704	0.1235	0.0412	0.0137	0.0046	0.0015	0	
CPD1, r3		10	3.3333	1.1111	0.3704	0.1235	0.0412	0.0137	0.0046	0.0015	0	
CPD2, r1		10	3.3333	1.1111	0.3704	0.1235	0.0412	0.0137	0.0046	0.0015	0	
CPD2, r2		10	3.3333	1.1111	0.3704	0.1235	0.0412	0.0137	0.0046	0.0015	0	
CPD2, r3		10	3.3333	1.1111	0.3704	0.1235	0.0412	0.0137	0.0046	0.0015	0	
H												

(c)

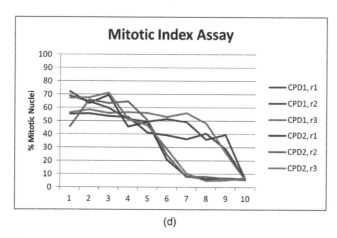

(d)

FIGURE T.17 *Mitotic index assay: data views and analysis.* (a) Heat map view of Nuclei count per image. (b) Heat map view of % mitotic cells. (c) Platemap showing dose titration of two compounds in triplicate. (d) Raw data graph showing different potencies of two compounds.

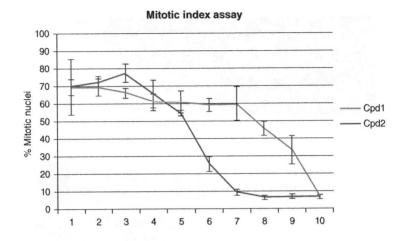

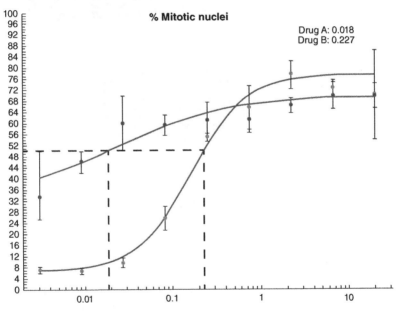

FIGURE T.18 *Mitotic index assay: data analysis and IC-50 determination.* (a) Average of triplicates for each compound. (b) Graph

experiment contains six compounds that are titrated left to right: high dose to low dose (Figure T.19c).

From these heat maps, one can observe the cell plating density across the plate ranges from 47 cells to 181 cells per well. Keep in mind that this is a single site per well—in a real experiment one would acquire multiple sites per well to capture a more representative distribution. One can also observe the different potencies of the six compounds, with the last compound being inactive in this assay.

135.0	99.0	119.0	105.0	57.0	94.0	151.0	164.0	123.0	151.0
109.0	107.0	152.0	113.0	110.0	113.0	91.0	121.0	150.0	89.0
106.0	113.0	103.0	125.0	47.0	63.0	135.0	132.0	127.0	97.0
95.0	87.0	99.0	96.0	91.0	102.0	139.0	96.0	116.0	63.0
181.0	169.0	87.0	100.0	84.0	109.0	140.0	98.0	67.0	126.0
116.0	104.0	79.0	83.0	97.0	90.0	94.0	99.0	137.0	149.0

(a)

1.4	1.4	1.4	1.5	2.3	2.7	3.1	3.7	5.0	4.5
2.0	2.9	3.2	3.9	4.9	5.0	4.5	4.3	4.7	5.6
1.5	1.5	1.9	2.6	3.9	4.4	3.8	4.5	4.2	4.1
2.9	3.5	4.7	4.2	5.0	4.7	4.7	4.9	4.8	1.4
1.4	1.7	2.3	3.5	4.5	4.4	4.1	4.6	6.1	1.3
3.5	4.6	5.1	4.8	4.7	5.0	5.6	4.7	3.5	1.3

(b)

	1	2	3	4	5	6	7	8	9	10	11	12
A												
CPD1		10	3.3333	1.1111	0.3704	0.1235	0.0412	0.0137	0.0046	0.0015	0	
CPD2		10	3.3333	1.1111	0.3704	0.1235	0.0412	0.0137	0.0046	0.0015	0	
CPD3		10	3.3333	1.1111	0.3704	0.1235	0.0412	0.0137	0.0046	0.0015	0	
CPD4		10	3.3333	1.1111	0.3704	0.1235	0.0412	0.0137	0.0046	0.0015	0	
CPD5		10	3.3333	1.1111	0.3704	0.1235	0.0412	0.0137	0.0046	0.0015	0	
CPD6		10	3.3333	1.1111	0.3704	0.1235	0.0412	0.0137	0.0046	0.0015	0	
H												

(c)

FIGURE T.19 *Translocation assay: data views and analysis.* (a) Heat map view of Nuclei count per image. (b) Heat map view of ratio: inner Intensity/outer intensity that is nuclear/cytoplasmic ratio. (c) Platemap showing dose titration of six compounds. As noted in both heat map views, compound 6 is inactive for specific pathway.

The Translocation-Enhanced Application Module generates a list of features other than the Total Nuclei and Inner/Outer Intensity Ratio shown above. As part of the assessment of this assay, one might want to look at other features.

Since the assay is a translocation of signal from the nucleus to the cytoplasm, we can visualize each feature separately (Figure T.20).

Congrats, you have generated an assay that can be used to rank compounds!

T.7 LESSON 4: QUALITY CONTROL IN A CELL-BASED ASSAY

This lesson will work through a number of suggested experiments for basic plate-based QC. This can flag the plate for a number of potential problems that include cell plating density, instrument errors, cell-health issues, and so on.

(a) Checking cell density (segmentation accuracy)
(b) Checking the number of cells/well (in the inner vs. the outer wells of plate)

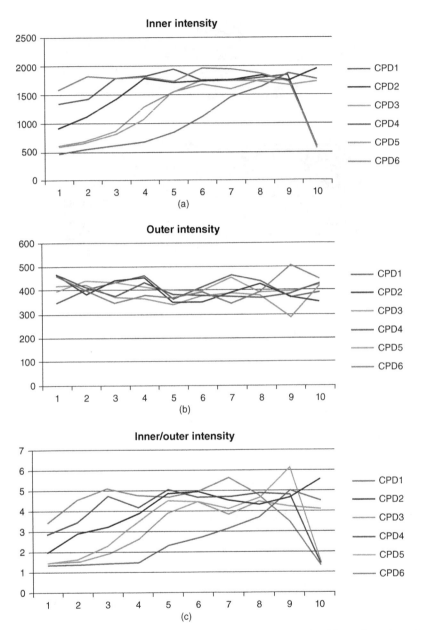

FIGURE T.20 *Translocation assay: different features.* (a) Graph of inner intensity only, representing the average intensity within nuclear compartment. (b) Graph of outer intensity only, representing the average intensity within cytoplasm compartment. (c) Graph of ratio of inner and outer intensity, representing nuclear/cytoplasm ratio.

(c) How many sites and/or wells do you need for a robust assay?

(d) The impact of signal intensity and signal-to-noise ratio on assay quality

(e) The impact of transfection efficiency on assay quality

Now that we can run the data analysis for the entire plate, there are a number of experiments that can be run for assay optimization and quality control.

- **Plate visualization to qualitatively assess image quality.** One should regularly view images from a number of wells across each plate to assess acquisition quality (e.g., auto-focus errors).

- **Identification of optimal magnification.** As previously mentioned, selecting the optimal magnification is important for accurate segmentation and quality data. Increasing the magnification may result in better pictures, but it also results in longer acquisition times. There may be cases where decreasing magnification results in similar quality data but significantly higher throughput. You can generate dose–response curves using data collected at different magnifications, and assess the impact of the change. In the case of the Mitotic Index assay, or similar "counting" assay, you may get similar data at 4×.

- **Checking number of cells per well and cell density.** One can easily look at nuclei count to assess the impact of a number of aspects of the process:
  - Different cell plating densities and its impact on segmentation accuracy
  - The variation of automated plating hardware on cell density across plate. For example, you may have a difference in the number of cells per well in one row compared to another row.

- **Determine optimal number of sites per well to capture.** Depending on the cell plating density, one must choose how many images to acquire in each well to capture enough cells to minimize noise in dose–response curves calculations. If you initially collect a large number of sites, one can chose to generate dose–response curve using different numbers of sites. Perhaps acquiring four sites will generate similar data as eight sites while decreasing collection time in half?

And you are done!

We hope that this virtual tutorial has provided some insight in the typical workflow for the image and data analysis aspects of an HCS assay. These relatively simple examples have stepped through the process from acquisition, image analysis, and basic data analysis. We have highlighted many of the specific parameters involved with each step, as well as some of the visualization tools available to you for protocol optimization and data assessment.

INDEX

*An Introduction to High Content Screening: Imaging Technology, Assay Development, and Data Analysis
in Biology and Drug Discovery*, First Edition. Edited by Steven A. Haney, Douglas Bowman, and Arijit Chakravarty.
© 2015 John Wiley & Sons, Inc. Published 2015 by John Wiley & Sons, Inc.

Printed and bound by CPI Group (UK) Ltd, Croydon, CR0 4YY

16/04/2025

14658523-0005